Wind Turbine Fundamentals for Technicians

Richard Skiba

AF882196

AFTER MIDNIGHT
PUBLISHING

Skiba, Richard (author)

Wind Turbine Fundamentals for Technicians

ISBN 978-1-7641699-4-3 (Paperback) 978-1-7641699-5-0 (eBook) 978-1-7641699-6-7 (Hardcover)

Non-fiction

Contents

Introduction

Wind energy has become one of the fastest-growing and most transformative industries in the world, reshaping how societies generate power and paving the way toward a cleaner, more sustainable future. As countries work to reduce carbon emissions and transition to renewable energy sources, wind farms—both onshore and offshore—are emerging as critical pillars of global energy infrastructure. At the heart of this transition are the wind turbine technicians: the highly skilled professionals who install, operate, maintain, and repair the machines that turn wind into electricity.

This book has been developed as a comprehensive guide for current and aspiring wind turbine technicians, industry trainers, and technical educators. It provides a complete overview of the wind energy sector and the competencies required to build a successful and sustainable career within it. Covering topics from the fundamentals of turbine operation to advanced diagnostic and maintenance procedures, the book bridges the gap between theory and practical fieldwork, ensuring readers are equipped with the technical knowledge, safety awareness, and professional insight necessary to thrive in this demanding and rewarding industry.

The content is structured to progress logically through the lifecycle of a wind energy project—from design and installation to operation, maintenance, and eventual decommissioning. Early chapters introduce the global context of wind energy, explaining how turbines generate electricity, how they are classified, and how technicians fit into this evolving workforce. Later sections explore core technical systems, including mechanical, electrical, and structural components, before advancing into specialist areas such as offshore operations, hybrid renewable systems, and emerging technologies like floating wind turbines and AI-based predictive maintenance.

A major focus of this book is safety, standards, and compliance—areas that underpin every aspect of wind turbine work. Detailed guidance is provided on risk assessment, personal protective equipment (PPE), emergency procedures, and adherence to international standards such as those set by the Global Wind Organisation (GWO). Readers will learn how to recognise and mitigate hazards, respond to incidents, and maintain the high levels of safety performance required in modern wind operations.

The book also addresses one of the most critical challenges faced by the global wind industry: developing a skilled, ethical, and future-ready workforce. In addition to technical topics, it explores professional practice, communication, documentation, leadership, and continuous professional development. These sections emphasise that a successful wind turbine technician must be more than a skilled tradesperson—they must also be an effective communicator, a problem solver, and a responsible environmental steward.

By combining technical rigour with industry relevance, this book aims to solve several key problems facing both learners and employers in the renewable energy sector. It provides a structured learning pathway aligned with international training standards, helping new entrants build competence and confidence, while also serving as a reference manual for experienced technicians seeking to deepen their expertise or transition into supervisory and leadership roles. For educators and training providers, it offers an up-to-date framework for curriculum design and assessment in vocational wind energy programs.

This book is written for a global audience, reflecting the international nature of the wind industry. Whether you are working on a wind farm in Australia, Europe, North America, or Asia, the principles of safety, efficiency, and sustainability remain the same. It is relevant to apprentices beginning their first placement, technicians upgrading their qualifications, engineers overseeing maintenance teams, and managers responsible for compliance and performance reporting.

Ultimately, this text aims to contribute to a safer, smarter, and more sustainable wind energy workforce. It recognises that every technician—through their hands-on work at great heights and in challenging conditions—plays a vital role in achieving the world's renewable energy goals. By equipping readers with the knowledge, standards, and professional values needed to excel, this book supports not only individual career growth but also the continued advancement and integrity of the global wind energy industry.

Part 1

Introduction to the Wind Energy Industry

Chapter 1

The Role of a Wind Turbine Technician

Wind power has moved from the margins to the mainstream. In just a few decades, wind has become one of the world's largest sources of new electricity, and with that rise has come an urgent need for people who can build, commission, operate, and maintain ever-larger, ever-smarter machines. This chapter opens the book by placing you— the wind turbine technician— at the centre of that story. It explains why technicians are the linchpin of reliable, affordable, and safe wind energy, and how the job is evolving as turbines scale up and digital systems become standard.

We begin with the global picture: how and why wind has grown so quickly on every continent, and what that means for workforce demand on both onshore ridges and offshore platforms. You'll see how policy, technology, and economics have converged to create a durable market for skilled technicians, and where the biggest opportunities lie over the next decade.

From there, we zoom in on the work itself. The chapter defines the technician's core responsibilities across the turbine lifecycle—installation support, commissioning, preventive maintenance, diagnostics, corrective repair, and documentation—and shows how these tasks interact with safety rules, quality standards, and grid requirements. You'll learn what a "typical day" really looks like, what tools and data streams you'll rely on, and how field decisions directly affect availability, energy yield, warranties, and lifecycle costs.

Finally, we map the career paths that technicians can follow, from entry-level roles to advanced specializations (blades, high-voltage, SCADA/CMS, offshore) and into leadership or training. You'll find the certifications and capabilities that employers value most, and how to align your learning with where the industry is headed.

By the end of Chapter 1 you'll understand the scale and momentum of the sector, the practical realities of the job, and the skills that turn a good technician into an indispensable one. It's a clear starting point for the rest of the book—grounding the big picture in the day-to-day work that keeps wind power turning.

The Growth of Wind Energy Globally

Wind energy has significantly transformed from a niche renewable source in the 1980s to a major player within the global power industry. As reported by the Global Wind Energy Council (GWEC), the world's cumulative installed wind capacity has surged over 100-fold since 1990, surpassing 1 terawatt (1,000 GW) in 2023, making it the second-largest renewable energy source following hydropower [1, 2]. This remarkable growth trajectory underscores wind energy's vital role in addressing the urgent need for decarbonization in light of climate agreements like the Paris Agreement [2].

Several key developments have propelled this exponential growth in wind energy. Firstly, there has been a significant reduction in technology costs, with the levelized cost of electricity (LCOE) for wind power expected to decline by 26% for onshore and 35% for offshore installations by 2025 [1]. Additionally, advancements in turbine efficiency have enhanced the viability and performance of wind farms [3]. Government policies aimed at fostering renewable energy, such as feed-in tariffs and tax incentives, have also played a crucial role in stimulating investment and ensuring the competitiveness of wind energy compared to fossil fuels [4].

Figure 1: An aerial view of of the Power County wind farm - Power County, Idaho with Nordex N100/2500 turbines. ENERGY.GOV, Public domain, via Wikimedia Commons.

According to the GWEC, as of 2021, over 80 countries are operationally harnessing wind power, with many more exploring new projects to meet both national energy demands and international climate obligations. The offshore wind sector, in particular, has been identified as possessing the greatest growth potential among renewable energy sources, with the goal of reaching nearly 2,000 GW of offshore wind capacity to limit global warming to below 1.5°C by 2050 [5]. Reports indicate that the offshore wind energy market is expected to expand rapidly, with notable contributions from countries like China, which accounted for the majority of new capacity installations in 2022 [6].

Figure 2: A windfarm and electricity pylons/power lines near Rye, East Sussex, England. Diliff, CC BY-SA 3.0, via Wikimedia Commons.

The projected trajectory of wind energy reflects broader economic growth patterns and an increasing recognition of sustainable energy solutions as necessary for combating climate change. With the rise of wind power installations, it is positioned to play a crucial role in the global energy transition, promising significant contributions toward achieving both national and international climate goals [2, 7].

Regional Trends

The global landscape of wind energy is witnessing significant regional trends that highlight the advancements and unique positions of various continents in harnessing both onshore and offshore winds.

Europe stands as a frontrunner in wind energy, particularly in onshore wind power, with countries such as Denmark, Germany, Spain, and the UK leading the way in capacity installations. The trend towards offshore wind energy has also culminated in Europe; specifically, the North Sea has become a hub for massive offshore wind projects. Innovative developments in the region, including integrated offshore grids, are expected to facilitate the growth of wind power installations substantially as the energy sector shifts toward greater reliance on renewable sources [8, 9]. Moreover, studies suggest that the offshore wind potential in this region is enhanced through advanced grid systems, which improve the reliability and efficiency of wind energy generation [8, 10].

Figure 3: Wind turbine under construction in Aalst, Flanders, Belgium, about 30 km (19 mi) northwest of Brussels. Aerial photo of the construction site 2020. MADe, CC BY-SA 4.0, via Wikimedia Commons.

In Asia, China dominates the wind energy sector, accounting for more than half of the new installations globally each year. Other nations like India, Vietnam, and South Korea are also experiencing rapid growth in their wind energy capabilities, driven by efforts to diversify energy supplies and reduce carbon emissions [11, 12].

Figure 4: Bangui Windmills are located in Bangui, Ilocos Norte, Philippines. Paolo Dala from Marikina City, Philippines, Philippines, CC BY-SA 2.0, via Wikimedia Commons.

The deployment strategies in these countries often involve leveraging reanalysis wind data to optimize offshore developments, particularly in areas where existing infrastructure might limit onshore growth [12]. This focus is critical as governments work to expand the renewable energy mix to meet increasing domestic demands while adhering to international climate commitments [11, 13].

Figure 5: Wind Turbines on Highway G312 Xinjiang, China. Matthew Summerton, CC BY-SA 3.0, via Wikimedia Commons.

North America, particularly the United States, showcases the second-largest installed capacity of wind energy, benefiting from advancements in both onshore and offshore wind projects. The Midwest and Texas are significant hubs of wind power generation, while the Atlantic coast is seeing a rapid acceleration in offshore projects. U.S. energy policies and incentives foster a competitive environment that encourages further expansion of wind energy infrastructures [14, 15]. Analyses predict that with strategic investments and technological advancements, the adaptability of existing grids to incorporate vast amounts of offshore wind will play a crucial role in meeting future energy requirements [15].

Figure 6: Wind turbine in a field in North Dakota. U.S. Fish and Wildlife Service - Midwest Region, Public domain, via Wikimedia Commons.

Emerging markets in Latin America and Africa, particularly Brazil, Chile, South Africa, and Egypt, are capitalizing on their strong wind resources. These countries see wind energy not just as a means to diversify their electricity supply but also as a potential driver for sustainable development due to its environmental benefits [16]. As these nations pursue energy diversification, the focus on efficient monitoring and maintaining biodiversity in wind farm planning becomes essential to mitigate potential negative impacts stemming from wind energy projects [17].

Figure 7: Albany Wind Farm, Western Australia. Nachoman-au, CC BY-SA 3.0, via Wikimedia Commons.

In Oceania, Australia has significantly expanded its onshore wind capacity and is establishing large offshore wind projects, such as Star of the South. The commitment to offshore developments highlights a broader policy shift aimed at achieving sustainable energy goals, reflecting the global trend of increasing offshore capacity [15]. Innovative solutions for optimizing wind farm installations, such as advanced simulation technologies, are being implemented to facilitate the deployment of offshore turbines in the region's complex marine environments [18].

Onshore vs Offshore Growth

Onshore wind energy has historically been the dominant form of wind power generation, providing the bulk of global capacity primarily due to its lower costs and simpler installation procedures. Technological advancements and economies of scale have significantly reduced the levelized cost of electricity (LCOE) for onshore wind energy, making it a highly competitive option for power generation. Stede et al. indicated that technological developments and improved financial conditions have led to a marked decrease in the LCOE for onshore wind, highlighting its role as a key player in renewable energy transitions globally [19]. Additionally, favourable policies in many regions have incentivized large-scale onshore wind projects, contributing to their prevalence [20].

In contrast, offshore wind energy is experiencing rapid growth, with significant annual increases in installed capacity. This growth is largely attributed to the higher capacity factors associated with offshore installations, which can utilize larger and more efficient turbines alongside more stable wind regimes in marine environments [21]. For instance, anticipated advancements in offshore turbine technology are projected to deliver potential capacities of up to 17 MW per turbine, compared to the typical onshore turbine capacity of around 5.5 MW [21]. Moreover, floating offshore wind technology presents a significant innovation, enabling the utilization of previously inaccessible deep-water sites where wind speeds tend to be higher, thereby enhancing energy generation potential [22].

Figure 8: The five Alstom Haliade 6 MW turbines of the Block Island Wind Farm. Ionna22, CC BY-SA 4.0, via Wikimedia Commons.

The differing growth patterns and capabilities between onshore and offshore wind can also be examined through their respective environmental impacts and land usage. Grimsrud et al. discussed how transitioning from onshore to offshore wind could reduce land requirements while offering substantial renewable energy potential [22]. It is noted that offshore wind farms typically occupy less land compared to their onshore counterparts, making them attractive options for densely populated areas where land availability is limited [13]. The competitive advantage of offshore wind is further highlighted by its higher energy density and lower turbulence, which contribute to more reliable energy production [14].

Onshore wind remains the dominant source of wind energy capacity globally due to its established infrastructure and cost benefits, offshore wind is positioned for significant expansion due to its favourable performance characteristics, innovative technologies, and ability to leverage oceanic areas for energy generation. As policies and technologies evolve, cooperation between onshore and offshore installations could play an essential role in meeting global energy demands sustainably.

Technology Improvements

The evolution of modern wind turbines is marked by significant advancements in size, efficiency, and technology compared to earlier models. One of the most notable developments has been in rotor diameter. In the 1990s, wind turbines typically featured rotor diameters around 80 meters, such as the Vestas V80 with a 2 MW power rating [23]. Today, particularly in offshore applications, turbines often boast rotor diameters exceeding 200 meters. This aerodynamic enhancement allows for more extensive wind capture over a larger swept area, substantially increasing energy generation capacity [24].

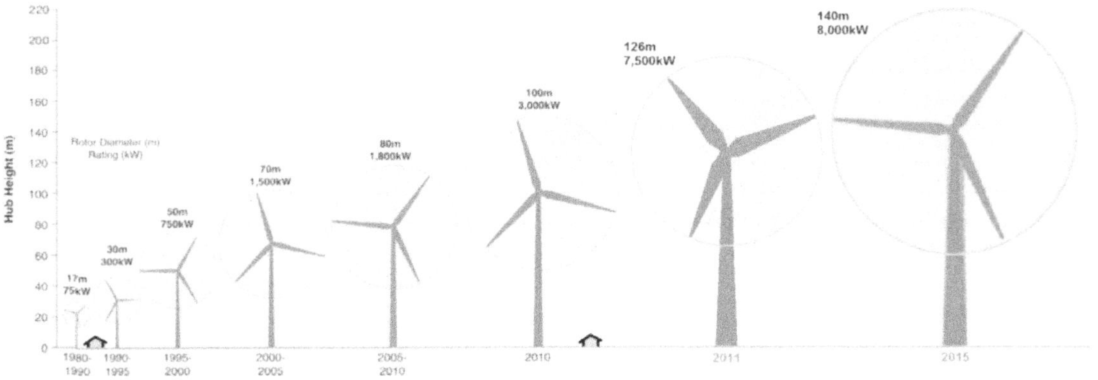

Figure 9: Wind turbine size increase 1980-2015, showing relative size of the swept area, as turbine size increased from 75 kW to 8 MW. Delphi234, CC0, via Wikimedia Commons.

Parallel to these advancements in rotor size, turbine power ratings have also escalated dramatically. Wind turbines from the 1990s commonly generated 1-2 megawatts (MW) of power; however, modern offshore units routinely deliver between 10-15 MW per turbine, with some prototypes anticipated to exceed these figures [25]. This substantial increase in power output means that fewer turbines are required to generate the same amount of energy, thereby reducing project costs inherently linked to infrastructure and maintenance [26]. For example, the NREL's offshore 5-MW reference wind turbine exemplifies this trend by showcasing the capabilities of modern designs, which optimize energy production while considering structural and control-system advancements [25].

The materials and technological components utilized in wind turbines have also progressed significantly. New blade designs leverage lightweight composites, which enhance not only aerodynamics but also durability [27]. Innovations in gearboxes and power electronics play a vital role in improving turbine reliability and efficiency, allowing for higher operational outputs and decreasing long-term maintenance costs [28]. As maintenance costs can consume a significant portion of wind farm revenues, integrating such advancements is critical for overall economic viability [29].

The integration of digital technologies represents a transformative shift in turbine operations and maintenance. Systems such as Supervisory Control and Data Acquisition (SCADA), combined with artificial intelligence and drones, revolutionize how turbine operators monitor performance and conduct maintenance. By enabling predictive maintenance, these technologies facilitate operational efficiency and reduce downtime associated with repairs [30]. For instance, predictive digital twin applications are emerging as valuable tools for assessing the remaining useful life of turbine structures, enhancing operational safety and cost-effectiveness in harsh offshore environments [29].

Through these developments, the wind energy sector has not only enhanced its power generation capabilities but has also positioned itself as a more reliable and economically feasible energy source in the long term. The convergence of larger rotor diameters, higher power ratings, advanced materials, and digital monitoring systems marks a pivotal transition in the evolution of wind turbines, establishing a robust foundation for future advancements in renewable energy technologies.

Understanding Megawatts (MW)

A watt (W) is a unit of power that measures the rate at which energy is produced or consumed. One megawatt (MW) equals one million watts. In electricity generation, a power rating indicates the maximum instantaneous output a generator can produce under ideal conditions. When a wind turbine is rated at 8 MW, it means that at full wind speeds and optimal conditions it can generate 8 million watts of electricity at that moment. This measurement is often called the "nameplate" or "rated" capacity.

By wind energy standards, this is a huge leap. In the 1990s, most onshore turbines produced only 1–2 MW each. Today's offshore turbines, at 8–15 MW, can generate four to ten times as much electricity per unit. This is why you now see fewer, much larger turbines at modern wind farms—each unit delivers far more power than its predecessors.

On average, one megawatt of continuous power can supply roughly 600–1,000 average Australian or U.S. homes, although the exact number varies by country and climate. A 2 MW turbine from the 1990s could power around 1,200–2,000 homes at full output, whereas a 15 MW modern offshore turbine could power approximately 9,000–15,000 homes at full output. This demonstrates how one large modern turbine can replace several older ones, dramatically increasing efficiency.

The common rule of thumb that "1 MW can power 600–1,000 homes" is based on average household electricity consumption in industrialised countries plus the capacity factor of the generator [31].

- In the U.S., the Energy Information Administration (EIA) reports average residential electricity use of about 10,500 kWh per home per year.
- In Australia, the figure is somewhat lower—around 6,000–7,000 kWh per home per year, depending on state and climate.

One megawatt of continuous power (i.e. 1 MW × 8,760 hours in a year) produces 8.76 million kWh per year. Dividing that by 10,500 kWh/home gives about 833 homes; dividing by 6,500 kWh/home gives about 1,346 homes.

Because wind doesn't blow at full speed all the time, actual delivered energy depends on the capacity factor (often 30–50% for wind turbines). So many public-facing sources (e.g. U.S. Department of Energy's *Wind Exchange*) use 600–1,000 homes per MW as a conservative range to reflect different consumption patterns and capacity factors.

Wind turbines can also be compared with other electricity sources to understand their scale. A typical residential rooftop solar system produces only 5–10 kilowatts (kW), or 0.005–0.01 MW. A small diesel generator typically generates 1–2 MW, which is about the same as an older onshore wind turbine from the 1990s. Modern onshore turbines now reach 3–6 MW, and modern offshore turbines produce 8–15 MW. By contrast, a single gas turbine peaker plant may deliver 50–200 MW, a large hydroelectric unit 100–700 MW, and a nuclear power reactor 900–1,600 MW. This comparison shows that although modern wind turbines are massive by renewable energy standards, they are still much smaller than a single large fossil fuel, hydroelectric, or nuclear power unit.

Economic and Environmental Impact

Wind power has emerged as a significant contributor to global job creation and environmental sustainability. The Global Wind Energy Council (GWEC) has reported estimates suggesting that wind power supports over 1.4 million jobs worldwide in sectors including construction, operation, and maintenance of wind energy systems [32]. This employment growth is indicative of the expanding wind energy sector, which not only creates direct jobs but also stimulates ancillary industries and services associated with renewable energy technologies [33]. In many regions, the transition to wind energy has led to considerable economic impacts, with policies specifically designed to enhance renewable energy production directly influencing job creation [34].

Additionally, wind energy plays a crucial role in mitigating climate change through effective reduction of carbon dioxide (CO_2) emissions and various air pollutants. By generating electricity without burning fossil fuels, wind power significantly displaces harmful emissions, thereby directly supporting net-zero targets adopted by various countries [35, 36]. The deployment of wind energy contributes to a decrease in greenhouse gas emissions, with studies indicating that increased capacity for wind power generation significantly reduces reliance on conventional energy sources [34, 37]. Thus, the dual benefits of economic stimulation through job growth and environmental enhancement through emission reductions mark wind power as a pillar in the transition toward sustainable energy systems.

Furthermore, the economic implications of wind energy extend beyond immediate job creation. Research has shown that policies promoting renewable energy technologies have long-term positive effects on overall economic growth and employment across sectors in various regions, particularly within the European Union [34, 38]. Similarly, in localized contexts, the establishment of wind farms has led to secondary employment effects that benefit surrounding areas through increased economic activities [33]. These findings align with the substantial environmental benefits that accrue from reduced reliance on carbon-intensive energy sources, reinforcing the interconnectivity between economic growth and sustainable practices in the energy sector.

Analysts are increasingly optimistic about the prospects of renewable energy, projecting a global capacity of approximately 2–3 terawatts (TW) by 2030, contingent on current growth trends. This development is facilitated by a combined push for offshore projects and hybrid renewable energy systems, which integrate wind, solar, and storage technologies. According to Gunn et al. [39], offshore renewable energy potential is substantial and could significantly contribute to this capacity. Their findings indicate that marine energy could provide around 2,300 terawatt-hours per year, illustrating the feasibility of achieving high output through maritime resources, complementary to wind and solar capabilities.

Moreover, the solar photovoltaic (PV) sector is expected to contribute significantly to this future capacity, driven by technological advancements and economic viability. Jan et al. Jan et al. [40] highlight solar PV's position as a key contributor, suggesting that it is well-suited for both large-scale and small-scale projects. This assertion is further supported by Victoria et al. ([41], who emphasize the cost competitiveness of solar energy technology and its potential to alleviate carbon emissions significantly. The growth forecast entails a sharp increase in the need for technicians skilled in installation, maintenance, and decommissioning as the fleet of renewable energy technology ages.

The hybrid systems that integrate various renewable sources, alongside storage capabilities, will require a skilled workforce for implementation and support. Specifically, offshore wind farms, discussed by Tusar and Sarker [42], are projected to increase in size and complexity, necessitating specialized technicians for maintenance and operational functions. The call for maintenance technicians is echoed in literature, with the expanding fleet indicating a tangible demand for skilled professionals capable of managing these technologies effectively as they mature and age over the coming decades.

The overall landscape implies that the push towards a greener energy future relies heavily on renewable technologies, including wind, solar, and hybrid systems, and will foster job creation through the necessity of a skilled workforce to maintain and support these systems. The growing trajectory of renewable energy installations signifies a transformative shift in the energy sector, moving towards sustainable practices and reliance on cleaner power sources.

Technician Opportunities

Wind Turbine Fundamentals for Technicians

Global wind capacity has expanded from a niche level in the 1980s to more than one terawatt in 2023, with analysts projecting between two and three terawatts by 2030. Every new turbine installed requires commissioning, regular inspections, servicing, and eventual decommissioning over its 20–30-year lifespan. This rapid expansion means a much larger fleet of assets that must be maintained, and technicians form the essential workforce that makes this possible. As global capacity doubles again by 2030, the demand for skilled wind-turbine technicians is expected to grow proportionally.

Offshore wind has emerged as the segment with the greatest growth potential, aiming for nearly 2,000 gigawatts of capacity by 2050 to meet the 1.5 °C climate targets. Offshore turbines are physically larger—currently rated at 8–15 MW with prototypes soon reaching 17 MW—more complex, and located in harsher environments than their onshore counterparts. These factors create opportunities for technicians with specialised skills in marine safety, high-voltage systems, rope access, and offshore logistics. Floating offshore wind will also open new roles in maintaining anchoring systems, moorings, and dynamic cables.

Advances in technology and digitalisation are transforming the nature of the job. Modern wind turbines feature larger rotors, advanced composite materials, improved gearboxes and power electronics, and sophisticated monitoring through SCADA systems, drones, artificial intelligence, and predictive maintenance. Technicians are increasingly required to install and calibrate digital systems, interpret data for predictive maintenance, repair composite blades, and work safely at height with integrated lifts and evacuation systems. This shift creates opportunities for upskilling and specialisation, moving the role beyond purely mechanical work into electrical, electronic, and data-driven tasks.

Figure 10: Meg White, from Master Point Rope Access Solutions, does a final check of the end of a wind turbine blade after sensors were attached which will sense impact, such as that from birds or bats, at the Foothills Campus of the National Renewable Energy Laboratory. U.S. Department of Energy from United States, Public domain, via Wikimedia Commons.

The sector is already a major employer, with the Global Wind Energy Council reporting 1.4 million jobs worldwide in 2022. Many governments are linking renewable-energy targets with local-content rules, apprenticeship programs, and training subsidies. This policy environment is creating new pathways for technicians to enter or retrain into wind power, especially in regions ramping up offshore projects such as China, the U.S. Atlantic coast, the North Sea, and Australia's Star of the South.

Hybrid systems that combine wind, solar, and storage are also creating demand for technicians who can work across technologies. Those trained initially on wind turbines may find opportunities in battery-storage maintenance and diagnostics, solar-array servicing alongside wind turbines, and high-voltage substation and grid-integration tasks. This diversification broadens the career prospects of wind-energy technicians and makes their skills more transferable.

As the first generation of large-scale wind farms reaches the end of its life, a new market is emerging in repowering, decommissioning, and recycling blades and components. Technicians with knowledge of safe dismantling and material handling will be in demand in this developing field.

The U.S. Bureau of Labor Statistics (BLS) projects that employment of wind turbine service technicians (SOC code 49-9081) will grow from 13,600 workers in 2024 to 20,500 workers in 2034 [43]. This represents an increase of about 6,800 jobs—a 50 percent growth rate—which is far higher than the average for all occupations. Most roles in the U.S. economy grow only a few percent per decade, so a 50 percent increase signals a very strong demand trend [43].

In addition to growth, there will be a steady stream of job openings. The BLS estimates roughly 2,300 positions will become available each year for wind turbine technicians over the next decade. These openings are not all new roles; many are driven by turnover as workers leave the occupation, move to different careers, or retire. This means that even if the growth rate slows later, consistent hiring will continue because the industry needs to replace experienced technicians exiting the workforce.

This rapid employment growth is driven by the expansion of wind power to meet rising electricity demand and decarbonisation targets. More turbines—particularly large onshore and new offshore units—are being installed across the U.S. Every turbine must be installed, maintained, inspected, and eventually repaired, creating direct employment for service technicians.

Wind-turbine service technicians already hold one of the fastest-growing jobs in the U.S. according to BLS occupational outlooks [43]. Because turbines are often located in remote or rural areas, there is a constant need for trained local personnel to ensure they operate safely and efficiently. Their work includes scheduled maintenance, troubleshooting faults, and upgrading components—tasks that cannot easily be automated or outsourced.

For someone considering this career, these figures point to strong job security and consistent hiring over the next decade. Training programs and certifications—such as those offered by the Global Wind Organisation (GWO) or community colleges working with wind companies—can position a worker to enter a field with high demand and excellent long-term prospects. Because a significant share of openings will result from retirements and career changes, there are opportunities for new entrants as well as for experienced technicians moving into supervisory or specialised roles.

Recent reports from the GWO and the Global Wind Energy Council (GWEC) further reinforce this trend. Their publication, *Global Wind Workforce Outlook 2023–2027*, forecasts that nearly 600,000 technicians will be needed over the next five years to construct, install, operate, and maintain the global wind fleet. These insights reveal promising career developments, with growing demand for new technicians both from full-time education and from other sectors looking to transition into renewable energy [44].

The *Outlook* highlights workforce needs in 10 countries in particular: Australia, the USA, Brazil, China, Colombia, Egypt, India, Japan, Kenya, and South Korea [44]. The ambitious wind power targets in these nations must be supported by a strong safety culture and a well-trained workforce. This trend underscores not only the scale of the workforce required but also the opportunities for local and national governments to leverage renewable energy expansion as a catalyst for job creation, training, and reskilling.

By 2027, it is projected that more than 574,000 technicians will be required globally for the construction, installation, operation, and maintenance of wind projects . This surge in demand is tied to the rapid expansion of the offshore wind market and a doubling of annual wind energy installations from 78 GW in 2022 to 155 GW in 2027 [44]. The forecast reflects a transformative moment for the industry and a wealth of opportunities for skilled workers entering or advancing within the wind-energy sector.

Key Responsibilities of Technicians

Wind turbine service technicians, sometimes called *windtechs*, are responsible for maintaining and repairing the complex machinery that turns wind into electricity. Although the salaries and the appeal of working outdoors can be attractive, not everyone is suited for this career. Windtechs must meet physical requirements such as climbing tall towers, lifting heavy parts, and sometimes working in confined spaces. They also need strong technical, communication, and problem-solving skills to manage the work effectively [43].

Like many trades, the role of the wind turbine technician is evolving to keep pace with rapid technological change. Advanced monitoring systems, diagnostic software, drones, and even robotic tools are now common. These technologies handle many inspection tasks at height that previously required a human presence, which improves safety and efficiency. However, technicians remain indispensable for interpreting the data that drones and robots collect and for conducting visual inspections and hands-on repairs that cannot be automated. As the technology improves, the job is expected to become less physically demanding and safer overall, but technicians will need to continually update their knowledge to remain proficient with new equipment and diagnostic tools.

Successful windtechs combine physical capability with mental agility. They need to lift and carry heavy items, crawl into tight spaces, and work comfortably at heights or offshore. Rope access training is often required, as is the ability to gather and interpret data from sensors and monitoring systems. As drones and robots take on more maintenance tasks, technicians will increasingly be expected to learn how to control and pilot these devices. They must also be willing to work outdoors in remote locations, often under challenging weather conditions, and maintain an open-minded attitude with a strong aptitude for learning as the job evolves.

Wind Turbine Fundamentals for Technicians

The day-to-day duties of windtechs cover the entire lifecycle of a turbine. They assist engineers and ironworkers in installing new wind turbines, inspect the exterior and physical integrity of towers, climb towers to repair or service equipment, perform routine maintenance, and test and troubleshoot electrical, mechanical, and hydraulic systems. They replace worn or malfunctioning components, collect turbine data for analysis, and may also service underground transmission systems, substations, and fibre-optic control networks. Although turbines have electronic monitoring equipment—usually located in the nacelle (pronounced as nuh-SELL /nəˈsɛl/) —to alert operators to problems, technicians still need to travel to the site to carry out physical inspections and repairs.

Most of a windtech's work focuses on the nacelle, the housing at the top of the tower that contains the generator and other critical systems, as well as on the blades attached to it. Maintenance schedules are typically based on operating hours but may vary by manufacturer. For turbines that operate year-round, routine maintenance is usually carried out one to three times annually. To reach the nacelle, which may be more than 200 feet high, windtechs use safety harnesses, climb assists, and sometimes elevators, while carrying the necessary hand and power tools. They also use computers and diagnostic equipment such as multimeters, infrared testers, and oscilloscopes to identify faults [43].

In 2024, about 13,600 people worked as wind turbine service technicians in the United States. The largest share (34 percent) was employed by wind electric power generation companies, but other technicians were self-employed or worked in construction, repair, or equipment supply businesses. Regardless of the employer, most windtechs work outdoors, often in extreme temperatures, and must be comfortable with heights. Repairing blades may involve rappelling from the nacelle down to the blade itself. Inside the nacelle, technicians work in confined spaces alongside other specialists, such as electricians, especially during major repairs.

Windtechs generally work full time and may be on call in the evenings or on weekends. They frequently travel to wind farms in rural areas or offshore, and offshore work may require staying away from home for days or weeks at a time. This lifestyle suits people who enjoy travel, adventure, and working in smaller, more remote communities rather than large urban centres.

O*NET lists repairing and troubleshooting, equipment maintenance, critical thinking, and operation monitoring as top skills for windtechs [45]. They also need to diagnose, test, and analyse systems using various software platforms. Typical tasks include diagnosing problems with turbine generators or control systems, climbing towers for inspections, testing electrical components, starting or restarting generator systems, and troubleshooting mechanical, hydraulic, and electrical malfunctions. Tools range from hot sticks and oil sampling kits to digital multimeters and stripping tools, while software can include analytical programs, Microsoft Outlook, enterprise resource planning (ERP) systems like SAP, computerised maintenance management systems (CMMS), and industrial control software such as SCADA [45].

A tower technician's routine work involves greasing bearings, changing filters, replacing brushes, torquing bolts, and running multiple electrical checks using industrial multimeters. Safety is paramount: harnesses, lanyards, climb assists, and elevators are standard, but technicians still need significant physical strength for repeated climbs. Many technicians look after about ten towers each and develop a strong familiarity with their assigned equipment. Complex problems can be time-consuming to diagnose, and resource technicians may be called in for particularly difficult faults.

Because wind turbines are usually located in rural or remote areas where wind speeds and land availability are favourable, technicians must be willing to work in isolated environments. Many travel as contract workers from site to site, which can be ideal for those who enjoy seeing new places. For those who prefer the outdoors and a sense of adventure, the lifestyle can be a great fit. Increasing offshore wind development also means more technicians will be needed who are comfortable working in marine environments, as some of the world's largest turbines are now offshore [46].

Safety is a critical focus of the wind industry. In 2019, the Global Wind Organisation more than doubled safety training requirements in North America, and in the UK there was only one fatal accident reported in 2021. Although U.S. statistics are harder to obtain, the track record of safety appears strong. OSHA highlights four main hazards—slips and falls, electrocution, being struck by objects, and being caught in or between equipment—but stresses that proper training significantly reduces these risks. Advances in personal protective equipment (PPE), automation, drones, and robotics continue to make the job safer than ever [46].

Automation and robotics are also transforming the work. Drones, infrared photography, and artificial intelligence now perform inspections that used to require human presence at height. This improves safety and provides higher-quality data but also places more emphasis on continuous learning for technicians, who must stay up to date with new systems and tools. The job of a wind turbine technician is therefore becoming a blend of hands-on mechanical work and high-tech data management—a role that rewards physical skill, adaptability, and a commitment to ongoing training [46].

Technician Role

Wind turbine technicians have a multi-layered role that spans the entire operational life of a turbine. Beyond day-to-day maintenance, they are also integral to the initial build and commissioning stages. At new sites they coordinate with engineers and installation teams, ensuring that structural, mechanical and electrical components are correctly assembled, aligned and tested before a machine goes online. This requires a solid grasp of technical drawings, lifting and rigging procedures and the safe handling of heavy equipment such as cranes and hoists.

Figure 11: Wind turbine construction, Pougny (Nièvre, France), rotor hub fairing of Enercon wind turbine No. 10 (E10). Cjp24, CC BY-SA 4.0, via Wikimedia Commons.

Once turbines are operational, technicians take responsibility for keeping them at peak performance. They follow manufacturer-defined service schedules to minimise wear and tear and avoid costly downtime. This includes preventive tasks on moving parts, electrical connections and control systems as well as visual checks for corrosion or damage across all major assemblies. Their aim is to intervene before small issues become major failures, thereby extending the life of each machine and protecting the operator's investment.

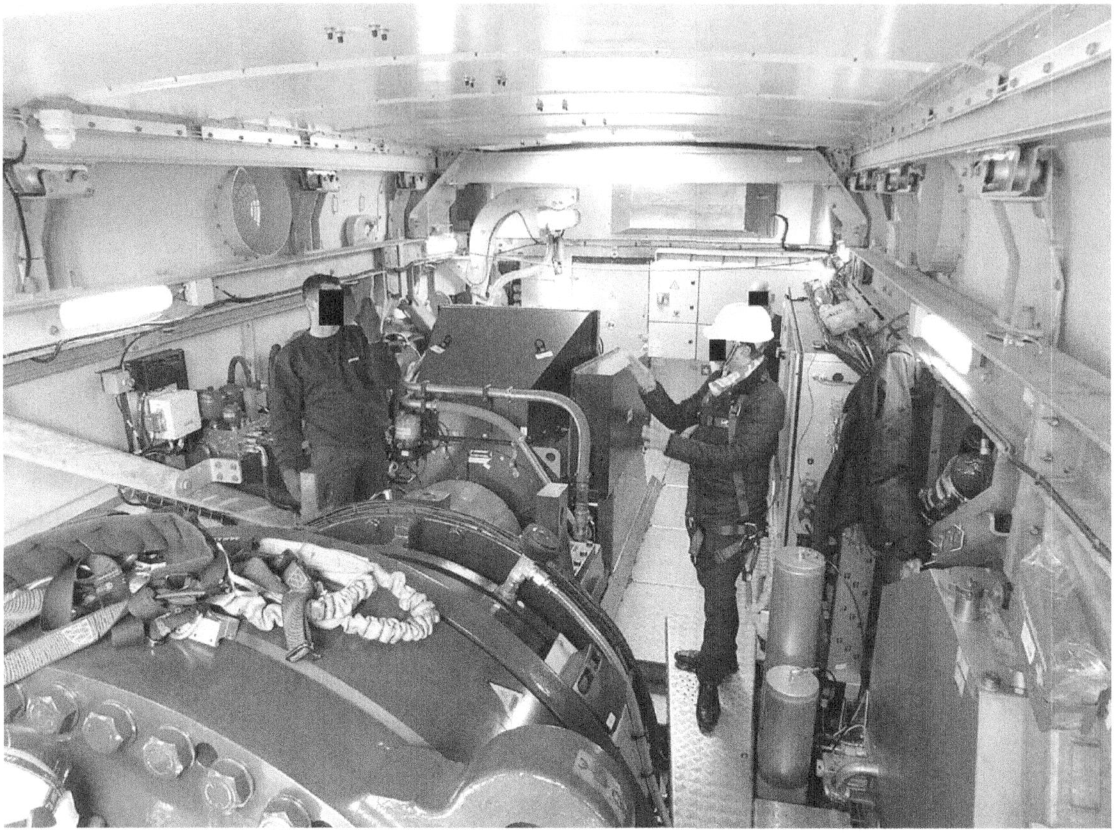

Figure 12: Vestas wind turbine generator house inside. Bundesstefan, CC BY-SA 4.0, via Wikimedia Commons.

An important part of the job is monitoring and diagnostics. Rather than relying only on scheduled checks, technicians analyse live data from sensors and control systems to spot anomalies. They use a range of specialised instruments and software to verify the condition of electrical, hydraulic and structural components and to record their findings for asset managers and regulators. This work underpins warranty compliance and supports long-term performance analysis.

When faults occur, windtechs move from monitoring to problem-solving. They isolate and correct failures in mechanical drives, electrical networks or hydraulic systems, calling in additional experts when a problem is especially complex. Because turbines are tall, enclosed and often located in harsh environments, much of this work takes place under demanding conditions and requires strict adherence to safety procedures.

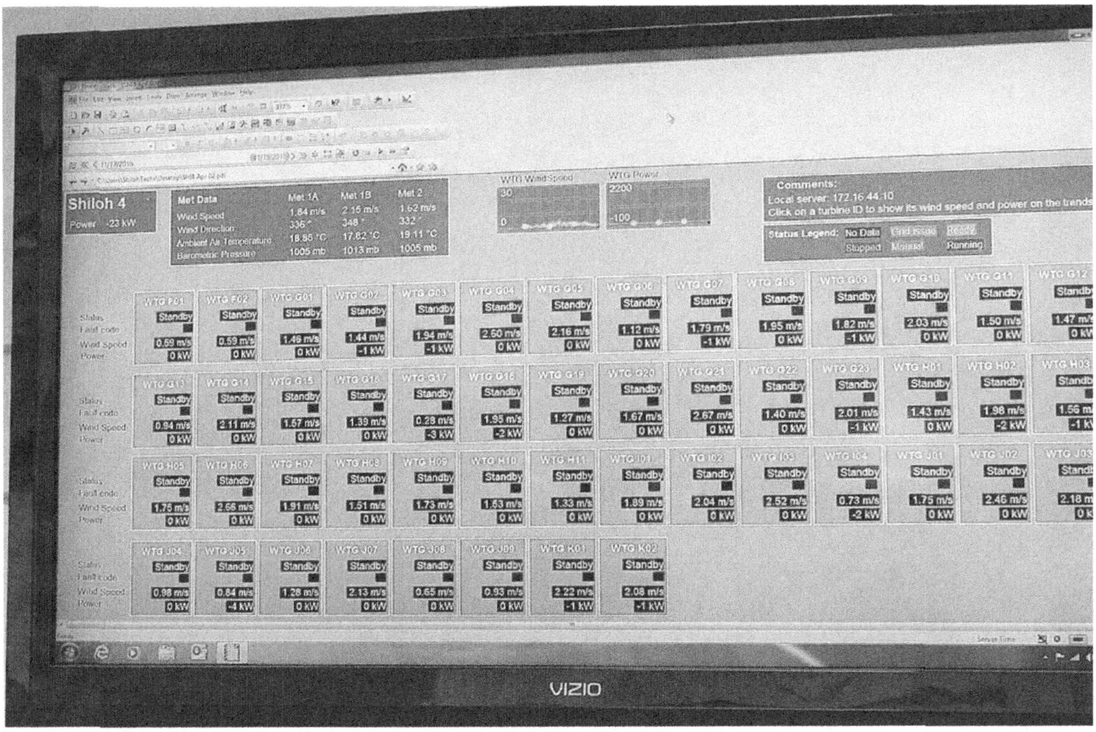

Figure 13: Video displays monitor energy output at turbine status at the Shiloh IV wind energy project in Solano County, Calif. Pacific Southwest Region USFWS from Sacramento, US, Public domain, via Wikimedia Commons.

Blades and towers also need ongoing attention. Technicians check the aerodynamic surfaces for erosion, cracking or lightning damage and verify that the tower structure remains sound. Depending on the severity of the issue, they may carry out minor repairs themselves or bring in specialist crews. Rope access and other advanced climbing techniques are often used to reach otherwise inaccessible areas safely.

Modern turbines are highly digital, and this has shifted the job from purely mechanical tasks to a mix of hands-on and data-driven work. Technicians now routinely interpret information from SCADA systems, deploy drones or robotic tools to inspect hard-to-reach components, and translate these findings into actionable maintenance plans. This trend towards predictive maintenance makes the role more analytical and less reactive than it once was.

Figure 14: Turbine Tower No 25 undergoing final testing. Technicians from the turbine manufacturer Nordex connected a laptop for the monitoring of more than 300 different parameters like temperature sensors, hydraulic sensors, pitch parameters, wind sensor set and power output. Paul Anderson, CC BY-SA 2.0, via Wikimedia Commons.

Because of the inherent hazards, every aspect of the job is underpinned by rigorous safety practice. Technicians maintain certifications, participate in rescue drills, and use protective equipment correctly, ensuring compliance with national and international standards. Equally important is accurate documentation: each inspection, repair or adjustment is logged in maintenance systems to support future planning and demonstrate compliance with contractual and regulatory requirements.

Mobility and adaptability are also key. Technicians may work across several wind farms or offshore platforms, adjusting schedules to match weather conditions and coordinating logistics to ensure that spare parts and tools are on hand. This mobility brings variety to the job but also requires resilience and good planning.

Finally, continual professional development is essential. As turbine designs evolve and new diagnostic technologies are introduced, technicians are expected to update their skills, gain additional certifications and, increasingly, learn to operate or oversee robotic devices. This ongoing learning not only keeps them effective in their current role but also opens pathways to supervisory, specialist or training positions within the industry.

Typical Day

While some wind turbine technicians contribute to the construction of new machines, the bulk of their work centres on keeping existing turbines in top condition. Their focus is on maintaining, troubleshooting, and repairing the sophisticated systems housed mainly in the nacelle—the compact unit at the top of the tower that contains the generator and related equipment [47].

In practice, this means technicians carry out a wide range of tasks. They examine the outside of towers to ensure structural integrity, climb to elevated work areas to investigate, diagnose, or fix faults, and collect operational data for testing, research, and analysis [47]. They carry out routine servicing, verify the performance of electrical, mechanical, and hydraulic components, and resolve malfunctions when they occur. Their remit also extends to underground cables, substations, and fibre-optic control systems, as well as replacing worn parts before they fail.

The timing and scope of maintenance depend mainly on operating hours and manufacturer specifications. Many turbine makers now recommend annual servicing that includes thorough visual inspections, torque checks on fasteners, and lubrication of moving parts. For machines running continuously, this level of work may be needed several times a year. Although turbines are monitored around the clock from central control rooms, technicians must still travel to site whenever a fault alert is raised to diagnose, repair, and restore normal operation.

A typical day involves scaling and examining multiple towers. Using computerised diagnostic tools, technicians identify electrical issues and schedule any required repairs. Because most maintenance work takes place within the nacelle—where drive-train components and sensitive electronics are tightly packed—they must be comfortable working in very confined spaces, including hubs and rotors [47]. Cleaning and lubricating shafts, bearings, and gears is routine, and so is using handheld power tools and measuring instruments to pinpoint electrical faults in control cabinets, generators, and other onboard systems [47].

Much of the job also happens outside on top of the nacelle. Technicians may need to replace instruments that measure wind speed and direction or coordinate with crane operators to swap out large components. Working hundreds of feet above the ground demands a heightened awareness of safety. Full-body harnesses are attached to rated anchor points, and movement is deliberate and cautious to prevent falls [47].

In addition to hands-on technical work, wind turbine technicians have administrative responsibilities. They may oversee a handful of turbines or a large section of a wind farm, order and track spare parts, manage inventory for future repairs, and coordinate day-to-day operations. This blend of physical, technical, and organisational duties keeps wind farms running smoothly and ensures that renewable energy keeps flowing into the grid [47].

Career Pathways and Certifications

The rapid expansion of wind energy worldwide has created a structured and increasingly professionalised career path for wind turbine technicians. What was once an emerging trade is now a global occupation with clear entry routes, specialist roles, and leadership opportunities. In every major wind market—from the United States and Australia to the UK, France, Germany, and Canada—technicians begin with vocational or technical training and build their expertise through internationally recognised safety and technical certifications. These credentials, such as those developed by the Global Wind Organisation (GWO), combined with national standards and apprenticeships, provide a solid foundation for advancement from entry-level maintenance work to specialist, supervisory, and even managerial positions in both onshore and offshore wind projects.

Becoming a wind turbine technician involves following a structured pathway of education, training, and skill development. It is a demanding career that calls for both specialised knowledge and practical competence, as technicians are responsible for maintaining and repairing complex, high-value equipment in often challenging conditions.

The first step is educational planning. Aspiring technicians need a strong academic base that combines theoretical understanding with hands-on skills. This usually involves choosing a field closely related to wind technology, such as electrical engineering, mechanical engineering, or renewable energy technology. While there is flexibility in specialisation, it is important to select a course of study that aligns with the needs of the renewable energy sector, as this will increase employability and ensure the skills gained are relevant to industry requirements.

Next comes enrolling in and completing formal training. Many institutions now offer courses tailored specifically to renewable energy and wind technology. These programs provide a rapid introduction to the technical, regulatory, and safety aspects of the industry, helping students build a foundation that prepares them for entry-level roles.

Hands-on experience is then essential. Wind turbine technicians work with tools and equipment every day, so practical experience is crucial. This might be gained through apprenticeships, internships, or on-the-job training at wind farms or renewable energy companies. Exposure to real-world maintenance and repair tasks allows future technicians to apply their classroom learning and develop confidence at height, in confined spaces, and in varied weather conditions.

Finally, aspiring windtechs must focus on developing key technical and professional skills. They need a strong grasp of electrical and mechanical systems, including hydraulics, because much of the job involves diagnosing and repairing faults. Problem-solving abilities are vital, as issues often arise unexpectedly and must be resolved quickly to keep turbines operational. Health and safety expertise is also critical, both for personal protection and to safeguard colleagues in high-risk environments. In addition, technicians must be comfortable with documentation—completing maintenance logs, writing reports, and accurately recording technical data—even if it is less glamorous than the hands-on work.

General Career Pathways in Wind Technology

Wind turbine technicians typically begin with technical or vocational training and progress through a structured career ladder. Common stages include:

- **Entry-level Technician** – performs basic inspections, preventive maintenance, and assists with installations.
- **Experienced/Lead Technician** – takes on more complex troubleshooting, trains new staff, oversees safety compliance.
- **Specialist Technician** – focuses on blade repair, high-voltage systems, SCADA/data analysis, or offshore operations.
- **Site Supervisor/Operations Manager** – manages a team of technicians, coordinates maintenance schedules, and liaises with asset owners.
- **Technical Instructor/Trainer** – delivers industry certification courses and safety training.
- **Transition to Engineering or Project Management** – experienced techs sometimes move into turbine commissioning, quality assurance, or renewable-energy project coordination roles.

At every stage, formal certification and ongoing training are essential for progression.

United States

Career Entry: Most U.S. wind technicians complete a 1–2 year Wind Energy Technology associate degree or certificate at community or technical colleges (e.g., Texas State Technical College, Iowa Lakes Community College). These programs teach mechanical, electrical, hydraulic systems, and climbing/rescue skills.

Certifications:

- Global Wind Organisation (GWO) Basic Safety Training (BST) – increasingly required by major operators even though it originated in Europe.
- OSHA training in fall protection, electrical safety, and confined spaces.
- First Aid/CPR certifications.
- Some technicians also earn NCCER (National Center for Construction Education and Research) credentials in related trades.

Career Pathway: Entry-level techs join as "Wind Turbine Service Technicians" and can advance to Lead Tech or Field Service Supervisor, and then into Wind Site Manager or O&M

(operations and maintenance) management roles. Some move laterally into commissioning, blade repair companies, or SCADA data analysis.

Australia

Career Entry: There is no single national "wind technician" qualification, but technicians usually hold:

- Certificate III in Electrotechnology Electrician (UEE30820) plus licensing through state electrical boards.
- Certificate IV or Diploma in Renewable Energy Engineering or Wind Energy Technology offered by TAFE or private RTOs.
- Many enter from related trades such as mechanical fitting or rigging.

Certifications:

- GWO Basic Safety Training (BST) and Basic Technical Training (BTT) are now widely adopted on Australian wind farms.
- Working at Heights, High Voltage Switching, and Confined Space courses accredited under Australian WHS standards.
- First Aid (HLTAID011) and emergency rescue training.

Career Pathway: Progression can lead to Senior Service Technician, Operations Coordinator, or Site Manager roles. Offshore wind is emerging (e.g., Star of the South), so Australian technicians with GWO offshore modules (Sea Survival, Manual Handling) will be in demand.

United Kingdom

Career Entry: Many UK wind techs come through:

- Apprenticeships in Wind Turbine Maintenance (e.g., at East Coast College or Grimsby Institute).
- NVQ/SVQ qualifications in mechanical/electrical engineering plus specialised wind modules.

Certifications:

- GWO BST, BTT, and Advanced Rescue Training (ART) – mandatory for most offshore work.
- IRATA Rope Access Certification for blade and tower work.

- Offshore work also requires BOSIET/FOET (Basic Offshore Safety Induction and Emergency Training) for helicopter transfer survival.

Career Pathway: Technicians start onshore or offshore as trainees, progress to Authorised Technician or Team Leader, and then to Wind Turbine Engineer, Offshore Coordinator, or Site Supervisor. Some move into health and safety or training roles within the UK's growing offshore sector.

France

Career Entry: French technicians often complete:

- BTS (Brevet de Technicien Supérieur) Maintenance des Systèmes or Électrotechnique.
- Licence Professionnelle Énergies Renouvelables with a wind specialisation.
- Industry-run training centres, e.g., GRETA or AFPA programmes for éolien (wind).

Certifications:

- GWO BST and BTT are now common for both onshore and offshore projects.
- CACES (Certificat d'Aptitude à la Conduite En Sécurité) for hoisting/lifting equipment.
- Travail en Hauteur (Working at Height) and Habilitation Électrique (electrical safety clearances).

Career Pathway: With France's growing offshore wind market (e.g., Saint-Nazaire, Fécamp), technicians can move from onshore maintenance to offshore O&M, supervisory positions, or project commissioning roles.

Germany

Career Entry: Germany has a well-established dual vocational training system. Prospective windtechs can take:

- Ausbildung zum Mechatroniker für Windenergieanlagen (Mechatronics Technician for Wind Turbines).
- Techniker für Windenergietechnik at specialised technical schools.

Certifications:

- GWO BST/BTT widely adopted.
- DGUV (German Social Accident Insurance) safety courses for working at height, high-voltage systems, and rescue.

- FISAT rope access certification.

Career Pathway: Germany's offshore sector (North Sea, Baltic Sea) offers clear routes from Service Technician to Lead Technician to Offshore Operations Manager. Some also specialise in SCADA diagnostics or grid-connection systems.

Canada

Career Entry: Canadian colleges offer programs such as:

- Wind Turbine Technician Diploma (Lambton College, Fanshawe College, Northern Lights College).
- Many electricians and millwrights transition into wind via additional training.

Certifications:

- GWO BST is increasingly recognised, especially for offshore projects in the Atlantic provinces.
- Working at Heights and Fall Arrest under Canadian OH&S regulations.
- Confined Space Entry, First Aid/CPR, and sometimes Rope Access.

Career Pathway: Technicians can progress to Lead Hand, Supervisor, or O&M Manager positions. Canada's large onshore wind base plus planned offshore developments will create demand for technicians with GWO offshore modules.

Chapter 2

Fundamentals of Wind Energy

Before a technician can service or operate a wind turbine, they must first understand the science that makes it work. Wind energy may appear simple—blades turning in the breeze—but beneath this motion lies a highly engineered process of aerodynamic design, mechanical transformation, and electrical generation. Chapter 2, *Fundamentals of Wind Energy*, explores this process from the ground up, explaining how moving air is converted into reliable electrical power through the integration of physics, materials, and control systems.

This chapter begins by breaking down the physical principles that govern wind behaviour and the mechanisms by which turbines extract energy from it. It traces the journey of energy conversion—from the kinetic energy of air currents to the mechanical torque of rotating blades, and finally to the production of electricity by sophisticated generators and control systems housed within the nacelle. Readers will gain a clear understanding of how aerodynamic lift and drag interact to turn a rotor, how turbines regulate speed and power output, and how design choices such as rotor diameter, tower height, and control systems influence performance and safety.

From there, the discussion moves into the two main turbine types—onshore and offshore—and compares their environments, structures, and maintenance requirements. This distinction is vital for technicians, as it determines the operational conditions, safety protocols, and skill sets required in each setting. Offshore work, for example, involves harsher weather, marine corrosion, and more complex logistics, while onshore installations demand adaptability to terrain and accessibility challenges.

The final section of the chapter examines the major components of a wind power plant, including the foundation, tower, nacelle, drivetrain, electrical systems, and blades. Each part plays a critical role in the overall performance, safety, and longevity of the turbine. Understanding how these systems function together allows technicians to diagnose problems, perform preventive maintenance, and ensure that turbines operate at peak efficiency.

By the end of this chapter, readers will have a foundational grasp of how wind energy systems work—knowledge that underpins every technical, safety, and maintenance practice covered in later chapters. It sets the stage for moving from theory to applied skill, transforming an understanding of wind into the ability to keep the world's clean energy machines turning.

How Wind Energy is Converted into Electricity

Wind turbines generate electricity by capturing the kinetic energy of moving air and converting it into mechanical rotation and then into electrical energy. The principle is the reverse of a fan. Instead of using electricity to turn blades and create wind, a wind turbine uses natural wind to turn blades and produce electricity. The movement of air results from three main factors acting together: uneven heating of the atmosphere by the sun, variations in the Earth's surface such as mountains, vegetation, and bodies of water, and the rotation of the planet itself. This motion energy has been harnessed for centuries for sailing ships or pumping water; today, it powers turbines to produce electricity.

Wind turbines are engineered to convert wind energy into electrical energy by harnessing the aerodynamic forces generated by their rotor blades. These blades function similarly to airplane wings, with air flowing over them creating a difference in air pressure that results in lift. This lift, which overcomes the drag force, causes the rotor to spin, ultimately driving a generator to produce electricity. Depending on the design, this connection can either be direct, as seen in direct drive turbines, or through a gearbox that increases the rotor's rotation speed to drive a smaller generator [48, 49].

Modern wind turbines are carefully designed to capture this energy efficiently. Most are horizontal-axis machines with three blades mounted on a tower between 50 and over 200 metres high. The blades are shaped like an aircraft wing so that when wind flows over them it creates a pressure difference—lower on one side and higher on the other. This produces lift, which is stronger than the drag acting on the blade, and causes the rotor to spin. Vertical-axis turbines also exist (for example, the eggbeater-shaped Darrieus), and although less common, they can accept wind from any direction without needing to be pointed into it.

Figure 15: Darrieus wind turbine near Heroldstatt-Ennabeuren, district Alb-Donau-Kreis, Baden-Württemberg, Germany. Dietrich Krieger, CC BY-SA 3.0, via Wikimedia Commons.

Wind turbines then, are categorized primarily into Horizontal Axis Wind Turbines (HAWTs) and Vertical Axis Wind Turbines (VAWTs), each with distinct advantages and design characteristics suited to specific environmental conditions and energy requirements [50, 51]. HAWTs, for example, are commonly used due to their efficiency at higher wind speeds, whereas VAWTs are often more adaptable for various local wind conditions [51]. Continuous advancements in turbine technology and designs seek to enhance efficiency further, making wind energy a critical component of the global energy landscape [52].

Figure 16: Gearbox , Rotor Shaft and Disk Brake Assembly, horizontal access wind turbine. Paul Anderson, CC BY-SA 2.0, via Wikimedia Commons.

As the blades rotate, they turn a central hub connected to a main shaft inside the nacelle—the streamlined housing at the top of the tower. This is where the drive train, gearbox, generator, and control systems are located. Because the rotor spins relatively slowly—usually 5 to 25 revolutions per minute for large turbines—the shaft connects to a gearbox that multiplies this speed to the 1,000–2,000 rpm needed by a conventional generator. This allows a physically smaller generator to produce significant power. Newer "direct-drive" designs avoid the gearbox altogether, using a large low-speed generator that can operate at variable speeds.

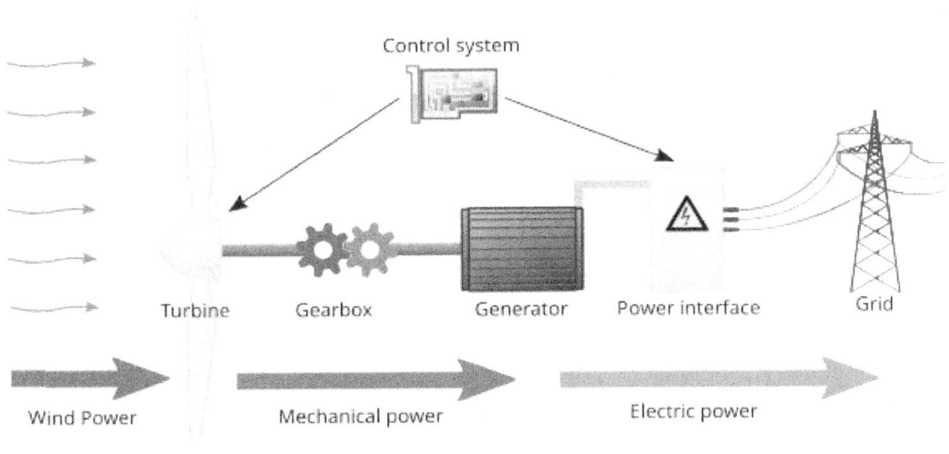

Figure 17: Functional schematic of the structure of a wind turbine. Jalonsom, CC BY-SA 3.0, via Wikimedia Commons.

The operational effectiveness of wind turbines is largely influenced by the wind conditions and geographic placement of wind power plants. Essential factors include wind speed, terrain, and proximity to power transmission networks, all of which play critical roles in optimizing energy capture and generation. Utility-scale wind farms comprise arrays of turbines that collectively feed into substations, where the harnessed electricity is then transferred to the power grid to supply communities [53, 54].

Inside the generator, the high-speed shaft moves magnets past coils of wire, creating a changing electromagnetic field. This induces an electric current in the coils, converting the mechanical energy from the rotor into electrical energy. The electricity produced is alternating current (AC). It travels down the tower through cables to a transformer at the base, which steps up the voltage for efficient transmission. From there it is sent by underground cables to a substation and integrated into the larger grid to power homes, businesses, and industries.

Several control systems ensure the turbine operates safely and efficiently. Motors and yaw systems at the top of the tower rotate the nacelle so the blades always face into the wind. Pitch systems adjust the angle of the blades to regulate rotor speed and optimise power output. Turbines typically begin generating at about 10 km/h wind speed and shut down automatically at very high winds (above about 90 km/h) to prevent damage.

Figure 18: Pitch system for controlling the angle of the three blades; this consists of three independent pitch gears driven by the blue electrical motors. Paul Anderson / Interior of the hub of Turbine No 3, CC BY-SA 2.0, via Wikimedia Commons.

The size of a turbine directly affects its capacity. Modern machines have blade diameters of 80 to over 200 metres, with taller towers accessing stronger, steadier winds at higher altitudes. Longer blades sweep a larger area, capturing more energy and producing more electricity per rotation. Offshore turbines, which benefit from stronger and more consistent winds, can now reach capacities of 8–15 megawatts per unit.

A wind turbine's actual electricity production depends on two key factors: the speed of the wind at the site and the rated power of the turbine. The rated power is the maximum output the turbine can produce under ideal conditions. For most large machines, this "sweet spot" occurs at about 15 metres per second (33 mph). At this speed, the rotor spins at its optimal rate, the generator runs at its designed capacity, and the turbine delivers its nameplate (rated) output. If the wind is slower, the turbine produces less power because the energy available in the wind drops with the cube of wind speed. If the wind is faster, protective systems limit output or shut the turbine down to prevent damage.

The diameter of the rotor—the circle swept by the blades—largely determines how much energy a turbine can capture. This is because the area of the circle increases with the square of the radius, so a small increase in blade length yields a large increase in swept area and therefore potential energy capture. The table you provided illustrates this scaling effect: a rotor diameter of 10 metres produces only about 25 kW, while a diameter of 80 metres can produce 2,500 kW (2.5 MW). As rotor size increases, towers also become taller to support the longer blades. The extra height allows access to stronger, steadier winds at higher altitudes, further boosting energy output.

Turbines are designed to operate safely within a specific wind-speed range. Most large turbines achieve their rated power at about 33 mph but will automatically shut down at around 45 mph (20 m/s) to avoid mechanical stress or structural damage. Several safety systems can trigger a shutdown. One simple device uses a metal ball attached to a chain perched on a small pedestal; excessive vibration dislodges the ball, which pulls the chain and signals the controller to stop the turbine.

The most common safety mechanism is the braking or power-control system. When wind speeds exceed the safe threshold, the controller "applies the brakes," reducing the rotor's speed and therefore the power output. When the wind subsides below the threshold, the brakes are released and normal operation resumes. Modern turbines typically use one of three braking or power-control strategies:

- **Pitch Control:** The controller constantly monitors power output and, at high wind speeds, commands the blades to change their pitch (the angle at which they meet the wind). By pitching the blades out of alignment with the wind, the rotor slows down and power output drops to a safe level. This requires blades that can rotate about their longitudinal axis.
- **Passive Stall Control:** In this simpler system, the blades are fixed but shaped so that above a certain wind speed the airflow over them becomes turbulent on the upwind side. This aerodynamic "stall" greatly reduces lift and slows the rotor automatically without moving parts.
- **Active Stall Control:** This hybrid system uses pitchable blades like a pitch-control system but, instead of pitching them out of the wind, the controller pitches them to deliberately induce stall. This provides more precise control over how much the rotor slows and how much power is produced at high wind speeds.

These design features ensure that a turbine captures as much energy as possible when conditions are optimal but automatically protects itself when conditions become extreme. Rotor size, tower height, wind speed, and control systems all work together to maximise output while safeguarding the structure and extending its service life.

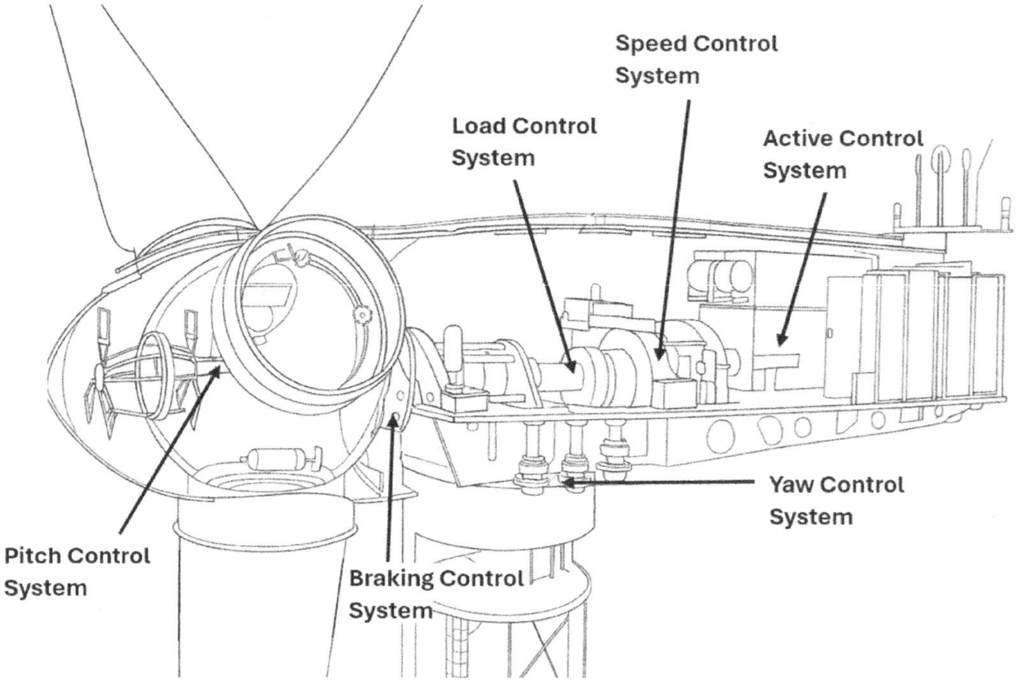

Figure 19: Wind Turbine Control Systems.

In essence, a wind turbine is an elegant chain of energy conversions: moving air creates aerodynamic lift; lift turns the rotor; the rotor spins a shaft; the shaft drives a generator; and the generator produces electricity that flows through transformers and cables into the grid. This combination of aerodynamic design, mechanical engineering, and electrical generation allows a single modern turbine to supply thousands of homes with clean energy.

Types of Wind Turbines (Onshore Vs Offshore)

The deployment of wind turbines is fundamentally categorized into two major types based on their installation location: onshore and offshore wind turbines. Each type operates within distinct environmental contexts, thus shaping their design, operational parameters, and technological requirements.

1. Location and Environment: Onshore wind turbines are typically installed on land, often in elevated or open areas where wind resources are abundant and easily accessible. These locations may include coastal regions, benefiting from strengthened wind patterns. In contrast, offshore wind turbines are positioned in bodies of water, including oceans, seas, and large lakes, and can be fixed to the seabed or mounted on floating platforms in deeper waters. This geographic distinction is crucial, as it influences the turbines' performance capabilities and the challenges they face in their respective environments. Offshore installations are increasingly pursued due to their potential to harness stronger and more consistent winds, making them a promising avenue for the future of wind energy development [14, 55].

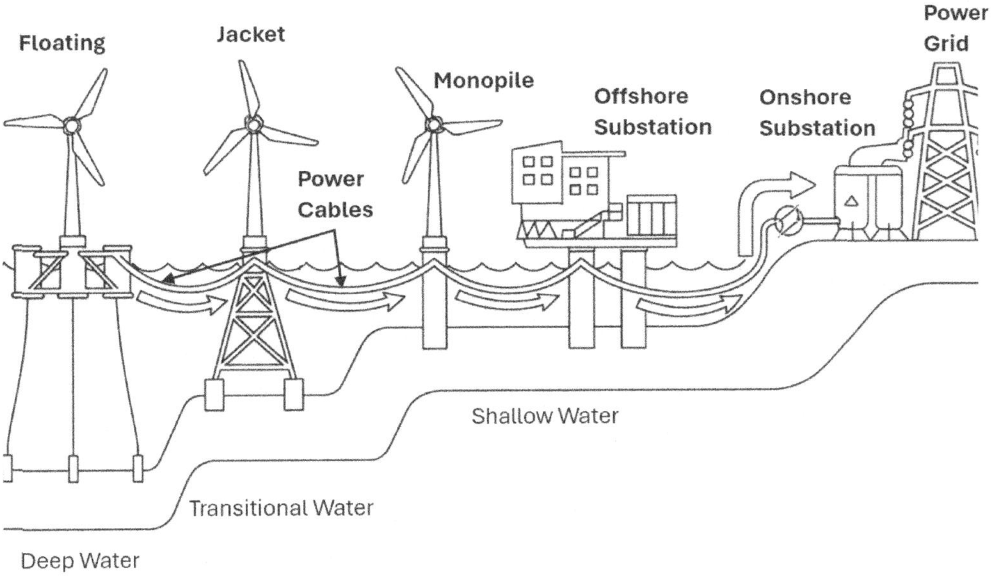

Figure 20: Offshore wind farm with turbine stations at sea or ocean.

2. Design and Size: The design specifications of onshore and offshore turbines differ significantly, particularly in rotor size and electrical capacity. Onshore turbines typically feature rotor diameters ranging from 80 to 140 meters, with a rated capacity of about 2 to 6 megawatts (MW) per unit, primarily constrained by transport limitations and local regulations regarding noise and visual impact [56]. Offshore turbines are built to a much larger scale, often exceeding rotor diameters of 200 meters and capacities reaching up to 15 MW or more per unit—a capacity made feasible by maritime logistics, alleviating some of the constraints faced by onshore turbines [57, 58]. This larger design captures more wind energy and positions offshore farms as pivotal assets for renewable energy strategies aimed at high-output generation.

3. Structural Variations: The structural components of onshore and offshore wind turbines also diverge substantially. Onshore turbines typically consist of tubular steel towers ranging from 80 to 120 meters tall, and sometimes utilize hybrid concrete-steel materials for enhanced durability [59]. The foundations for onshore turbines are primarily reinforced concrete pads or piles suited to local soil conditions. Conversely, offshore structures must contend with harsher marine environments (including high winds and saltwater corrosion) and often exceed 150 meters in hub height. They employ foundations like monopiles, gravity bases, or floating systems tailored for deeper waters, designed to endure the combined effects of waves, currents, and wind loads [60, 61].

4. Operation and Maintenance: Operational challenges faced by onshore versus offshore turbines underscore their differences. Onshore turbines allow for straightforward access, facilitating routine maintenance due to the ease of reaching the sites via vehicle [62]. Offshore maintenance requires specialized vessels or helicopters and is heavily influenced by marine weather conditions, complicating scheduling and increasing operational costs [63]. The remoteness of offshore sites necessitates advanced supervisory control and data acquisition (SCADA) systems to monitor conditions and manage maintenance needs efficiently [64].

5. Performance and Economic Factors: In terms of performance, onshore turbines typically achieve capacity factors of around 25–40%, influenced by variable wind speeds and local climatic conditions, while offshore turbines can reach factors of 40–60% due to more favourable and consistent wind conditions [65]. Economically, although onshore wind turbines are generally less expensive to install per megawatt, offshore turbines tend to provide significantly higher energy output, justifying their higher installation and maintenance costs. This balance of cost and output efficiency highlights the complementary roles of onshore and offshore technologies within the broader context of renewable energy generation [66, 67].

Table 1: Onshore and Offshore Wind Turbine Comparison Summary.

Feature	Onshore Wind Turbines	Offshore Wind Turbines
Typical Capacity	2–6 MW	8–15 MW (17 MW prototypes)
Rotor Diameter	80–140 m	150–220 m+
Tower Height	80–120 m	120–150 m+
Foundation Type	Concrete pad or piles	Monopile, jacket, gravity-base, floating
Environment	Land-based winds, temperature swings	Saltwater, waves, higher, steadier winds
Grid Connection	Direct to local substation	Offshore substation, HVAC or HVDC to shore
Access for O&M	Road vehicles	Vessels, helicopters, marine safety training
Capacity Factor	25–40 %	40–60 %

The differences between onshore and offshore turbines shape the daily work and career pathways of wind turbine technicians in significant ways. The environment in which a technician operates determines not only how they access the turbines but also the skills, certifications, and maintenance strategies they need to master.

Onshore and offshore settings create very different day-to-day conditions for technicians. Onshore windtechs usually drive to turbines by road and can plan routine maintenance with relative flexibility. Tools, spare parts, and crew support are readily available, and adverse weather rarely makes a site inaccessible for extended periods. By contrast, offshore technicians work in a marine environment where access is limited by sea state, tides, and weather windows. They often travel by crew-transfer vessels, service operation vessels, or helicopters and may live offshore for days or weeks. This demands additional training in marine survival, offshore emergency response, and rope access, as well as a higher tolerance for working in isolated and unpredictable conditions.

The much larger size of offshore turbines also affects the scope of maintenance tasks. Offshore units can exceed 200 metres in rotor diameter and 150 metres in hub height, meaning technicians work with larger components, higher voltages, and more complex systems such as dynamic cables, offshore substations, and high-voltage direct current (HVDC) export links. Onshore turbines, with rotor diameters of 80–140 metres and capacities of 2–6 MW, generally have simpler grid connections and less demanding lifting operations. This difference means offshore work often involves heavy-lift coordination, working with large cranes, or using specialised blade- or gearbox-replacement tools at sea, whereas onshore maintenance remains more straightforward and routine.

Because of these differences, offshore technicians require a broader skill set and more credentials than their onshore counterparts. Onshore wind techs typically hold electrical, mechanical, or renewable energy qualifications plus safety training in working at heights, first aid, and confined space entry. Offshore technicians, however, must complete Global Wind Organisation (GWO) Basic Safety Training and Sea Survival modules, obtain marine-safety certificates such as BOSIET/FOET, and often hold rope access certification (IRATA or FISAT) for blade work. They also need familiarity with offshore-specific SCADA systems and remote monitoring technology. This creates a career pathway in which technicians may start onshore and later upgrade training and certifications to qualify for offshore work, which usually commands higher pay but involves greater risk and responsibility.

Maintenance strategies also diverge between the two environments. Because offshore access is difficult and costly, maintenance is more data-driven and proactive. Offshore turbines rely heavily on SCADA, condition monitoring, and predictive maintenance software to identify faults before they escalate, requiring technicians to interpret digital diagnostics and plan interventions efficiently. Onshore turbines also use SCADA but can tolerate more reactive maintenance since crews can reach them quickly and cheaply. This shifts offshore technicians' roles toward remote monitoring, pre-mission planning, and precision troubleshooting to minimise time spent on site.

Safety demands are elevated offshore as well. In addition to fall protection, offshore technicians must be trained for helicopter transfers, sea rescue, and evacuation procedures, and they must be comfortable working at greater heights and in confined nacelles exposed to salt spray, humidity, and vibration. Emergency response options are more limited at sea, so technicians must be self-reliant and strictly follow protocols to prevent incidents. Onshore technicians still face hazards from heights, electricity, and heavy components but have faster access to ground-based rescue services.

These contrasts also shape career development. Many technicians begin onshore, gaining experience with mechanical, electrical, and hydraulic systems, and then progress to offshore roles once they accumulate experience and offshore-specific certifications. Offshore positions often offer higher salaries and faster advancement to supervisory roles but demand longer shifts away from home. Over time, technicians can specialise in blade repair, high-voltage systems, SCADA/data analysis, or move into site management, training, or engineering support roles.

In practice, the onshore/offshore distinction means that wind turbine technicians are not doing one uniform job. Onshore work is more accessible, less logistically complex, and focused on routine maintenance and troubleshooting. Offshore work involves larger machines, stricter safety and certification requirements, advanced digital monitoring, and working in remote, challenging conditions. These factors shape both the skills technicians must develop and the way their careers progress within the rapidly expanding wind energy industry.

Major Components of a Wind Power Plant

Modern wind turbines are complex machines that integrate mechanical, electrical, and digital systems to convert the kinetic energy of wind into usable electricity. Although they look simple from a distance, each turbine contains dozens of specialised components working together to ensure safe, efficient, and reliable operation. At the base, the foundation anchors the structure to the ground or seabed, while the tower elevates the rotor and nacelle into stronger, steadier winds. Inside the tower are ladders, lifts, cables, and control equipment that give technicians access to the upper components.

At the top of the tower sits the nacelle, a streamlined housing that contains the turbine's drivetrain and control systems. Here, the main shaft, gearbox, generator, and braking systems convert the slow rotation of the blades into high-speed electrical energy, while sensors and monitoring equipment constantly feed data to the control system. The rotor blades, attached to a central hub, are shaped like aircraft wings to capture the wind efficiently. Their pitch control systems adjust blade angles to regulate power output and protect the turbine in strong winds.

Around these core elements are other essential systems: the yaw mechanism turns the nacelle to face the wind, anemometers and wind vanes measure wind speed and direction, and transformers and cables carry electricity down the tower to the grid. Together, these components form a highly engineered system designed to maximise energy capture, withstand harsh environments, and provide decades of safe, clean power generation.

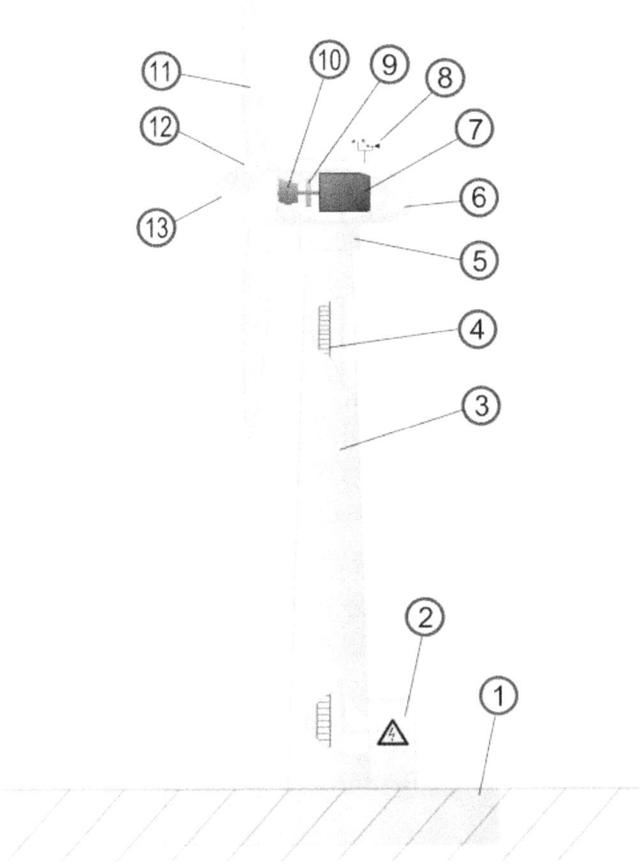

Figure 21: Components of a typical Land-based Horizontal Axis Wind Turbine Arne Nordmann (norro), CC BY-SA 3.0, via Wikimedia Commons.

Illustration of a wind turbine.

1. Foundation
2. Connection to electric grid
3. Tower
4. Access ladder
5. Wind orientation control
6. Nacelle
7. Generator
8. Anemometer
9. Brake
10. Gearbox
11. Rotor blade
12. Blade pitch control

13. Rotor hub

1. Foundation

The foundation anchors the entire turbine to the ground or seabed and transmits structural and dynamic loads safely into the supporting medium.

- Onshore foundations are typically reinforced concrete gravity pads or piles. Gravity pads spread the load over a large footprint; piles transfer loads into deeper, stronger soil or rock strata.
- Offshore foundations include monopiles (large steel cylinders driven deep into the seabed), jacket/tripod lattice structures for deeper waters, and gravity bases. For water depths beyond about 60–70 m, floating platforms such as semi-submersible, spar-buoy, or tension-leg platforms are used with mooring lines.
- Foundations are designed to resist not only the static weight of the tower and nacelle but also wind thrust, rotor torque, dynamic vibrations, and—in offshore installations—wave and current forces.

Figure 22: Wind Turbine Foundation, Goshen North Wind Farm, Idaho Falls. Idaho National Laboratory, CC BY 2.0, via Wikimedia Commons.

2. Connection to the Electric Grid

Electric power generated in the nacelle is stepped up and transmitted to the grid.

- High-voltage cables run from the generator and control cabinets in the nacelle down the tower to a transformer, usually at the base or in an external substation.
- Onshore turbines connect via underground or overhead cables to a local substation, where voltage is further increased for regional transmission.
- Offshore turbines transmit power to an offshore substation. For export distances up to ~60–80 km, high-voltage alternating current (HVAC) cables are used. Beyond that, high-voltage direct current (HVDC) systems are preferred to reduce losses. Offshore substations also house transformers, switchgear, and reactive power compensation equipment.
- Modern turbines include power-electronic converters to match grid frequency and comply with grid codes.

3. Tower

The tower elevates the rotor and nacelle into stronger, less turbulent winds.

- Most utility-scale towers are tapered tubular steel segments bolted together on-site, ranging from 80–150 m tall for onshore and up to 170 m for offshore.
- Some designs use hybrid concrete-steel towers for extra stiffness or lattice towers for smaller applications.
- Towers are designed to handle axial, bending, and torsional loads and often include internal platforms, lifting rails, and vibration dampers.
- They also house cables, control panels, and sometimes transformers.

A wind turbine's tower is the tall structure that elevates the nacelle and rotor to the height where wind is faster and less turbulent. Because wind speed increases with altitude (due to reduced ground friction), the rotor must be placed high enough to achieve a relatively stable wind flow. Without sufficient height, the turbine's energy yield drops dramatically.

A general design guideline is that the tower height should be roughly 1 to 1.5 times the rotor diameter. For example, a rotor 100 m across would typically be mounted on a tower 100–150 m tall. Small or micro turbines often have proportionally higher towers relative to their rotor size to compensate for ground-level turbulence and obstacles such as trees or buildings.

The tower must support not only the weight of the nacelle, gearbox, generator, and rotor but also resist dynamic loads from wind gusts and rotor torque. This demands high-strength materials and careful structural design. However, towers are also a major cost element in a wind turbine, so engineers must balance stiffness, fatigue resistance, transportability, and price.

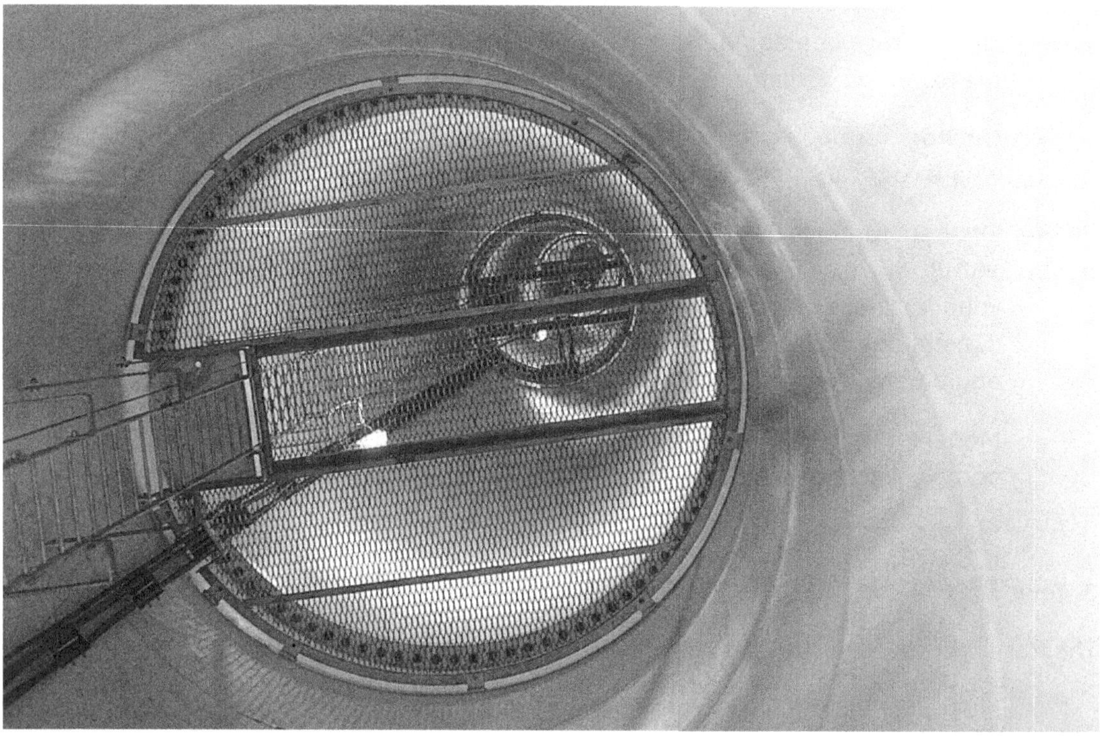

Figure 23: Looking Up The Wind Turbine Tower, Byron, Wisconsin. Michael Pereckas, CC BY 2.0, via Wikimedia Commons.

Different tower designs are used depending on turbine size:

- Micro Wind Turbines: Typically use a single iron or steel pipe mast stabilized by guy cables (tensioned wires anchored to the ground). This simple design is inexpensive and easy to erect but unsuitable for large loads.
- Small and Medium Turbines: Often use truss towers (also called lattice towers) made of angle steel or other profiles. These are low-cost, lighter to transport, and can be climbed externally. However, they are visually less appealing and can present safety issues for the public because of exposed ladders.
- Large Utility-Scale Turbines: Almost exclusively use tubular towers—cylindrical steel or concrete columns. Tubular towers provide an enclosed space for internal access and wiring, reduce aerodynamic drag, and look cleaner on the landscape. Steel tubular towers are prefabricated in sections and bolted together on site, but transporting very large steel sections is challenging. This has led to the increasing use of reinforced concrete towers, which can be cast or assembled on site to achieve greater heights without oversized loads.

Figure 24: The base of a wind turbine tower in the Causeymire Wind Farm. david glass / Wind Turbine Base, CC BY 2.0, via Wikimedia Commons.

4. Access Ladder

Inside the tower is an access ladder system allowing technicians to climb from the base to the nacelle.

- Ladders are fitted with vertical lifelines, fall-arrest systems, and safety rails.
- Many turbines now incorporate climb-assist devices—motorised hoists that reduce effort—or full service lifts for personnel and small parts.
- Ladders serve as both primary and emergency access.

Figure 25: Inside Turbine 2, Westmill Wind Farm, Watchfield, showing access ladder. Brian Robert Marshall, CC BY-SA 2.0, via Wikimedia Commons.

5. Wind Orientation Control (Yaw System)

The yaw system keeps the rotor facing into the wind for maximum energy capture.

- It consists of electric yaw motors, gear reducers, and a large yaw bearing between the nacelle and tower flange.

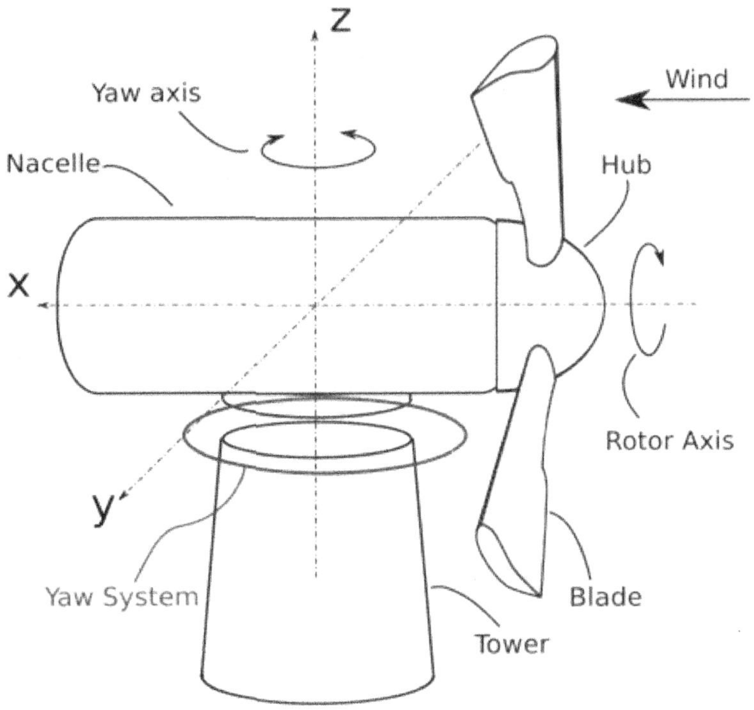

Figure 26: Yaw system location, Hanuman Wind ,CC BY-SA 3.0, via Wikimedia Commons.

- A wind vane and anemometer on top of the nacelle provide direction data to the turbine controller, which commands the yaw motors to rotate the nacelle horizontally.
- Once aligned, yaw brakes lock the nacelle in position to prevent unwanted rotation and reduce bearing wear.

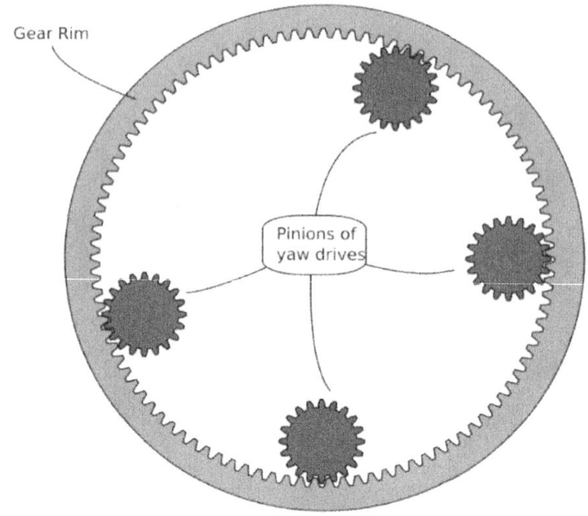

Figure 27: Gear rim with yaw drives. Hanuman Wind at en.wikipedia, CC BY-SA 3.0, via Wikimedia Commons.

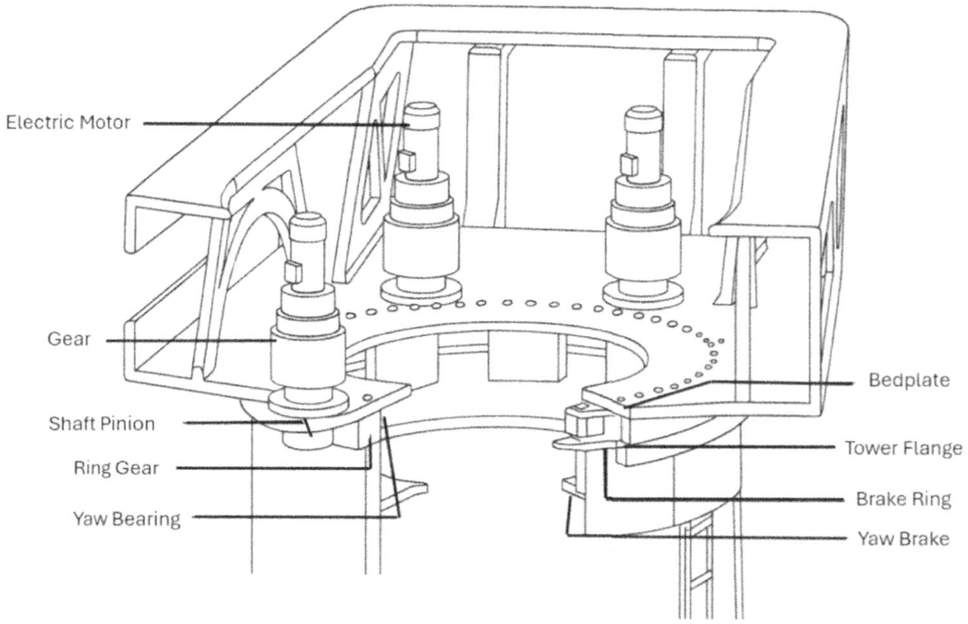

Figure 28: Yaw System Example.

Wind Turbine Fundamentals for Technicians

Modern wind turbines look sleek and simple from a distance, but behind the turning blades lies a complex set of control systems designed to extract as much energy as possible from constantly changing winds. One of the most critical of these systems is the yaw control system, which acts much like a steering mechanism for the turbine. If you imagine a traditional windmill turning its sails to face the breeze, you already understand the basic idea. However, modern turbines take this principle to a much higher level of sophistication because of their scale, cost, and the demands of grid-connected electricity generation.

"Yaw" is the technical term for rotation of the entire nacelle (and the rotor attached to it) around the turbine's vertical axis. Because the wind direction shifts frequently, a turbine that stays fixed in one position would quickly become misaligned with the oncoming flow. Misalignment reduces aerodynamic lift on the blades, cuts power output, and places uneven loads on the rotor. The yaw control system prevents this by constantly adjusting the nacelle so that the rotor faces directly into the wind. In effect, it is the turbine's "steering wheel," ensuring that the blades present their full swept area to the oncoming air and maximising energy capture.

This active alignment is economically as well as technically significant. Even small yaw errors—five or ten degrees—can cause measurable drops in power production and accelerated wear on structural components. Over the life of a wind farm, poor yaw control translates directly into lost revenue and reduced return on investment. Reliable, precise yaw control is therefore a cornerstone of both the technical performance and the business case for modern wind energy.

A typical yaw system comprises several interacting components. Wind direction sensors—usually anemometers and wind vanes mounted on top of the nacelle—continuously measure the incoming wind speed and direction. Their signals feed into a central control unit that acts as the system's brain. When a change in wind direction is detected, the controller commands the yaw drive, a set of electric or sometimes hydraulic motors coupled to reduction gearboxes, to rotate the nacelle about the yaw bearing, a large-diameter slewing bearing designed to carry the combined weight of the nacelle and rotor and resist enormous wind loads. This bearing allows smooth rotation and distributes forces evenly so that the yaw motors can make fine adjustments without excessive wear. The interplay of sensors, controller, drive, and bearing enables the turbine to track wind shifts dynamically.

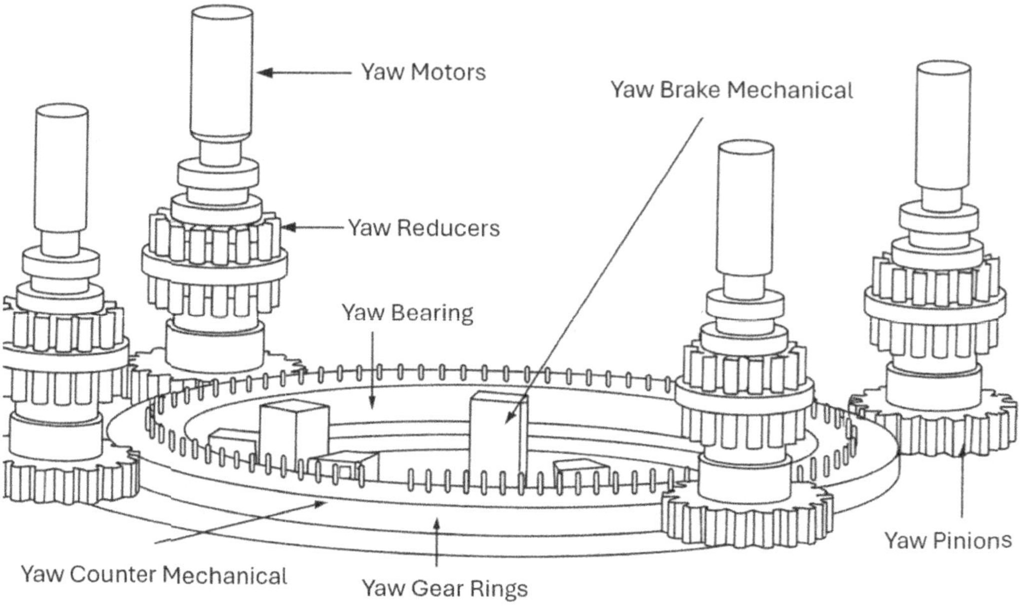

Figure 29: Wind Turbine Yaw System.

The car-on-a-winding-road analogy is apt: just as a driver constantly makes small steering inputs to stay in lane, the yaw control system continually "steers" the nacelle to follow the changing wind. Large, grid-scale turbines use active yaw control, which means powered motors actively rotate the nacelle based on sensor input. Some small or older machines use passive (free) yaw, where the rotor and nacelle align themselves aerodynamically without motors. Passive yaw is mechanically simpler but much less precise and therefore unsuitable for today's multi-megawatt machines where maximising yield is critical.

At a deeper engineering level, yaw systems are highly specialised. The drives must produce very high torque to rotate hundreds of tonnes of nacelle and rotor against wind loads, yet do so smoothly to avoid shock loading. Large turbines may use multiple yaw motors working together or even hydraulic systems for greater torque capacity, although hydraulic designs bring their own maintenance considerations such as potential fluid leaks. Yaw bearings are equally demanding, often several metres in diameter with hardened raceways and high-capacity rollers, designed to cope with both axial and radial loads over decades of service. Even small increases in bearing friction can lead to higher energy losses and faster wear, so material choice, surface treatment, and precision manufacturing are critical.

Control strategies have also evolved beyond simple feedback loops. Basic proportional–integral–derivative (PID) controllers work well in steady winds but can oscillate or lag in turbulent conditions. Advanced systems incorporate predictive algorithms, sometimes drawing on wind forecasting models or using machine learning to anticipate directional changes and pre-emptively adjust yaw angle. These approaches improve energy capture and reduce the frequency of yaw movements, which in turn lowers mechanical stress and extends component life.

Understanding yaw misalignment highlights why this matters. Persistent misalignment not only reduces output but can create asymmetric loading on blades, hubs, and main bearings, accelerating fatigue and increasing maintenance costs. For operators, optimising yaw control therefore improves both immediate revenue and long-term reliability.

The yaw control system is far more than a simple turning mechanism. It is an integrated suite of sensors, heavy-duty mechanics, and intelligent control software that constantly aligns a wind turbine with the wind, maximising aerodynamic efficiency and protecting the machine from unnecessary stress. Active yaw control has become the standard for modern turbines because it enables high-precision tracking of variable winds, ensuring that wind energy remains economically viable and technically reliable as a cornerstone of the global shift to sustainable power.

6. Nacelle

The nacelle is the streamlined housing at the top of the tower that contains most of the drivetrain and control systems.

- Inside are the main shaft, gearbox, generator, brake system, cooling systems, and control electronics.
- It includes internal walkways, lifting equipment, and sealed enclosures to protect components from dust, moisture, and—in offshore cases—salt spray.
- The nacelle rotates with the yaw system and is often pressurised or dehumidified to safeguard sensitive electronics.

Figure 30: Nordtank NTK 1500 wind turbine nacelle in Gammelby (Germany). Windkraftfan SH, CC BY-SA 4.0, via Wikimedia Commons.

The frame or chassis forms the structural backbone of the nacelle. Built from heavy-duty transverse beams and longitudinal girders, it creates a rigid platform on which all major components—including the gearbox, generator, brake system, and auxiliary equipment—are mounted. This design must withstand vibration, torsion, and the combined loads transmitted by the rotor and drivetrain, ensuring that forces are evenly distributed and safely transferred to the tower below.

At the front of the nacelle, the main shaft (or spindle) links the rotor hub to the gearbox, transmitting the rotational torque produced by the blades. This shaft is supported by large bearings designed to absorb axial thrust from the wind pushing on the rotor, radial loads from the weight of the rotor and gearbox, and torsional forces from the rotor torque itself. Most turbines use a front bearing at the hub to take the majority of these forces and, in some cases, a rear bearing to support the tail end of the shaft. Spherical roller bearings are common because they tolerate misalignment and reduce stress concentrations. In other designs, the low-speed gearbox input couples directly to the shaft without a rear bearing.

Most utility-scale turbines include a gearbox that steps up the slow rotation of the main shaft—typically 10–25 rpm—to the much higher speed required by the generator, usually between 1,000 and 1,800 rpm. The gearbox's high-speed shaft connects to the generator through a flexible coupling, which absorbs small misalignments and dampens torsional vibrations.

The generator, mounted on the same frame as the gearbox, converts mechanical energy into electricity. To stop the rotor safely during maintenance, faults, or high winds, brake discs are mounted on the generator-side shaft. Brake callipers clamp onto these discs to bring the rotating shaft to a halt. This braking system works in tandem with aerodynamic braking—achieved by feathering the blades—to ensure safe shutdown of the turbine.

Running through the main shaft is a hollow passage that allows hydraulic lines, power cables, and signal cables for the pitch system to pass from the nacelle controller to the hub's pitch actuators. These lines are routed through a rotary slip-ring assembly that transmits both power and data between the stationary nacelle and the rotating hub. Some turbine designs instead use mechanical connecting rods to actuate blade pitch directly.

The main shaft is usually installed with a slight upward tilt, just a few degrees above horizontal. This tilt increases blade clearance over the tower to prevent blade strikes and reduces the overhang length of the shaft. This subtle design feature improves safety and lowers bending stresses on the drivetrain.

Because gearboxes and generators operate under heavy loads, the nacelle houses dedicated lubrication and cooling systems. Gearboxes use pumps, filters, and heat exchangers to circulate and cool oil, while large generators rely on air-to-air or liquid-to-air cooling systems to prevent overheating. These systems are essential for reliability and to extend the life of major components.

Beneath the nacelle sits the yaw system, a ring of gears powered by electric motors or hydraulic drives that rotates the nacelle so the rotor always faces the wind. At the very top of the nacelle, an anemometer and wind vane continuously measure wind speed and direction. Their signals go to the control cabinet inside the nacelle, which commands the yaw motors to adjust the turbine's orientation and the pitch actuators to adjust blade angles for optimal performance.

Some turbines use a direct-drive configuration, eliminating the gearbox entirely. In these machines, the rotor shaft connects directly to a large, low-speed generator with many poles. This reduces mechanical complexity, eliminates gearbox-related losses, and lowers maintenance requirements, although it produces a heavier nacelle and a larger-diameter generator.

Inside the nacelle, all of these systems function as an integrated unit. The frame secures the drivetrain, the bearings support massive loads, the gearbox and generator transform slow rotor motion into usable electricity, and the braking, yaw, and pitch systems maintain safe and efficient operation in changing wind conditions. Lubrication, cooling, and slip-ring assemblies provide the support needed for continuous, high-reliability power generation.

7. Generator

The generator converts the mechanical rotation of the rotor into electrical energy.

- Gearbox turbines use high-speed generators (1,000–1,800 rpm), typically doubly-fed induction or synchronous machines with power converters.
- Direct-drive turbines use low-speed, large-diameter permanent-magnet generators that eliminate the gearbox.
- Generators are equipped with cooling (air or liquid), temperature and vibration sensors, and power-electronic interfaces to deliver grid-compliant output.

Most modern wind turbines generate electricity using three-phase alternators (AC generators). Three-phase AC is the industry standard because it delivers smoother power, higher efficiency, and easier integration with the electrical grid than single-phase systems. For miniature and small wind turbines, where space and cost are critical, manufacturers often use claw-pole generators (a compact form of alternator also known as a Lundell generator). These machines are lightweight, have a small footprint, and can produce usable power at relatively low speeds—ideal for off-grid or battery-charging applications.

Depending on the size of the turbine and the control strategy, both synchronous generators (including permanent-magnet types) and induction generators (especially squirrel-cage or wound-rotor designs) are used.

- Synchronous generators produce power at a frequency tied to their rotational speed; when paired with power electronics (rectifiers and inverters) they can supply stable grid-quality power even if rotor speed varies widely.
- Induction generators can connect directly to the grid without extensive electronics but need the rotor speed to remain close to synchronous speed, which limits how much it can vary.

The choice between these generator types is closely linked to the control mode of the wind turbine.

In this mode, the rotor blades have a fixed pitch angle, and the generator speed varies with the wind. Because the blades cannot adjust to control power, the output fluctuates as the wind changes. This simple arrangement is common for small, stand-alone turbines not connected to the grid—such as those charging batteries on remote sites. Typically, permanent-magnet synchronous generators are used here because they can produce power efficiently at variable speeds without the need for external excitation. The raw output is often rectified to DC to charge batteries or passed through an inverter to produce stable AC for appliances.

Large and medium wind turbines traditionally operated at nearly constant rotational speed to supply grid power at 50 Hz (or 60 Hz). Because the wind speed changes constantly, the turbine uses a pitch control system to adjust blade angle, attempting to keep the rotor speed and generator output frequency steady. Most turbines in this category employ cage-type (squirrel-cage) induction generators, which can output a stable 50 Hz AC when their rotational speed changes only slightly. This allows direct grid connection without inverters. However, the limitation of this setup is that it cannot always maintain optimal aerodynamic efficiency at varying wind speeds—especially under "fixed pitch" operation—so energy capture is lower compared with newer variable-speed designs.

With advances in power electronics and computer control, the industry has moved to variable-speed/constant-frequency (VSCF) operation. Here, the rotor is allowed to spin over a wide range of speeds to extract maximum aerodynamic efficiency from the wind. The generator output is first rectified to DC, then passed through an inverter to produce AC at the grid frequency (50 or 60 Hz). This decouples rotor speed from grid frequency, so the turbine can always operate at its optimal tip-speed ratio.

In this setup, the pitch system is used mainly for power regulation above rated wind speed (to limit loads and protect the drivetrain), while speed control is achieved through generator torque and converter settings. Two main types of generators dominate:

- Doubly-fed induction generators (DFIGs), where only the rotor circuit is connected to a power converter while the stator is connected directly to the grid. This allows variable-speed operation with smaller converters.
- Direct-drive permanent-magnet generators, which eliminate the gearbox entirely and produce variable-frequency AC directly at low speeds.

Both approaches support variable speed/constant frequency operation, which is now standard for large onshore and offshore turbines.

8. Anemometer

An anemometer measures wind speed, and is usually paired with a wind vane for direction.

- These sensors are mounted atop the nacelle, away from blade turbulence.
- Data informs the control system when to start or stop the turbine, adjust blade pitch, or yaw the nacelle.
- Many modern turbines use ultrasonic (no moving parts) anemometers for reliability and redundancy.

Wind is fundamentally the horizontal movement of air across the Earth's surface. It originates from solar heating, which creates uneven air pressure: warm air rises, cool air sinks, and these pressure differences drive airflow. Terrain, vegetation, and bodies of water further modify this flow. Classic examples of large-scale winds include land and sea breezes, valley winds, monsoons, and trade winds.

For engineers and operators of wind turbines, two parameters are crucial:

- Wind magnitude (speed) – how fast the air is moving
- Wind direction – the angle from which the air is flowing

Accurate measurement of these parameters ensures optimal turbine yaw and pitch control, directly affecting power output and structural safety.

Wind speed and direction can be measured using several types of sensors, classified by their measurement principle:

- Mechanical (rotating cups, propellers, vanes)
- Ultrasonic (time-of-flight of sound waves)
- Acoustic oscillation
- Pressure and hot-wire anemometers
- Laser radar (LIDAR) and microwave radar

Among these, mechanical and ultrasonic transducers are the most common in wind turbines because they offer a good balance of reliability, cost, and accuracy.

Mechanical anemometers are among the oldest and most widely used tools for measuring wind speed and direction. The simplest and most common design for wind speed measurement is the three-cup anemometer. In this device, three small cups are mounted symmetrically on a rotating arm. As wind speed increases, the cups spin faster, and a built-in speed transducer converts this rotational motion into an electrical signal that can be read by a controller or data logger. This setup is simple, inexpensive, and effective, making it a standard choice for many applications.

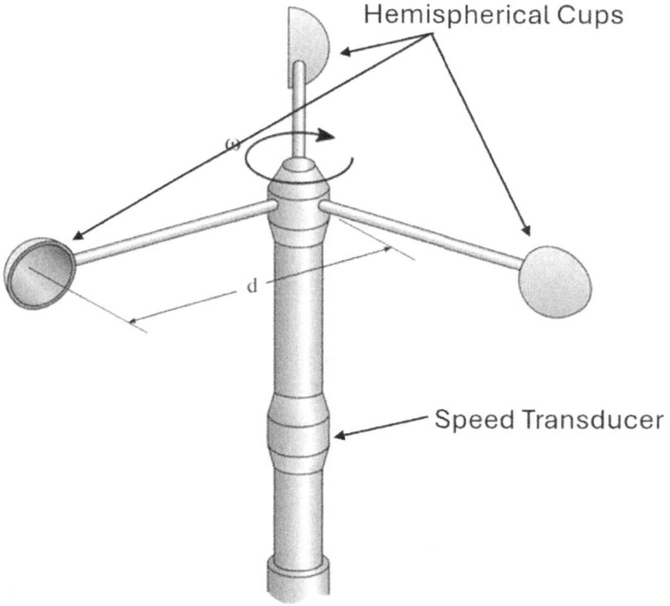

Figure 31: Three-cup anemometer-wind speed transducer. Adapted from Brina Schenk with UBC and Douglas College, CC BY 4.0, via Wikimedia Commons.

For measuring wind direction, a wind vane is typically used. This instrument consists of a fin-shaped blade mounted on a shaft that aligns itself with the wind. Inside the vane, an angle transducer converts the physical position of the blade into an electrical signal, allowing for precise measurement of the wind's direction.

Some instruments combine both speed and direction measurement into a single device. These combined wind sensors often have a small propeller at the front to measure wind speed and a fin at the rear to indicate direction. Each component has its own transducer, producing separate electrical outputs for wind speed and direction, which can then be processed together by a controller.

The main advantages of mechanical anemometers are their simplicity, low cost, and widespread familiarity. However, because they rely on moving parts, they are vulnerable to wear and tear from sand, ice, or salt in the air. This makes them prone to degradation over time, requiring regular maintenance and periodic calibration to ensure accuracy.

Ultrasonic anemometers measure wind speed and direction without any moving parts, making them fundamentally different from mechanical devices. These instruments, often called ultrasonic wind speed and direction transducers, work by transmitting ultrasonic pulses between pairs of transducers. The key principle is the time it takes for sound to travel between the transducers. In a tailwind, the sound travels faster; in a headwind, it travels slower. By comparing these time differences in opposite directions, the device can precisely calculate both wind speed and wind direction.

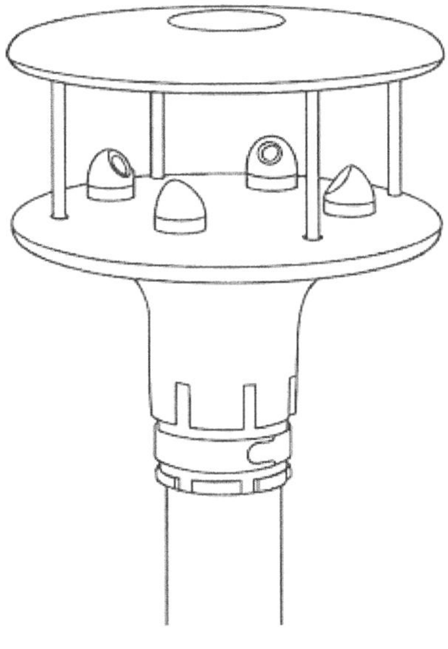

Figure 32: Ultrasonic Anemometer.

Each pair of ultrasonic probes measures wind movement along a single axis. By combining the data from two perpendicular pairs of probes, the instrument can reconstruct the full wind vector, giving a comprehensive measurement of both speed and direction simultaneously. This setup allows for continuous, highly accurate readings.

The primary advantages of ultrasonic anemometers are their lack of mechanical wear and high precision, making them particularly well suited for long-term deployments in demanding environments. However, they do have limitations. Their larger size makes them more conspicuous, and their performance can be affected by rain, snow, fog, or dust, which can scatter or slow the ultrasonic pulses and introduce errors. To mitigate these issues, many modern ultrasonic anemometers incorporate probe-heating systems to prevent frost or ice accumulation on the transducers, ensuring reliable operation even in severe weather conditions.

Laser Wind Radar, more commonly known as LIDAR (Light Detection and Ranging), represents one of the most advanced technologies for wind measurement. Unlike mechanical or ultrasonic anemometers mounted directly on the turbine, LIDAR systems measure wind conditions remotely. They operate by emitting laser pulses into the atmosphere and analysing the Doppler shift in the backscattered light from airborne particles such as dust, aerosols, or water droplets. This shift reveals both the speed and direction of the moving air.

Because LIDAR can detect wind conditions hundreds of metres ahead of a turbine, it provides a detailed profile of the approaching wind field rather than just the conditions at the nacelle. This capability enables far more predictive control strategies. For example, the turbine's yaw system can begin turning to face a changing wind direction before it actually arrives, and the pitch control system can pre-adjust blade angles to handle gusts or shifts in wind speed. This advanced information helps optimise energy capture, reduce mechanical stress on components, and improve the overall reliability and efficiency of modern wind turbines.

The integration of wind speed and direction transducers within modern wind turbines plays a crucial role in optimizing their operational efficiency and longevity. These transducers, typically mounted on the nacelle, provide real-time data that informs two key control systems: the yaw controller and the turbine controller. The yaw controller utilizes wind direction data to adjust the turbine's orientation, ensuring that the rotor is aligned optimally with prevailing wind conditions to minimize aerodynamic drag and maximize energy capture [68, 69]. In contrast, the turbine controller leverages wind speed data to manage the pitch angle of the blades and the generator's loading, thereby maintaining optimal power output and efficiency [70].

Real-time feedback from these sensors is vital, as it allows the turbine to dynamically respond to fluctuations in wind conditions, which is essential for enhancing both efficiency and safety [71]. For example, the implementation of active yaw control has been shown to alleviate wake losses for downstream turbines, thus improving overall wind farm performance [71]. The optimization of yaw and pitch control systems not only enhances immediate energy capture but also contributes to the long-term reliability and lifespan of the turbine by mitigating excessive mechanical stresses caused by misalignment with the wind [72, 73].

Moreover, advances in technology, such as using Machine Learning (ML) and Internet of Things (IoT) approaches, further augment the capabilities of these control systems. By employing sensors to collect extensive real-time data, ML algorithms analyse this data to forecast wind conditions and predict potential failures, thereby allowing for pre-emptive maintenance scheduling that can further improve operational safety and efficiency [74, 75]. As wind turbines operate in increasingly complex and variable environments, integrating these advanced control methodologies becomes essential in mitigating risks associated with wind-induced stresses and optimizing performance [76].

9. Brake

The brake system stops or slows the rotor for maintenance, emergencies, or overspeed conditions.

- Mechanical brakes (disc or caliper) are mounted on the high-speed shaft or main shaft to provide a positive stop.
- Aerodynamic braking is achieved by pitching blades out of the wind (feathering) or into stall to reduce lift.
- Brakes are designed fail-safe: if hydraulic or electric power is lost, springs apply braking force automatically.

Wind turbines operate under constantly changing wind speeds, which means the rotor's speed can vary significantly. Braking systems are critical for controlling this motion to prevent dangerous overspeed conditions, facilitate safe maintenance, and protect the structure and its components from damage. Without effective braking, the turbine could experience excessive torque or runaway rotor speeds during high winds, resulting in catastrophic failures such as blade loss, gearbox damage, or tower collapse.

Braking systems also stabilize the rotor during grid faults (sudden power loss) or emergency stops, preventing uncontrolled rotation. This is especially important for large utility-scale turbines, where the rotor diameter can exceed 150 m and the stored rotational energy is enormous.

Modern turbines combine several braking approaches—mechanical, aerodynamic, and yaw brakes—into an integrated system. Each serves a unique purpose but together they provide redundancy, precise control, and safe operation.

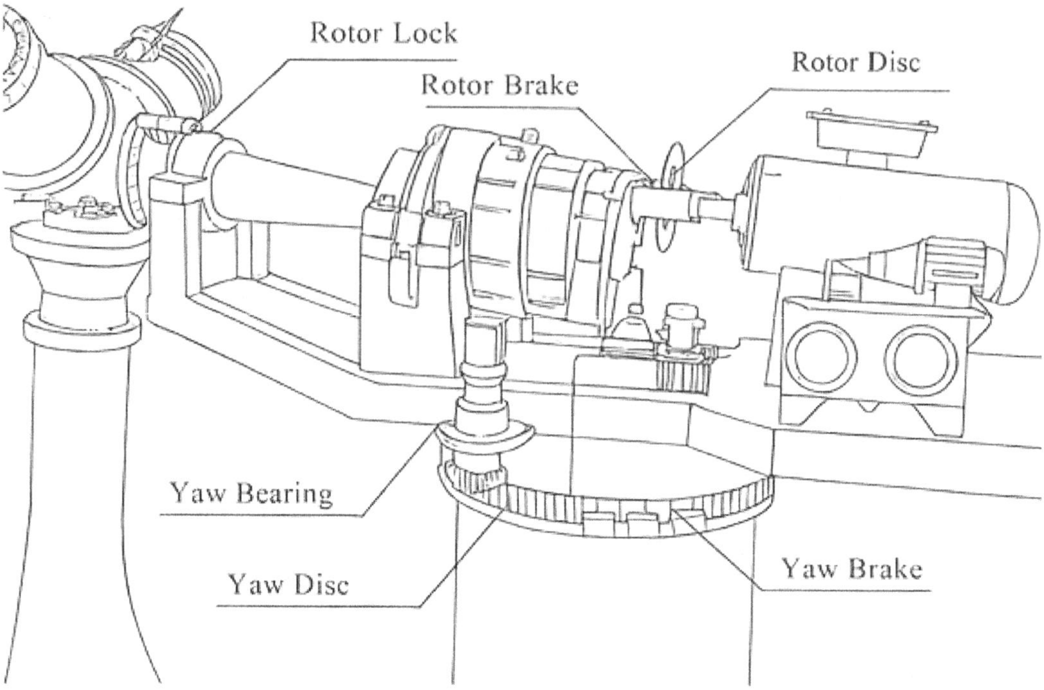

Figure 33: Braking system of wind turbine.

Rotor brakes act directly on the rotating components of the drivetrain to slow or stop rotation. They are especially important during planned maintenance or in emergencies.

- **Mechanical Rotor Brakes:** These are disc-brake systems mounted on the high-speed shaft (near the generator) or on a separate brake disc. Calipers clamp onto the brake disc, creating friction to slow the shaft. Because the high-speed shaft rotates much faster than the rotor, the brake can be smaller and more manageable while still dissipating significant energy. Mechanical brakes must be robust enough to handle high torque and environmental extremes such as cold, humidity, or salt spray. Regular inspection and replacement of pads and seals are essential to prevent wear-induced failures.
- **Aerodynamic Rotor Brakes:** This type of braking doesn't use friction at all. Instead, it adjusts the pitch of the blades—turning them out of the wind—to reduce lift and increase drag. In high winds, blades can even be "feathered" to almost parallel with the wind flow, drastically slowing the rotor. Aerodynamic braking is especially effective for gradual deceleration and is always used as the primary braking method before mechanical brakes engage.

Most turbines use a combination: first pitch the blades to slow the rotor (aerodynamic braking), then apply mechanical brakes to bring it to a complete stop.

Yaw Control Brakes: Yaw brakes aren't used to stop rotation of the rotor but to lock the nacelle (the housing on top of the tower) in position. They work by applying clamping pressure to the yaw bearing or friction plates, preventing the nacelle from swinging under wind forces. This ensures the rotor faces the correct wind direction for maximum efficiency and also prevents oscillations or misalignment during high winds.

Modern systems combine yaw motors (to rotate the nacelle) with hydraulic or spring-applied friction brakes (to hold it steady once aligned). Without yaw braking, the nacelle could rotate uncontrollably, damaging cables and components.

Blade Pitch Control: Although technically part of the rotor system rather than a brake itself, pitch control is essential for regulating rotor speed. By dynamically adjusting blade angle, it allows the turbine to optimize energy capture at low and medium winds and limit power at high winds. When combined with aerodynamic braking, it provides a seamless way to reduce rotational speed before a full stop.

Pitch systems (hydraulic or electric) also feature emergency feathering capability: if grid power is lost, backup accumulators or batteries drive the pitch actuators to feather the blades automatically, preventing runaway rotation.

When wind speeds exceed safe limits or an emergency stop is required, the wind turbine's control system initiates a coordinated sequence of braking actions. First, the blade pitch control automatically rotates the blades into a feathered position—nearly parallel to the wind flow. This aerodynamic braking greatly reduces the lift force on the blades and therefore lowers the torque being transmitted through the drivetrain. By slowing the rotor gradually, the turbine avoids abrupt mechanical stress.

Once the rotor has slowed to a manageable speed, the mechanical brake engages. This friction-based brake—usually mounted on the high-speed shaft or on a dedicated disc—clamps onto the brake disc and brings the drivetrain to a complete stop. This is especially important for maintenance or grid faults where the rotor must be fully stationary.

Meanwhile, yaw brakes lock the nacelle (the housing at the top of the tower) in place to prevent it from swinging under wind forces. Keeping the nacelle stable ensures that all loads are safely transferred to the tower and prevents twisting or misalignment of internal cables and systems during the stop.

Finally, backup power systems, such as hydraulic accumulators or battery packs, ensure that braking can still occur if grid power is lost. These systems provide stored energy to operate the pitch actuators and release mechanisms so that the blades feather and the brake clamps automatically, even during a complete power outage. This multi-layered braking approach minimizes component stress, increases reliability, and provides critical redundancy for safety.

Modern wind turbine braking systems are evolving to be more efficient, intelligent, and durable. One major innovation is regenerative braking, which allows part of the kinetic energy from deceleration to be converted into electrical energy for auxiliary systems, reducing energy waste.

Another improvement is the use of advanced materials for brake pads, discs, and seals. High-performance composites resist heat, moisture, and corrosion, extending the service life of the brake and reducing maintenance requirements in harsh environments.

In addition, the rise of predictive maintenance is transforming how braking systems are managed. Sensors embedded in brakes, yaw drives, and pitch actuators continuously feed data to the turbine's SCADA (Supervisory Control and Data Acquisition) system. By analyzing this data, operators can forecast wear, schedule maintenance proactively, and minimize downtime.

Finally, integrated safety algorithms in the turbine's control software now coordinate aerodynamic and mechanical braking automatically. This ensures smooth deceleration, avoids abrupt loads, and optimizes the timing of each step in the braking sequence. Together, these advancements make braking systems more reliable, safer, and better suited for the ever-larger turbines being deployed worldwide.

10. Gearbox

Most turbines use a gearbox to step up the slow rotation of the rotor to the higher speed needed by the generator.

- Typical blade speeds are 5–25 rpm; the gearbox multiplies this to 1,000–2,000 rpm for the generator.
- It usually combines planetary and parallel helical gears to handle high torque.
- Gearboxes have oil-lubrication, filtration, and cooling systems and are instrumented with vibration and wear-particle sensors for condition monitoring.
- Some turbines are direct drive and omit the gearbox entirely.

A wind turbine's rotor turns very slowly compared with a conventional electrical generator. A two-pole three-phase alternator needs to rotate at roughly 3,000 rpm to produce 50 Hz electricity; a four-pole alternator at about 1,500 rpm; and a six-pole alternator at about 1,000 rpm. By contrast, a small wind turbine rotor may only reach a few hundred revolutions per minute at most, while a large or medium wind turbine rotor typically spins at only tens of revolutions per minute—sometimes as low as 10 rpm.

This huge speed mismatch means the slow-turning rotor cannot drive a standard generator directly. In geared wind turbines, a gearbox sits between the low-speed rotor shaft and the high-speed generator shaft to step up the rotational speed. The gearbox multiplies the rotor speed by tens or even hundreds of times, so the generator runs at its rated rpm.

Figure 34: Preparing Gearbox, Rotor Shaft and Disk Brake for Lifting. Paul Anderson, CC BY-SA 2.0, via Wikimedia Commons.

Wind turbine gearboxes must deliver very high speed increases in a compact space. Two basic gear arrangements are used:

- Spur gear stages, where gear teeth are cut straight and shafts are offset (input and output shafts are on different axes).
- Planetary gear stages, where planet gears rotate around a central sun gear inside a ring gear (input and output shafts are coaxial).

Planetary gears are especially suited to wind turbines because they can achieve large ratios with small size, distribute loads across multiple gear meshes, and keep input/output shafts aligned.

A typical utility-scale turbine uses either:

- Two stages of planetary gears, or

- One planetary stage plus one spur stage for a very large ratio, so that the final high-speed shaft can drive the generator at thousands of rpm. The exact configuration depends on the turbine's rated power and rotor speed.

The main input shaft of most wind-turbine gearboxes is tubular with a central through-hole. This hole allows hydraulic lines, electrical cables, and signal wiring for the hub pitch system to pass from the stationary nacelle into the rotating hub without interference. It is a unique feature of wind turbine gearboxes not usually seen in industrial gearboxes.

Because the input torque from the rotor is huge and often fluctuating, the gearbox must be extremely robust. Housing castings, gears, and bearings are made from high-strength, case-hardened steels. Precision machining and strict quality control are essential to avoid premature failure.

Wind turbines operate in harsh outdoor conditions—temperature swings, vibration, dust, salt spray—and are expected to run for 20–30 years with minimal downtime. As a result, their gearboxes have high material and process requirements and rely on sophisticated lubrication systems.

These systems include:

- Forced-oil lubrication with pumps and filters to ensure constant oil flow to gear meshes and bearings.
- Oil coolers (air-oil or water-oil heat exchangers) to dissipate the heat generated by high-speed gear operation.
- Condition monitoring for oil temperature, pressure, and contamination to prevent failures.

The gearbox transforms slow, high-torque rotor rotation into high-speed, low-torque rotation suitable for the generator. Planetary stages keep the package compact and coaxial; spur stages can be added for extra ratio. The tubular input shaft doubles as a conduit for pitch-control lines. Lubrication and cooling are critical to reliability. Together these design choices allow a modern wind turbine to produce grid-quality electricity continuously despite the very low speed of its rotor.

11. Rotor Blade

Rotor blades capture the wind's kinetic energy using aerodynamic lift.

- Blades are made from composite materials (fiberglass, carbon fibre, epoxy) for high strength-to-weight ratio.

Figure 35: Wind turbine blades are unloaded in Lerwick, Shetland. Ronnie Robertson, CC BY-SA 2.0, via Wikimedia Commons.

- Their airfoil cross-section and twist vary along the span to optimise lift-to-drag ratio at different wind speeds.
- Modern blades can exceed 100 m in length offshore.
- They are bolted to the hub with massive blade bearings to accommodate loads and pitch movement.

Figure 36: Attaching a Turbine Blade. Paul Anderson, CC BY-SA 2.0, via Wikimedia Commons.

A wind-turbine blade is essentially a long, twisted airfoil—a shape designed to create lift from moving air just like an aircraft wing. Along the blade's length the profile (cross-section) gradually changes from a thicker, more robust root section near the hub to a thinner, more aerodynamic tip section. This variation allows the blade to carry structural loads at the root while maintaining high aerodynamic efficiency at the tip [77].

Figure 37: The blades of a wind turbine are affected by four forces: drag, lift, centrifugal, and gravitational forces. Wind Turbine professor, CC BY-SA 4.0, via Wikimedia Commons.

In aerodynamics, the key geometric reference is the chord line of the airfoil (a straight line connecting the leading and trailing edges). The angle of attack (α) is defined as the angle between the chord line and the incoming airflow. Adjusting this angle directly controls how much lift and drag the blade generates at any moment [77].

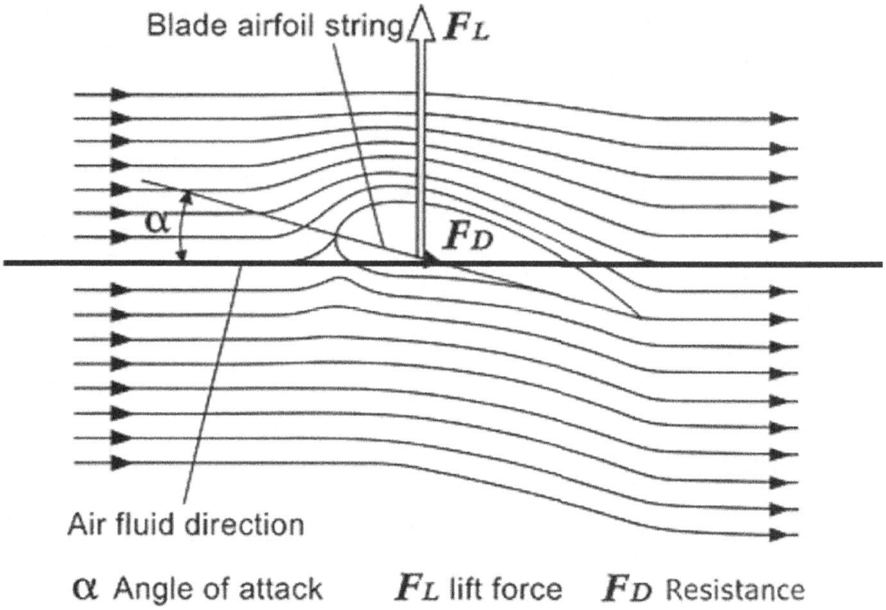

Figure 38: Attack angle of the airfoil.

When the blade is correctly pitched, air flows smoothly over both sides of the airfoil. According to the Bernoulli principle, air moving faster over the curved upper surface has lower pressure than air moving slower underneath. This pressure difference creates an upward force perpendicular to the flow known as lift (Fl). At the same time, the blade also experiences drag (Fd)—a force parallel to the flow resisting motion [77].

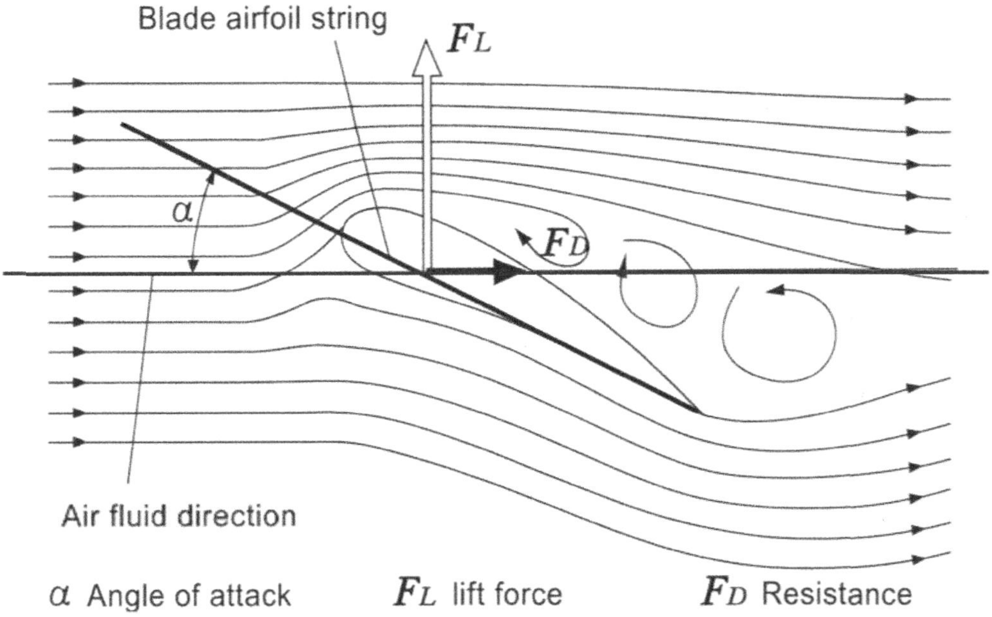

Figure 39: The airfoil starts to stall after the angle of attack exceeds the stall angle of attack.

For a wind turbine, lift is not vertical but acts tangentially to the rotor plane, driving the blade's rotation and producing torque. Drag, on the other hand, wastes energy and slows rotation. Good blade design therefore aims for a high lift-to-drag ratio, so the rotor converts as much wind energy as possible into rotational energy [77].

However, the blade does not always produce high lift. If the angle of attack α becomes too large, the smooth flow of air separates from the upper surface of the blade. This separation causes turbulent eddies and a dramatic increase in drag while lift drops sharply. This condition is called stall. The critical angle at which stall begins is known as the stall angle of attack.

In most thin airfoil sections used on wind-turbine blades, the lift coefficient increases steadily with α up to about 15°. Beyond this, stall occurs: the lift curve suddenly collapses while the drag curve spikes. By about 45°, lift and drag forces become roughly equal, meaning the blade is no longer producing useful torque but mostly resisting the wind [77].

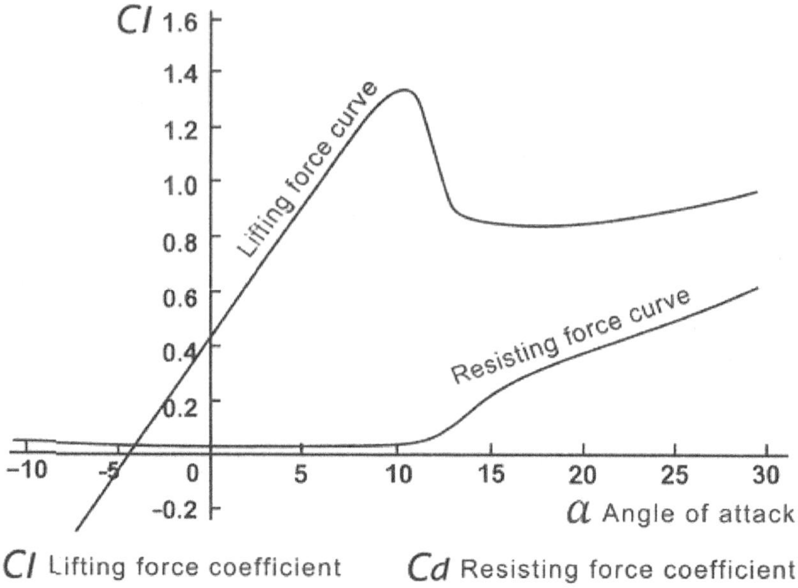

Figure 40: Lift curve and drag curve of the airfoil [77].

The stall curves described above represent static stall, measured when airflow is steady and the angle of attack changes slowly. Under real operating conditions, especially in gusty or turbulent winds, the situation is more complex. Rapid changes in wind direction or blade pitch can cause dynamic stall, where the lift and drag curves deviate significantly from the ideal. Dynamic stall can create brief but extreme loads on the blade, tower, and drivetrain, making it a critical factor in structural design and control strategies [77].

Understanding lift, drag, and stall is important for designing and operating wind-turbine blades. Blade pitch control systems are programmed to keep the angle of attack below the stall threshold in normal operation to maximise energy capture. During high winds, they deliberately increase pitch angle (reducing α) to spill excess wind and prevent overloading. The airfoil's geometry, twist distribution, and material strength all work together with the control system to manage these aerodynamic forces safely and efficiently.

Early commercial wind turbines adopted existing NACA (National Advisory Committee for Aeronautics) airfoil shapes, which had been developed originally for aircraft wings. These profiles offered good basic lift and drag characteristics but were not specifically optimised for the unique demands of large rotating machines. Unlike an aircraft wing—which has a uniform profile and operates at relatively constant angles of attack—a wind-turbine blade experiences a wide range of speeds, angles, and loads from root to tip and from one revolution to the next.

A key distinction is that a wind-turbine blade must support very high bending and torsional loads, particularly near the hub where torque is greatest. To achieve the required structural stiffness, the blade's root section is much thicker (often 35–50% of chord length). While this improves strength and space for bolted connections, it also increases aerodynamic drag if not carefully designed. As a result, wind-turbine airfoils must balance structural needs (thickness) with aerodynamic efficiency (low drag).

Another difference is that turbine blades rotate, so each section experiences different relative wind speeds and angles of attack across the span. This creates complex flow conditions including turbulence, tower shadow effects, and dynamic stall—all of which the airfoil must handle without sudden loss of lift or large fluctuations in loads.

Specialised wind-turbine airfoils are designed to meet several critical criteria:

- **High Maximum Lift Coefficient (Clmax):** To extract as much energy as possible, especially at low wind speeds.
- **High Lift-to-Drag Ratio (Cl/Cd):** Maximising torque while minimising drag-induced losses.
- **Gentle, Predictable Stall Behaviour:** Stable aerodynamic performance when approaching stall to avoid abrupt load changes and vibration.
- **Low Sensitivity to Surface Roughness:** In real operation, insects, salt spray, or ice can roughen the leading edge. An ideal airfoil maintains good performance despite such contamination.
- **Thickness Variability:** Thicker profiles near the root for strength and thinner profiles near the tip for aerodynamic efficiency, while maintaining smooth transition along the blade.

Recognising these unique requirements, several countries and research institutions began developing dedicated wind-turbine airfoil families in the mid-1980s:

- **NREL S-Series (United States):** Developed by the National Renewable Energy Laboratory for small to large turbines, these profiles combine high lift, low drag, and robust stall characteristics.
- **Risø Airfoil Series (Denmark):** Optimised for offshore and onshore turbines, focusing on thick root sections and insensitivity to roughness.
- **FFA-W Series (Sweden):** Designed for stable performance under variable conditions and ease of manufacturing.
- **DU (Delft University) Series (Netherlands):** Emphasises high lift and thick sections for large blades.
- **WA Series (China):** A domestic development tailored to Chinese wind conditions and manufacturing capabilities.

Each series offers a range of thicknesses so blade designers can select a root profile and transition to tip profiles that best suit the desired rotor size and operational conditions.

The move from generic NACA profiles to custom wind-turbine airfoils has been one of the key enablers of modern high-efficiency, multi-megawatt turbines. Purpose-designed airfoils allow blades to be longer and lighter yet still structurally sound, improving annual energy production and reducing cost per kilowatt-hour. They also make turbines more tolerant of rough operating environments, which is critical for offshore and remote onshore installations.

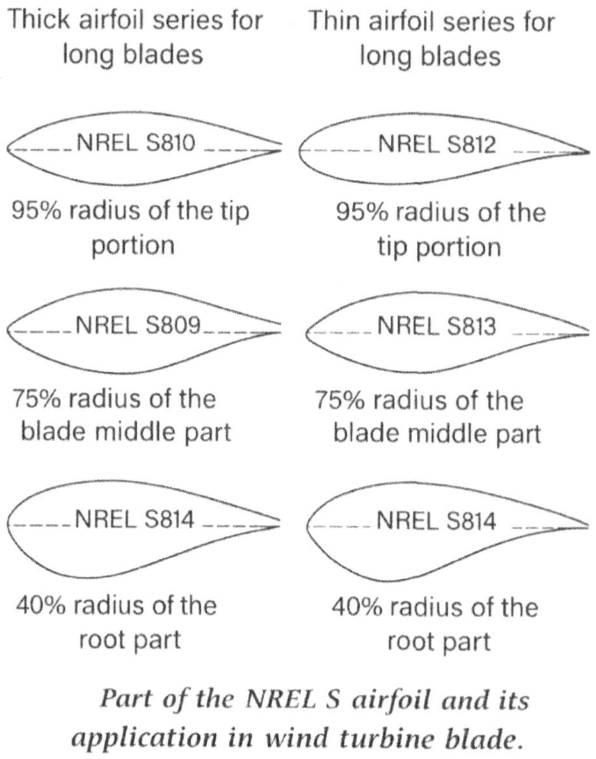

Figure 41: Wind turbine dedicated airfoil [77].

Every section of a wind-turbine blade acts like an airfoil. The pitch angle (β) is the fixed angle between the blade's chord line (or string line) and the plane of rotation. As the rotor spins, each blade section sees a relative wind speed made up of two components:

- The free wind speed (v) approaching the turbine.
- The linear speed (u) of that blade section caused by the blade's rotation.

Combining these two vectors produces the relative wind velocity (w). The angle between this relative wind and the blade's chord line is called the angle of attack (α). For efficient operation, each blade section should run at an angle of attack just below its stall point—this yields maximum lift and minimal drag. For a given wind speed and blade rotational speed, designers select the pitch angle β that will create the optimum angle of attack α.

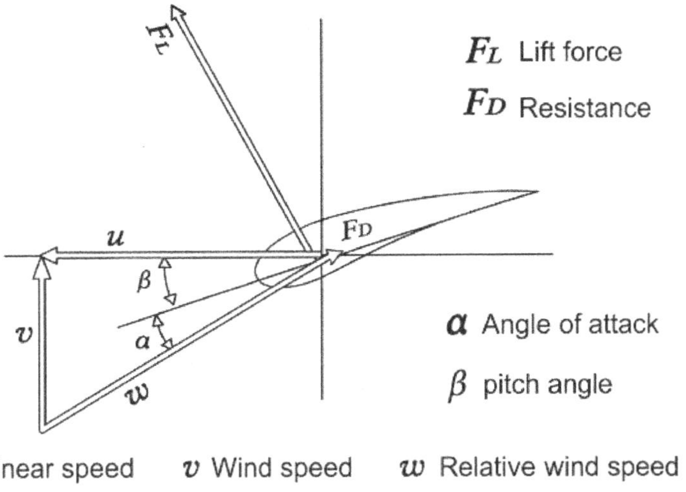

Figure 42: Airfoil pitch angle [77].

A rotating blade does not experience the same conditions along its length. The linear velocity (u) increases from root to tip because the tip travels a larger circumference in the same time. For the same incoming wind speed (v), this means:

- Near the tip, the relative wind is more "edge-on" and the angle between the relative wind and the rotor plane is smaller.
- Near the root, the relative wind appears "steeper" and the angle between the relative wind and the rotor plane is larger.

If the blade geometry were constant, each section would experience a different, often suboptimal, angle of attack. This would lower the total aerodynamic efficiency.

To compensate for these changing conditions, wind-turbine blades are manufactured with a gradual twist along their length. This twist continuously changes the local pitch angle so that each section of the blade meets the relative wind at its most efficient angle of attack. Typically, the root has the largest pitch angle relative to the rotor plane, while the tip has the smallest pitch angle. This ensures that all sections contribute effectively to torque generation.

The blade also changes in airfoil shape and thickness from root to tip. Near the root, the airfoil is thicker to provide structural strength and house bolts or bearings for mounting. Towards the tip, the profile becomes thinner and more streamlined to minimise drag and noise. This combination of twist and profile variation allows modern blades to achieve both structural integrity and high aerodynamic efficiency.

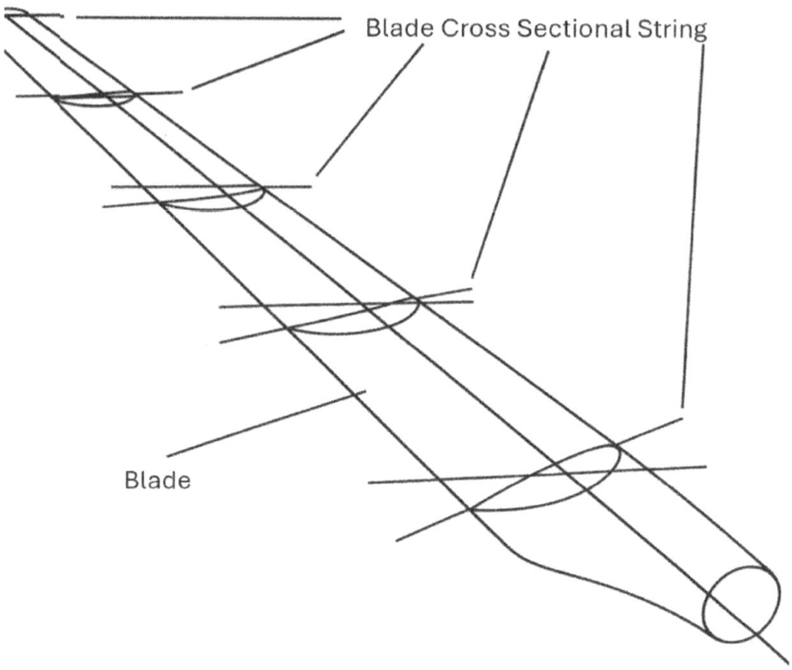

Figure 43: Wind turbine cross section of different parts of the blade.

The amount of twist and the exact profiles used are determined during the turbine's design process. Engineers use aerodynamic models to optimise blade geometry for a target wind speed (for example, 8 m/s for a medium-sized turbine). The resulting blade has a specific pitch-angle distribution from root to tip—often differing by 10–20 degrees or more—to achieve maximum energy capture under typical operating conditions.

Modern turbines rarely use a single generic airfoil along the whole blade. Instead, they combine several airfoil sections (for example, from the NREL S-series or Risø series) tailored to the loads and speeds at each radius. Thicker sections near the hub provide strength and space for internal components; thinner sections near the tip reduce drag and noise. All of these are integrated smoothly to produce a blade that runs at its optimal angle of attack over as much of the rotor swept area as possible.

This carefully engineered blade structure—with variable pitch, twist, and airfoil shapes—is crucial to maximising a turbine's power output, minimising structural loads, and ensuring stable behaviour under varying wind conditions. Without it, large portions of the rotor would operate inefficiently or be subject to excessive stresses, reducing both energy production and component life.

12. Blade Pitch Control

Blade pitch control regulates the angle of each blade relative to the wind to control power output and rotor speed.

- Pitch bearings at the blade root and electric or hydraulic pitch motors rotate each blade.
- At low to moderate winds, blades are pitched for maximum lift; at high winds, they feather out of the wind to reduce aerodynamic loads.
- The pitch system also provides emergency braking and load control to protect the drivetrain.
- Backup power (batteries or hydraulic accumulators) ensures blades can be feathered in case of power loss.

The ability to adjust each blade's angle precisely and in real time is one of the most important reasons wind turbines can generate so much electricity safely and reliably. This function is carried out by the pitch control system. To picture the basic idea, imagine a traditional windmill: its wooden sails had to be set at just the right angle to the wind to turn effectively. A wind turbine is the same concept at far greater scale, using aerodynamic blades hundreds of feet long to capture the wind's kinetic energy and convert it into electricity. The pitch control system is the mechanism that constantly tweaks each blade's "attack angle" to the oncoming wind so that the turbine produces maximum power without overstressing its components. In other words, it is both the "throttle" and the "safety valve" for the entire machine.

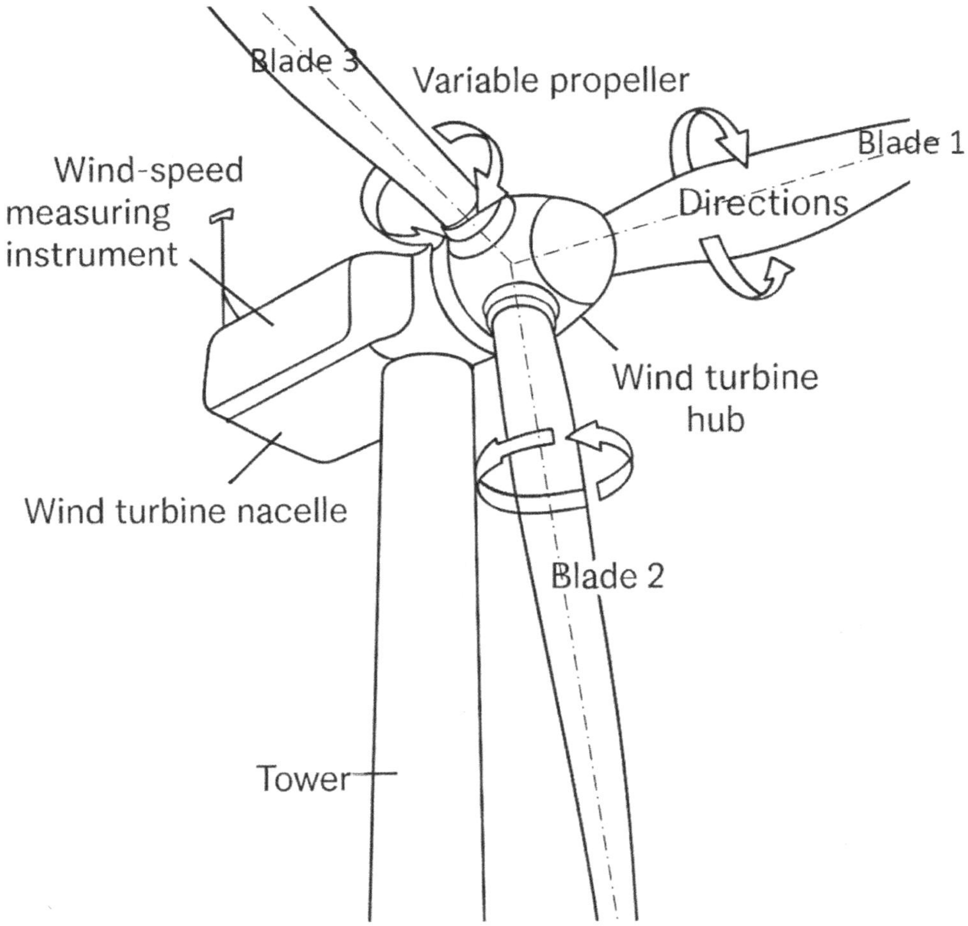

Figure 44: Wind Turbine Blade Pitch Control.

Technically, the pitch angle is the rotational position of a blade around its own longitudinal axis. This single parameter controls how much aerodynamic lift the blade generates, which in turn sets how fast the rotor spins and how much torque reaches the generator. In low winds, the controller orders the blades to pitch into a "high-lift" position, perpendicular to the flow, so that the rotor captures as much energy as possible. As winds strengthen, the system automatically fine-tunes the pitch to maintain the turbine at its rated power output, preventing overload of the generator or converter. When wind speeds become dangerously high, the blades are "feathered"—turned nearly parallel to the flow—so that the rotor slows dramatically and the turbine's structure is protected. All of this happens automatically and continuously, with the controller updating commands every second based on fresh sensor data.

The pitch control system can therefore be thought of as having a dual purpose. First, it optimises power output by adjusting the blades to suit constantly changing wind speeds. This is especially important in regions with variable or turbulent winds, where fixed-pitch blades would waste a great deal of energy. Second, it protects the turbine by reducing aerodynamic loads when winds exceed safe levels. Without such a system, a modern multi-megawatt turbine would either have to be massively over-engineered to survive storms or risk catastrophic damage.

A basic pitch control system consists of several integrated subsystems working together. On top of the nacelle sit wind speed and direction sensors—anemometers and wind vanes—that continuously measure the prevailing conditions. Their data flows to the pitch controller, the system's "brain," which runs algorithms to determine the optimal blade angle based on wind speed, generator speed, and power output. The controller then sends commands to the pitch actuators, the "muscles" of the system, which physically rotate the blades. Actuators are either hydraulic cylinders or electric motors. Hydraulic systems, once dominant because of their high force capacity and robustness, are still common in very large turbines, while electric systems are increasingly favoured for their higher efficiency, cleaner operation, and lower maintenance demands.

The actuators transmit their force through a blade pitch mechanism built into the hub. This is a heavy-duty bearing and drive assembly that allows each blade to rotate smoothly and precisely under high aerodynamic loads. The entire system is supported by power and communication infrastructure inside the hub and nacelle, which supplies electricity to the actuators, sensors, and controller, and ensures real-time feedback between them. Redundancy is built into critical parts so that a single failure does not leave a blade "stuck" at the wrong angle.

The operation of the pitch system is fully automated but highly sophisticated. Modern controllers use not just simple feedback (reacting to current wind) but also predictive strategies that anticipate gusts and directional shifts, reducing mechanical stress and smoothing power output. By combining sensors, algorithms, and actuators into a single integrated loop, pitch control systems allow turbines to produce more energy over their lifetime while suffering less wear and tear.

In short, the pitch control system is what makes a modern wind turbine both productive and resilient. It constantly balances aerodynamic efficiency against structural safety, adjusting each blade in real time to changing conditions. This ensures maximum energy capture in light winds, protects the machine in storms, and stabilises the electricity supplied to the grid. Without it, large-scale wind power would be far less viable as a dependable source of renewable energy.

Once you move beyond the basics, understanding pitch control systems becomes less about memorising a definition and more about appreciating the different designs, control strategies, and operating modes that exist in real turbines. The essential purpose—optimising energy capture while protecting the machine—remains the same, but the ways engineers achieve it vary significantly. This diversity in implementation is what gives modern wind turbines their extraordinary ability to generate electricity safely and efficiently under a wide range of conditions.

At the intermediate level, one of the first distinctions is between hydraulic and electric pitch control systems. Hydraulic systems have traditionally been the workhorse of large turbines because they can deliver the high forces needed to move multi-tonne blades quickly and reliably. They use pumps to pressurise fluid, which then drives cylinders to rotate the blade around its axis. This setup is robust and well-proven in harsh environments but carries maintenance and environmental risks, such as leaks. Electric pitch systems, by contrast, employ servo or gear motors to turn each blade. They are inherently more energy-efficient because they do not need to keep pumps running, eliminate the risk of hydraulic fluid spills, and are easier to monitor and maintain. Advances in motor technology and power electronics have made electric actuation the preferred choice for many new turbine designs, especially offshore.

A second layer of sophistication is the shift from collective pitch control (CPC) to individual pitch control (IPC). In CPC, all blades change angle together; this is simpler and effective for regulating rotor speed and power output but cannot compensate for asymmetric loads caused by turbulence, wind shear, or tower shadow. IPC takes a more granular approach by adjusting each blade independently based on the wind it encounters at that instant. This reduces uneven loads on the rotor and drive train, improves fatigue life, and can even increase energy capture by fine-tuning aerodynamics blade by blade. IPC requires more sensors, faster control loops, and more complex algorithms, but the payoff is smoother operation, higher power quality, and reduced mechanical stress.

The control strategy itself also changes depending on wind speed. Below the cut-in speed, the blades are generally fixed to maximise starting torque. In the intermediate “Region 2” of operation—above cut-in but below rated wind speed—the goal is to extract as much energy as possible, often by keeping the tip-speed ratio constant and adjusting generator torque rather than blade pitch. Near and above the rated wind speed (“Region 3”), the control objective flips: the pitch system becomes highly active to limit power output, regulate rotor speed, and maintain safe loads on the generator and gearbox. In extreme winds (“Region 4”), the system commands full feathering of the blades and engages the braking system to stop the rotor completely. This staged behaviour shows that pitch control is not static but dynamic, with different priorities in different wind regimes.

Inside the hub, the hardware of a pitch system reflects these demands. Wind sensors on the nacelle feed data to the pitch controller—the “brain”—which runs algorithms ranging from simple proportional–integral controllers to model-predictive control and even machine learning in some cutting-edge designs. The controller commands actuators that rotate each blade via a heavy-duty bearing assembly designed to withstand enormous axial and radial loads. Redundant power supplies and communication links ensure that even during grid disturbances or component failures, blades can still be pitched to a safe position.

The impact on turbine performance is profound. By optimising blade angle in Region 2, pitch control allows the rotor to operate at the maximum aerodynamic efficiency (highest Cp) for the current wind, boosting energy yield and improving the economic viability of a wind farm. In Region 3, precise pitch adjustments smooth out power output despite gusts, which is critical for grid integration. IPC, in particular, reduces power fluctuations and torque oscillations, improving power quality. Equally important, active pitch control significantly reduces mechanical loads and fatigue on blades, bearings, and gearboxes, extending turbine life and lowering maintenance costs. Modern turbines can even use pitch control to contribute to frequency regulation and voltage support on the grid, something fixed-pitch machines cannot do.

In essence, at the intermediate level the pitch control system emerges as more than just a mechanism for turning blades. It is a highly integrated electromechanical and control system whose design choices—hydraulic versus electric actuation, collective versus individual control, simple versus predictive algorithms—directly affect a turbine's energy capture, reliability, grid performance, and maintenance profile. Its sophistication is a key enabler of wind energy's transition from a niche resource to a mainstream, grid-friendly power source.

Independent Pitch Systems

Modern medium- and large-scale wind turbines place very high demands on blade pitch performance. The pitch system must keep each rotor blade at the correct angle in three situations: starting, normal operation, and shutdown/feathering. By continuously adjusting the blade's angle of attack, the pitch system maximises energy capture under varying winds while protecting the turbine from overspeed or structural overload.

An independent pitch system is one where each blade has its own dedicated drive unit—hydraulic or electric—so the three blades can be pitched individually rather than only collectively. This design allows more precise control of loads and performance, which is especially important for today's multi-megawatt machines.

Pitch control in different states:

- **Starting state:** When the turbine is at rest, blades are normally "feathered" at about 90° to the wind to minimise drag. As wind speed rises to the cut-in threshold, the blades rotate toward ~45° to generate starting torque. Once the rotor accelerates to operating speed, the blades transition to about 0° to capture wind efficiently.

- **Normal operation:** Below rated power, blades stay near 0° pitch to extract maximum energy. When power output reaches or exceeds the rated value, the controller incrementally increases pitch (up to ~30°) to limit aerodynamic loading and maintain generator power at the rated level.

- **Shutdown/feathering:** For planned stops, emergency stops, or grid faults, the blades rotate back to near 90°. This "feathered" position uses aerodynamic drag to slow the rotor to zero and prevents runaway. Large turbines are designed so the pitch system can automatically feather the blades even during a power outage, using stored hydraulic or battery energy.

An independent pitch system allows each turbine blade to be controlled by its own dedicated drive unit. Rather than moving all the blades in unison, this design lets each blade adopt a different pitch angle if necessary. Such independent control improves aerodynamic performance in turbulent or asymmetric wind conditions, reduces fatigue loads on blades and bearings, and enhances overall energy capture. Modern turbines achieve this using either hydraulic or electric drive technologies.

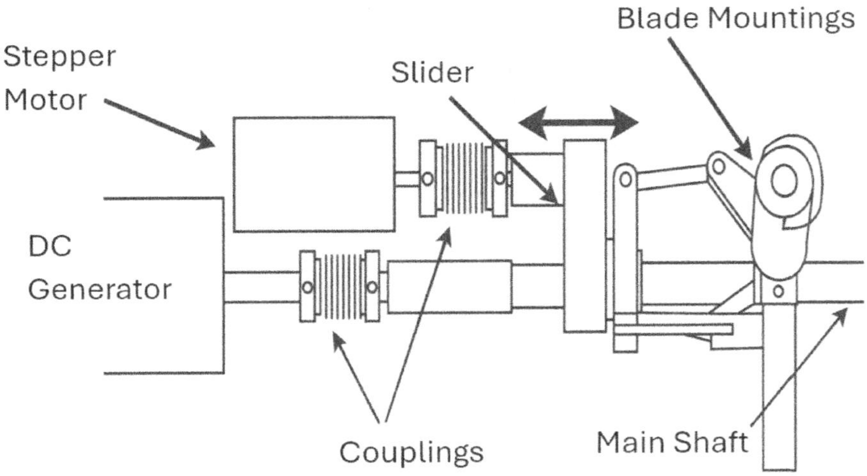

Figure 45: Working principle of the pitching mechanism.

In a hydraulic pitch system, each blade root is bolted into the inner ring of a large pitch bearing that is attached to a "three-forked" hub. The outer ring of the bearing is fixed to the hub, while the inner ring rotates with the blade. A hydraulic cylinder with an extendable piston rod connects to a pitching lever on the blade root plate. As the piston extends or retracts, it pushes or pulls this lever, causing the blade to rotate about its longitudinal axis. Because the pitching handle moves in an arc, the hydraulic cylinder itself must pivot on a swing shaft attached to the hub to follow the motion smoothly. Each cylinder contains a linear displacement sensor that measures piston travel so the controller can calculate the blade's pitch angle in real time. Hydraulic hoses and signal cables pass through the main shaft and gearbox to a rotary joint or slip ring, which transmits both hydraulic fluid and electrical signals between the rotating hub and the stationary nacelle controller. Under computer command, the hydraulic system continuously adjusts blade pitch to maintain rated speed, feathers the blades during high winds, and automatically feathers them if power is lost.

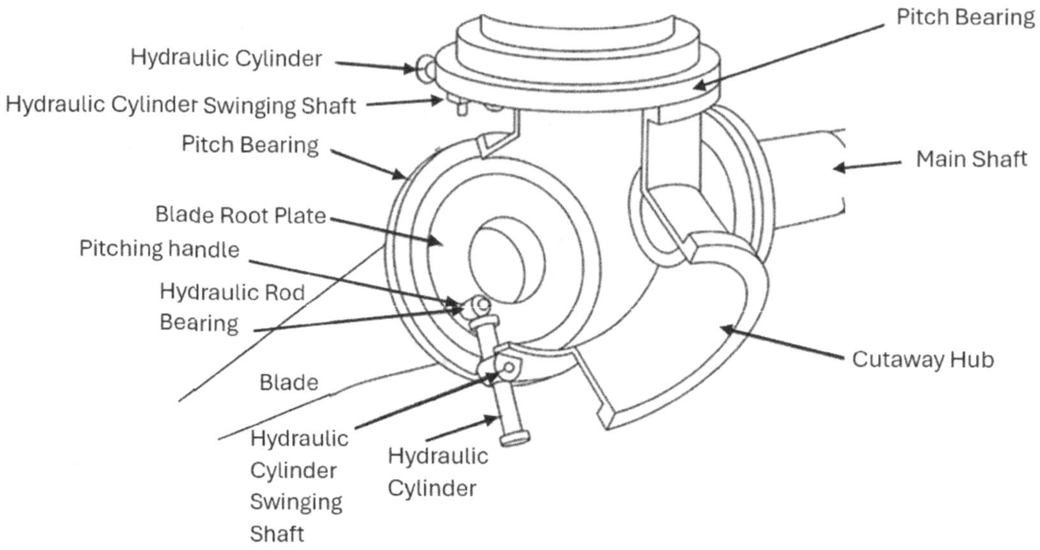

Figure 46: Hydraulic cylinder of hydraulic pitch system.

In an external gear electric pitch system, the pitch bearing's outer ring is fixed to the hub and integrates an external gear around its circumference. Each blade has its own electric pitch drive motor with a pinion gear that meshes with this external gear. The motor—whether DC, AC, or permanent magnet—drives the pinion through a gearbox, rotating the blade to change its pitch angle. Although each blade's motor operates independently, they are synchronised by the turbine's control computer. Power and control electronics are mounted inside the hub and connected to the nacelle via slip rings. Electric systems eliminate the need for hydraulic fluid, reducing leak risk and maintenance requirements, which makes them increasingly popular in new wind turbine designs.

An internal gear electric pitch system achieves the same function but reverses the gear arrangement. In this configuration, the pitch bearing's inner ring is integrated with a large internal gear, and the blade mounts to this inner ring. Three pitch motors are mounted around the hub interior, each with a pinion that meshes with its blade's internal gear. As the motor turns the pinion, it rotates the blade via the internal gear. This arrangement provides a compact layout, protects the gears from external contamination, and still allows each blade to be independently actuated but synchronised by the central controller.

Auxiliary Devices and Safety Features: Modern wind turbines rely on several auxiliary devices to ensure reliable operation and robust safety. Because the hub rotates continuously during operation, all power cables, sensor lines, and hydraulic tubes must pass between moving and stationary components without tangling or breaking. This is achieved through slip rings and rotating joints, which act as rotary interfaces. Slip rings allow electrical signals, control data, and even hydraulic fluid to transfer seamlessly from the stationary nacelle to the rotating hub.

This uninterrupted connection is essential for maintaining blade control and monitoring systems in real time.

Turbines also incorporate emergency backup systems to protect the machine during power interruptions. Accumulators in hydraulic systems or backup batteries in electric pitch systems provide stored energy to drive the pitch actuators in an emergency, enabling the blades to feather and slow the rotor even if grid power is lost. This automatic response prevents uncontrolled rotation and significantly reduces the risk of mechanical damage.

Accurate measurement of blade pitch is critical for precise control. Blade angle transducers—typically encoders and counters meshed with the pitch gear—feed real-time pitch-angle data to the turbine's controller. This feedback loop allows the control system to make rapid adjustments, optimising both performance and safety. Similarly, electric pitch motors are fitted with magnetic brakes to lock the blades in their current position when the motors are not energised. This prevents uncommanded movement of the blades due to aerodynamic or mechanical forces, further enhancing operational stability.

Independent pitch systems represent a major advance over older collective pitch mechanisms. Instead of adjusting all blades simultaneously to the same angle, an independent system allows each blade to respond individually to its own aerodynamic conditions. This finer control is especially valuable in reducing uneven stresses caused by wind shear, turbulence, or tower shadow effects. By actively balancing loads across the rotor, independent systems extend the life of critical components such as bearings, gearboxes, and blade roots.

The benefits extend beyond structural health. Independent pitch systems improve power quality by smoothing out fluctuations in electrical output during turbulent conditions. This steadier power delivery makes grid integration easier and enhances overall system reliability. They also provide greater safety and redundancy. If one actuator fails, the remaining blades can still feather, helping to slow the rotor and prevent catastrophic overspeed. This built-in redundancy is vital for turbines operating in remote or offshore locations where immediate manual intervention is difficult.

As blade lengths and rotor torque increase in today's utility-scale machines, precise control of each blade has become essential. While some medium-sized turbines still use unified pitch drives with a single actuator and linkages, nearly all new large-scale wind turbines now employ independent pitch systems—either hydraulic or electric—to meet modern standards of performance, safety, and reliability. This shift reflects the industry's move toward larger, more powerful turbines where individual blade control is no longer optional but fundamental.

13. Rotor Hub

The rotor hub is the central component that connects the blades to the main shaft.

- It contains blade bearings and pitch mechanisms, allowing each blade to rotate independently for pitch control.

- In most three-blade designs, the hub is bolted directly to the main shaft (or low-speed shaft) and transmits torque to the gearbox or generator.
- The hub must withstand massive bending moments and fatigue loads from changing wind conditions.
- Access hatches inside the hub allow technicians to service pitch motors, hydraulic lines, and blade connections.

Inside the nacelle, the gearbox and generator sit on a rigid frame. At the very front of the nacelle, the spindle—also called the main shaft—connects the rotating wind rotor to the gearbox. Its primary function is to transmit the huge torque produced by the rotor blades to the gearbox input shaft. At the same time, it must resist the bending, thrust, and torsional loads caused by wind gusts, turbulence, and rotor weight.

Because of these stresses, the spindle is manufactured from high-strength alloy or stainless steel, often forged and heat-treated to withstand fatigue and corrosion. The front end of the spindle carries a hub coupling flange, which bolts the hub securely to the shaft. The rear end couples to the low-speed gearbox input via a flexible or rigid coupling.

A distinctive feature of most modern spindles is the hollow centre (through-hole). This passage allows hydraulic lines, power cables, and signal cables—or in some designs a mechanical rod—for the pitch control system to pass from the stationary nacelle controller to the rotating hub. This design reduces clutter, protects lines from damage, and makes the hub's pitch actuators possible.

The hub is the large, cast-steel or cast-iron structure at the very front of the turbine that anchors the blades to the spindle. Together with the blades, the hub forms the wind rotor—the part of the turbine that captures wind energy.

Each blade root is bolted into a pitch bearing that sits inside the hub. The hub provides the mounting points for these bearings and houses the pitch drive mechanisms (hydraulic or electric) that rotate each blade around its longitudinal axis. By enclosing these systems inside a robust casting, the hub protects them from wind, rain, and debris.

Figure 47: Hub for Turbine Tower. Paul Anderson / Hub for Turbine Tower No 11, CC BY 2.0, via Wikimedia Commons.

Wind turbine hubs come in two main forms, which correspond to different pitch systems:

- **Spherical Hub**: This is a rounded, ball-like casting with three equally spaced flanges for pitch bearings. It can house a synchronous pitch drive (a single drive system controlling all three blades simultaneously) or independent drives. Spherical hubs are common in older or medium-sized turbines.

- **Three-Cylindrical or "Three-Pronged" Hub**: This design has three protruding cylinders or arms, each forming a blade root chamber. It is especially suited to independent pitch systems, where each blade has its own dedicated drive unit and electronics. The separate arms give technicians more space to install and service actuators, sensors, and hydraulic components for each blade.

Regardless of shape, the hub is bolted directly to the spindle flange. When assembled with three blades, it forms the complete rotor. A shroud or spinner (an aerodynamic nose cone) is often installed over the hub to streamline airflow and protect internal components from weather.

In operation, the spindle and hub work as a single assembly. The spindle provides a strong, aligned path for torque into the gearbox, while the hub supports the blades and integrates the pitch control hardware. Together, they transfer enormous aerodynamic loads into the nacelle frame and tower while allowing the blades to adjust their angle for maximum efficiency and safety.

This design—spindle with through-hole plus a hub accommodating pitch mechanisms—is fundamental to modern utility-scale wind turbines. It allows engineers to build larger rotors with sophisticated blade controls while keeping the drivetrain compact, reliable, and maintainable.

How These Components Work Together

Wind enters the rotor blades, which create lift and turn the hub and low-speed shaft. The gearbox (if present) steps up speed to drive the generator in the nacelle, producing electricity. Pitch control and the yaw system optimise blade angle and direction for maximum efficiency and safety. The tower and foundation provide structural stability, while anemometers and sensors feed real-time data to the control system. Finally, electricity flows down the tower to transformers and on to the grid connection, delivering clean power to homes and industries.

Part 2

Safety, Standards, and Compliance

Chapter 3

Occupational/Work Health and Safety Principles

Working on a wind turbine is one of the most technically demanding and high-risk occupations in the renewable energy sector. Every climb, lift, or electrical isolation task carries potential hazards that must be controlled through rigorous safety systems. *Chapter 3: Occupational/Work Health and Safety (WHS) Principles* introduces the essential frameworks, regulations, and practices that underpin safe operations for wind technicians worldwide.

This chapter establishes the foundation for understanding how health and safety principles are applied across the wind energy industry—from legislative requirements to the technical standards that govern turbine design, site operations, and workforce training. It begins by exploring the global structure of occupational health and safety regulation, comparing international and national approaches such as ISO 45001, IEC 61400, and the Global Wind Organisation (GWO) standards. Readers will see how these frameworks ensure that technicians receive consistent protection whether they are working on a remote Australian wind farm, an offshore platform in the North Sea, or a desert installation in the United States.

From there, the chapter examines how legislation and safety systems translate into day-to-day field practices. It outlines the core controls common to every wind site—working at height, electrical and high-voltage isolation, confined-space entry, heavy lifting, and lightning or weather risk management. Each of these areas is governed by strict procedures and verified through training, documentation, and supervision. Practical examples and real-world case studies show how technicians apply these systems in the field, reinforcing the link between written standards and safe behaviour.

A central theme in this chapter is **competence**—the idea that safety depends not only on compliance, but on the knowledge, skill, and judgement of every technician. Programs such as GWO's Basic Safety Training (BST) and Basic Technical Training (BTT) are discussed in detail, highlighting their role as the global minimum benchmark for entry into the wind industry. These certifications, along with national licensing and refresher training, ensure that technicians can recognise hazards, apply correct procedures, and respond effectively to emergencies.

The chapter also explains how risk assessments, Job Safety Analyses (JSA), and permit-to-work systems form the backbone of safe work processes. These tools encourage proactive hazard identification and structured decision-making before any job begins. Readers will learn how to implement safe work practices that comply with international standards, including lock-out/tag-out procedures, fall-protection systems, and emergency rescue planning.

Finally, *Occupational/Work Health and Safety Principles* emphasises the shared responsibility embedded in modern safety culture. From technicians and supervisors to site managers and equipment manufacturers, everyone has a duty of care to eliminate or minimise risks so far as is reasonably practicable. By understanding how legal frameworks, standards, and management systems integrate, readers gain the insight needed to create and sustain a culture of safety—protecting lives, assets, and the long-term integrity of the wind industry.

This chapter is therefore more than a checklist of rules—it is a practical guide to the mindset and systems that keep wind turbine technicians safe in one of the world's most dynamic and high-risk engineering environments.

Health and Safety Legislation and International Equivalents

Wind-turbine technicians operate under a framework of global and local regulations that hinge on three core safety pillars: Management Systems, Wind-Specific Engineering Standards, and Workforce Training. These elements are crucial for maintaining high safety standards among technicians operating onshore and offshore wind installations.

Management Systems: The ISO 45001 standard for occupational health and safety management systems has gained worldwide acceptance as a fundamental requirement imposed by wind turbine owners and operators. This standard emphasizes systematic approaches to risk assessment, safety training, incident investigation, and continuous improvement in workplace safety cultures. It is noted that insurers in Germany reference DGUV rules, aligning industry practices with recognized safety standards, thus enhancing the regulatory framework governing technicians' work environments [78, 79].

Wind-Specific Engineering Standards: The IEC 61400 series provides a comprehensive set of guidelines focusing on the design and safety of wind turbine technology. Key parts, particularly IEC 61400-1 and IEC 61400-24, detail the necessary structural and electrical safety measures that technicians must adhere to. These standards inform site operational protocols for managing diverse weather conditions, such as executing lock-out/tag-out procedures or conducting inspections after lightning strikes [79, 80]. Adherence to these standards ensures that safety practices are uniformly applied across different geographical locations while accommodating local environmental conditions.

Workforce Training: Workforce training is solidified through frameworks established by the Global Wind Organisation (GWO), encompassing modules that address various safety and technical competencies necessary for working in high-risk environments. The GWO's Basic Safety Training (BST) and Basic Technical Training (BTT) have become standard credentials that wind technicians need to acquire. For instance, industry bodies like Safe Work Australia recognize these training modules as benchmarks for ensuring high safety levels [79]. Continuous training is necessary not only for competency retention but also for adapting to new safety technologies and methods, underscoring the importance of regular refresher courses [80, 81].

The purpose of the GWO Basic Safety Training (BST) is to equip participants with the essential safety knowledge and skills needed to work safely in a wind turbine environment and to respond effectively to emergencies. This training forms the foundation for all technicians entering the wind energy sector.

BST covers several key modules. *Working at Heights* teaches fall prevention, rescue techniques, and the correct use of fall-arrest systems. *Manual Handling* focuses on safe lifting and carrying practices to reduce strain and injury. *Fire Awareness* introduces the science of fire, types of fires, prevention strategies, and basic firefighting techniques. *First Aid* equips participants to provide immediate care, with an emphasis on common wind-industry injuries such as electrical incidents. For offshore technicians, *Sea Survival* is included, covering safe transfer, emergency procedures, and survival at sea.

BST is regarded as an industry minimum requirement and is mandatory before any worker can enter a wind turbine tower. Certificates are typically valid for two years, after which refresher training is required to maintain certification.

The purpose of the GWO Basic Technical Training (BTT) is to provide participants with the fundamental technical skills necessary to perform specific tasks on wind turbine systems under the supervision of experienced personnel. It builds on the safety foundation provided by BST.

BTT consists of several key modules. The *Mechanical* module covers the principles of mechanical systems, gearboxes, brakes, and inspection tasks. The *Electrical* module focuses on reading and understanding electrical diagrams, assembling components, and performing safe measurement techniques. The *Hydraulics* module explains the basics of hydraulic systems, including pumps, actuators, and valves. Finally, the *Bolt Tightening* module provides practical training on the use of manual and power-operated tightening tools—essential skills for maintaining turbine integrity.

Unlike BST, BTT is considered an “enduring qualification.” This means the certificate does not have a fixed expiry date, although technicians are expected to keep their skills current through ongoing practice and site experience.

BST acts as a technician's passport to the wind energy sector. It provides the essential safety knowledge needed to protect themselves and others, making it the first step for anyone entering a turbine.

BTT builds on that foundation by giving participants practical technical skills to carry out tasks under supervision, demonstrating readiness for modern wind energy work. Together, these two programs ensure that technicians are not only safe but also technically competent to perform their roles effectively.

Europe and the United Kingdom

Within the European Union, the umbrella law is the Framework Directive 89/391/EEC, which requires employers to prevent risks, assess hazards, and train workers. Additional directives on work equipment (2009/104/EC) and PPE (89/656/EEC) govern MEWPs, hoists, harnesses, rescue kits, and tools. The Machinery Directive 2006/42/EC applies to turbines and access equipment placed on the market. Most turbines show compliance with IEC 61400-1/-24 for design loads and lightning, which in turn shapes site controls.

In the United Kingdom, which still mirrors the EU approach post-Brexit, daily work is governed by the Health and Safety at Work etc. Act 1974 and the Work at Height Regulations 2005—the primary reference for planning climbs, competence, rescue, and anchor points. Other key rules include PUWER (work equipment), LOLER (lifting operations and gear), the Electricity at Work Regulations 1989, MHSWR, RIDDOR, and CDM 2015 for construction or major component exchanges. The UK Health and Safety Executive (HSE) provides detailed guidance.

North America

In the United States, OSHA standards are split between construction and general industry. Typical wind work touches both:

- 29 CFR 1926 (construction) covers fall protection (§.501), cranes and rigging (Subpart CC), steel erection, and related hazards.
- 29 CFR 1910 (general industry) covers walking–working surfaces, lock-out/tag-out (§1910.147), confined spaces (§1910.146), hazard communication (§1910.1200), electrical safety (Subpart S), and power-generation work (§1910.269).

Sites also adopt consensus standards such as NFPA 70E (arc-flash risk categories, energized work permits) and ANSI fall-protection standards. Offshore U.S. projects add marine-safety oversight by BOEM and BSEE, with operators layering GWO Sea Survival and marine transfer procedures on top of OSHA rules.

In Canada, the Canada Labour Code Part II and each province's OHS Act set the legal duties. The Canadian Centre for Occupational Health and Safety (CCOHS) aggregates requirements for fall protection, confined space, electrical safety, and WHMIS. Common CSA standards include CSA Z259 for fall-protection systems and CSA Z462 for electrical safety and arc-flash. Atlantic offshore sites add marine authority requirements, and most mandate GWO plus marine survival training.

Australia

The Model Work Health and Safety (WHS) Act 2011 and WHS Regulations (adopted in most states and territories) establish duty of care, risk management, and competency requirements. Safe Work Australia codes of practice cover fall prevention, confined spaces, electrical work, hazardous manual tasks, and construction. State electrical regulators and AS/NZS standards govern high-voltage switching and installations. GWO BST/BTT is now widely used on Australian wind sites.

Germany and France

In Germany, the Arbeitsschutzgesetz (ArbSchG) and BetrSichV (Industrial Safety Ordinance) underpin risk assessments and the safe use of work equipment. The DGUV rules are highly practical for wind—DGUV Rule 113-004 on rope access and DGUV 112-198/-199 on fall-arrest are two examples—plus DGUV information sheets tailored to wind farms.

In France, the Code du travail sets legal duties. Working at height and habilitation électrique (H0/B0, BS/BE, B2V, etc.) are mandatory for electrical tasks. INRS publishes the technical guidance used by employers and training bodies.

What This Means Day to Day for Technicians

Everywhere in the world, core controls converge:

- **Work at Height.** Planned climbs, certified anchor systems, twin-lanyard use, fixed ladders or elevators, rescue plans, and regular practical rescue drills (GWO ART where required). UK WAH, EU directives, OSHA Subparts D & M, and DGUV rules all align on this.
- **Electrical and High Voltage.** Lock-out/tag-out, test-before-touch, arc-flash PPE, energized work permits, and competency or authorization records—framed by OSHA 1910, NFPA 70E, CSA Z462, and EU work-equipment rules.
- **Confined Spaces.** Nacelle, hub, and tower interiors often meet confined-space criteria and therefore require entry permits, atmospheric testing, attendants, and retrieval procedures.
- **Lifting and Major Component Exchanges.** LOLER/PUWER (UK/EU) or OSHA crane rules (US) apply, plus certified rigging plans and competent lifting supervisors.

- **Lightning and Weather.** Procedures follow IEC 61400-24 for strike inspections, earthing continuity, surge protection, and weather shutdown/wind-speed limits.
- **Marine/Offshore Transfers.** CTV, SOV, or helicopter transfers require marine survival training (GWO Sea Survival; UK often adds OPITO BOSIET/FOET), immersion suits, man-overboard drills, and permit-to-work interfaces with vessel masters.
- **Management Systems.** Documented risk assessments (JSA/JHA), toolbox talks, incident reporting, and continual improvement under ISO 45001 tie everything together.

Equipment and Documentation Commonly Audited

Technicians are typically asked to show valid GWO certificates (BST/BTT, ART for rescue, Sea Survival offshore) plus country-specific certifications such as UK Work at Height, French habilitation électrique, or German DGUV rope-access credentials. Equipment such as calibrated anemometers, gas meters, LOTO kits, arc-rated PPE, and rescue gear must have traceable inspection records. Emergency-response plans must cover medical care, trauma, hypothermia, lightning/ice procedures, helicopter winch plans offshore, and current first-aid/CPR certificates.

Work at Height

Wind turbines are tall structures—typically 60 m to 150 m hub height—so climbing and working at height is routine for technicians. This makes fall prevention, safe access, and emergency rescue planning some of the most tightly regulated aspects of the job worldwide. Work at height requirements sit on top of general occupational health and safety (OHS) laws and are reinforced by wind-specific training such as the Global Wind Organisation's Advanced Rescue Training (ART).

Climbs into a nacelle or up a tower are treated as planned high-risk activities. Before starting, technicians complete a risk assessment or job safety analysis (JSA), check weather and lightning conditions, verify that rescue equipment is available, and sign a permit-to-work. This process ensures that anchor points, ladders, and elevators are inspected and that rescue arrangements are in place.

Inside towers and nacelles there are fixed fall-arrest systems (rail or cable). Anchor points must be certified to national or EU standards (e.g., EN 795 or CSA Z259). Technicians clip on using full-body harnesses and twin-lanyard systems so they are always connected when moving past intermediate anchors. Many towers also have vertical lifeline systems or guided-type fall arresters for climbing ladders.

Where installed, service lifts or climb-assist devices reduce fatigue, but even then a fall-arrest harness and connection are still required.

The "100% tie-off" principle applies: before detaching from one anchor, the technician clips onto the next anchor with the second lanyard. Harnesses must be inspected before each use and recorded in a traceable inspection log. Self-retracting lifelines (SRLs) or inertia reels may also be used in nacelles and hubs.

Figure 48: Wind turbine work at height.

Rescue planning is mandatory. Employers must have written rescue and evacuation procedures for tower interiors, nacelles, hubs, and blades. This covers both self-rescue and assisted rescue, including scenarios such as an unconscious worker on a ladder or inside a confined space. Technicians practice these rescues regularly—at least annually—to keep their skills current. The GWO Advanced Rescue Training (ART) standard formalises these practical rescue drills for blade, hub, and nacelle scenarios, including stretcher use and casualty lowering or raising systems.

Figure 49: Working on the hub of a turbine. Paul Anderson/Working on the Hub of Turbine No 2 at Scout Moor, CC BY 2.0, via Wikimedia Commons.

Alignment of Laws and Standards:

- **United Kingdom** – *Work at Height Regulations 2005* require planning, competent persons, appropriate equipment, inspection of anchor systems, and a rescue plan before work begins.
- **European Union** – Framework Directive 89/391/EEC plus the Work Equipment and PPE directives mandate risk assessments, equipment certification, and training for working at height.
- **United States** – OSHA 29 CFR 1910 Subpart D (Walking-Working Surfaces) and 29 CFR 1926 Subpart M (Fall Protection) set fall-protection requirements. Turbine owners typically also adopt ANSI Z359 standards for harnesses, lanyards, and anchors.
- **Germany** – DGUV rules such as DGUV 112-198/-199 (use of personal fall-arrest equipment) and DGUV Rule 113-004 (rope access) provide detailed, practical guidance for wind-farm work at height.

Across all these jurisdictions, the core controls are essentially the same: plan the climb, use certified anchor systems, maintain continuous tie-off with twin lanyards or SRLs, inspect equipment regularly, and train/refresh practical rescue skills.

Electrical and High Voltage

Wind turbines contain multiple electrical hazards that technicians must manage. Common sources of shock and arc-flash risks include the generator stator (typically 400–690 V AC), high-voltage switchgear or transformers (5–36 kV onshore; 33–66 kV offshore; and up to 132 kV or more on export circuits), and auxiliary panels. Stored energy hazards arise from DC-link capacitors in converters, UPS and battery systems, spring-charged breakers, and induced or capacitive energy in long cables. These hazards are intensified because work often takes place in confined and elevated spaces—inside nacelles, hubs, or towers—where egress is limited and enclosures are metallic.

Before any work begins, technicians must plan the isolation using a single-line diagram and a written energy-control procedure. This includes identifying all energy sources such as grid HV, generator LV, auxiliary feeders, control power, batteries, and backfeed paths from cranes or temporary generators. Once identified, workers de-energize, isolate, and secure the equipment by opening or racking-out devices, applying personal locks, and tagging them with their name, contact details, time, and reason. Stored energy must be released—DC capacitors discharged (per OEM time constants), springs blocked or released, and pneumatics or hydraulics vented. A try-out is then performed to verify zero response locally and via SCADA. Where multiple trades are involved, a group LOTO lock box is used, with the supervisor's lock applied last and removed last. This practice aligns with OSHA 1910.147, CSA Z462, and EU "use of work equipment" rules.

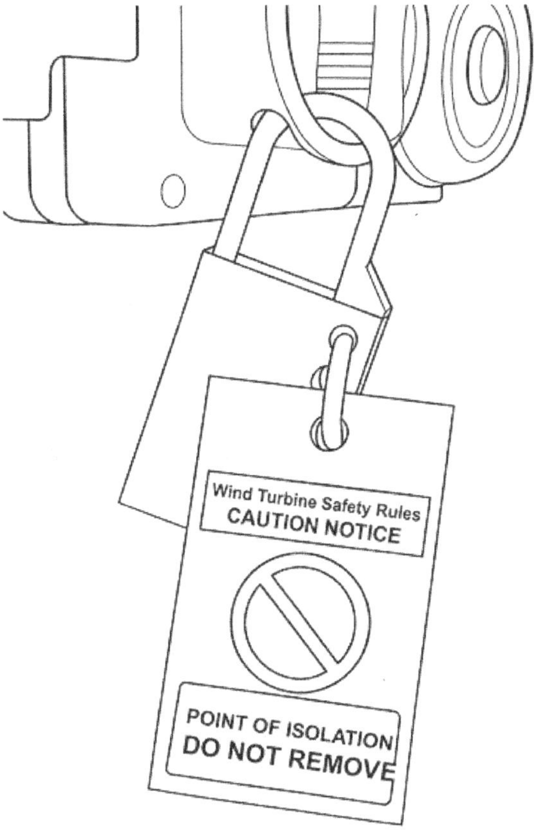

Figure 50: Lock-out/tag-out.

Technicians follow a Prove–Test–Prove sequence: first verifying their tester on a known live source, then testing all phases (including phase-to-ground), and finally re-verifying the tester. Where required, portable protective grounding and bonding is applied after testing to eliminate induced or capacitive charges—especially on long medium-voltage feeders and offshore arrays. Only rated test instruments (CAT III/IV) and insulated tools appropriate for the site's voltage class should be used.

The "Prove–Test–Prove" methodology is a critical three-step verification process applied in electrical safety management, particularly when establishing the absence of voltage before interfacing with electrical systems. This process is essential to adhere to safety standards, such as those set forth by OSHA 1910, NFPA 70E, and CSA Z462, ensuring that all conductors are properly de-energized before any maintenance or handling occurs.

<u>Step 1 – Prove Your Tester Works</u>

The first step involves verifying that the electrical testing instrument functions correctly by checking it against a known live source. This initial verification is crucial; if the tester malfunctions—due to a dead battery, damaged leads, or inherent faults—it may provide misleading readings, resulting in potentially hazardous situations. Effective risk management in electrical safety hinges on ensuring functional detection instruments, as highlighted by Neitzel, who emphasizes the importance of robust safety protocols and the ramifications of equipment failure in his review of electrical hazards and risk assessment practices [82].

Step 2 – Test the Circuit or Equipment

After confirming the tester's operational integrity, the next phase is to test all conductors and phases of the equipment needing work. This step aims to identify any present voltage, guaranteeing that the equipment is indeed de-energized. In complex systems, such as wind turbines, backfeed and induced voltages can complicate the verification process. The thoroughness of this step ensures that technicians do not assume systems are safe without adequate checks [83].

Step 3 – Prove Your Tester Again

The concluding step of re-proving the tester on a known live source reinforces confidence in the previous readings. Instruments can fail during use; thus, a second verification is vital to ascertain that the readings indicating "zero volts" are valid. This step minimizes the risk of human error associated with assuming the tester remains functional throughout the process [84, 85]. The importance of thorough and repeat testing is further corroborated by findings in electrical safety audits that advocate for regular checks and maintenance of equipment to sustain operational reliability [86].

The "Prove–Test–Prove" process is fundamental in preventing critical errors that could arise from faulty tester readings and unwarranted assumptions about equipment isolation. In environments laden with risks such as stored energy or induced voltage, this sequence is not merely procedural but a life-saving practice. The electrical power industry, including sectors like wind energy, relies on technicians adhering to this rigorously defined and executed protocol to mitigate hazards that could affect worker safety [87, 88].

In practice, technicians working near live electrical systems must deploy CAT III/IV rated instruments, ensuring not only that they are safe for the expected voltage classes but also that their integrity and functionality are consistently verified through every step of the work process [89].

Overall, the "Prove–Test–Prove" approach is critical for maintaining electrical safety standards and significantly mitigating risks associated with electrical hazards in various operational fields.

Standards such as NFPA 70E, CSA Z462, EN 50110-1, and IEC 61482-2 require either an incident-energy analysis or a PPE category approach. Equipment must be labelled with available fault current, protective device clearing times, arc-flash boundaries, and the required PPE. Technicians observe both shock approach boundaries (limited and restricted) and the arc-flash boundary (where incident energy drops to 1.2 cal/cm^2). PPE must match the label or category—typically arc-rated coveralls or layered systems, arc-rated face shield or hood, balaclava, leather and arc gloves over correctly classed rubber insulating gloves, hearing protection, EH-rated boots, and safety glasses. Safe work methods include using insulated mats, barriers, remote racking or switching, and a "no live work" policy unless justified under an energized work permit.

If de-energizing is infeasible—such as for diagnostics or when de-energizing poses a greater hazard—technicians must obtain an energized work permit. This includes documented justification, a shock and arc assessment, approach boundaries, PPE requirements, work procedures, a job briefing, and approval by an authorized manager. Only live-work tools (insulating sticks, covers, blankets) are used, and safe distances must be maintained at all times.

Sites maintain authorization records specifying who may operate HV/LV switchgear, perform testing, apply grounds, or issue and accept permits. Training for authorized workers typically covers LOTO, test-before-touch, arc-flash awareness, HV switching, first aid/CPR/AED, and GWO Advanced Rescue for electrical emergencies. Gloves and tools must be controlled: dielectric gloves tested at mandated intervals, torque tools and meters calibrated, and PPE inspected before each use.

Wind turbines present unique challenges. DC-link capacitors require adherence to OEM discharge times and verification of zero energy at designated test points before opening converter cabinets. Backfeed risks from cranes, temporary power, or auxiliary generators must be included in the energy-control plan. Offshore, induced voltages on long submarine cables demand portable grounds and checks on permanent earthing continuity, with attention to corrosion and salt exposure. Work in confined spaces such as nacelles or hubs requires ventilation, gas testing, and staging of rescue equipment. For transformers and MV switchgear, HV access rules, interlocks, key exchange systems, and test/earth switches apply, and only authorized persons may perform switching. SCADA coordination is essential—permits must be aligned with the control room, turbines placed in local or safe maintenance mode, and remote commands blocked.

A typical pre-job checklist includes reviewing single-line diagrams and LOTO steps to identify every source and backfeed; confirming permits, weather windows, communications, and rescue plans; inspecting PPE, tools, and test instruments; executing LOTO and stored-energy release; proving absence of voltage and applying grounds if needed; maintaining boundaries with posted attendants; and keeping only essential personnel in the arc-flash zone. Upon completion, workers remove grounds, clear tools, remove personal locks, function test the system, and document its return to service.

OSHA 1910 provides the legal baseline in the U.S. for controlling hazardous energy, walking-working surfaces, electrical safety, and power-generation work, and is often paired with NFPA 70E for arc-flash methods and PPE. CSA Z462 in Canada harmonizes with NFPA 70E and is enforced alongside provincial OHS acts and electrical codes; CSA Z259 covers fall-protection equipment. In the EU, employer duties fall under Framework Directive 89/391/EEC, with the Use of Work Equipment Directive 2009/104/EC and national transpositions governing isolation, LOTO, and competence. EN 50110-1 provides detailed guidance on the operation of electrical installations, while IEC 61482-2 covers arc-rated PPE.

While terminology differs between countries, the expectations are consistent: plan the isolation, verify absence of voltage, control arc-flash energy, justify any energized work with permits, and prove competence with training and records. Following these practices is how wind sites keep technicians safe around powerful electrical systems.

Confined Spaces

In wind-turbine environments, a confined space is any area large enough for a worker to enter and perform tasks but with limited or restricted entry and exit and not designed for continuous occupancy. Such spaces can present serious hazards, including oxygen deficiency, toxic or flammable atmospheres, or entrapment risks.

Typical examples on wind turbines include cramped nacelle interiors with only one or two hatches, long tower shafts with restricted ventilation, sealed hubs around blade roots and pitch mechanisms, and converter or transformer compartments at the tower base. Even when these areas are technically "open" at one end, hazardous atmospheres may still accumulate — such as ozone, carbon monoxide from welding, leaked hydraulic oil vapours, or low oxygen following a fire suppression discharge.

Figure 51: Turbine room at the top of a wind turbine This shot shows the turbine. Just to the left at the top is a ladder leading to an access hatch. Allan Macdonald, CC BY-SA 2.0, via Wikimedia Commons.

Working inside these areas exposes technicians to a range of risks. Atmospheric hazards include low oxygen levels, flammable vapours, refrigerants, cleaning solvents, or hydrogen and ozone from batteries and electrical gear. Heat stress is also common because equipment generates heat and ventilation is limited, and long climbs increase exertion.

Limited egress adds another challenge — narrow ladders, hatches, and vertical climbs make self-rescue difficult. There is also the danger of falling objects or moving machinery during maintenance work, as tools and components may be suspended overhead. Finally, electrical hazards and stored energy from switchgear, capacitors, and spring-loaded mechanisms in cramped cabinets pose additional threats.

Figure 52: Climbing Wind Turbine Ladder. Idaho National Laboratory, CC BY 2.0, via Wikimedia Commons.

Across most jurisdictions — including OSHA 1910.146 in the U.S., CSA Z1006 in Canada, EU Directive 89/391/EEC and national rules, the UK Confined Spaces Regulations 1997, Germany's DGUV and TRGS, and the French Code du Travail — similar controls are mandated for confined-space work.

Workers must obtain a written entry permit specifying hazards, controls, entrants, attendants, and the duration of work, which is reviewed and signed off before entry. Atmospheric testing using calibrated multi-gas meters must be done before and during entry to monitor oxygen, combustible gases, and toxins. Spaces should be ventilated and isolated by purging or ventilating and by locking out or tagging out all electrical, hydraulic, and mechanical systems feeding the space.

A trained attendant must remain outside to monitor entrants and summon rescue if needed. Two-way radios or intercom systems are essential where noise or distance makes voice communication impractical. A rescue plan and retrieval system must be staged in advance rather than relying solely on public emergency services. This includes using tripods, davits, inertia reels, and casualty-haul systems inspected before use, with technicians trained in GWO Advanced Rescue Training (ART) or equivalent rope/rescue courses. All personnel must also be competent and trained in confined-space hazards, atmospheric testing, emergency procedures, first aid/CPR, and proper PPE use, with refresher training provided at prescribed intervals.

Wind turbines add unique challenges to confined-space safety. The vertical distance of tower climbs means retrieval systems must be capable of lifting from high and tight spaces. Offshore turbines bring extra factors such as vessel transfer and marine survival, increasing the complexity of rescue plans.

Simultaneous operations — such as lifting heavy components, electrical switching, and confined-space entry — often occur together, requiring integrated permit-to-work coordination. Turbines must also be placed in a "maintenance" or "local control" mode through SCADA systems to prevent automatic start-up while technicians are inside.

Any time you enter a nacelle, hub, or tower interior, treat it as a potential confined space until hazards are fully assessed. Always obtain a valid entry permit, test the atmosphere before and during entry, and ensure a trained attendant and a robust rescue plan are in place. Verify that all energy sources are locked out or tagged out, and wear the correct PPE, including fall protection, respiratory protection if needed, proper lighting, and reliable communication devices.

Lifting and Major Component Exchanges

Wind turbines contain extremely large and heavy components—blades weighing several tonnes, gearboxes exceeding 50 t, and generators or transformers positioned hundreds of feet above ground. Replacing ("exchanging") these parts involves high-risk lifting operations with cranes, hoists, or winches in exposed environments. Even small mistakes can lead to dropped loads, structural damage, or serious injuries. This is why every jurisdiction requires formal lift planning, competent supervision, and the use of certified equipment.

In the United Kingdom and European Union, the Lifting Operations and Lifting Equipment Regulations 1998 (LOLER) require all lifting operations to be properly planned, supervised, and carried out by competent persons, with lifting equipment thoroughly examined and certified. The Provision and Use of Work Equipment Regulations 1998 (PUWER) govern the safe use and maintenance of equipment such as winches, hoists, MEWPs, and hydraulic tools. Both sets of rules stem from EU Directives 2009/104/EC and 89/655/EEC on work equipment safety.

Figure 53: Turbine Blade lift. Paul Anderson, CC BY-SA 2.0, via Wikimedia Commons.

In the United States, OSHA 29 CFR 1926 Subpart CC (Cranes and Derricks in Construction) and related sections in 1910 (general industry) govern crane use, rigging, signalling, and operator certification. OSHA also expects written lift plans, qualified signalpersons, and annual equipment inspections.

Other jurisdictions follow similar models. Canada applies provincial OHS Acts and CSA standards (such as CSA Z150 for cranes). Australia enforces Model WHS Regulations and state electrical/lifting codes. Germany uses BetrSichV and DGUV rules, including DGUV 100-500 and DGUV 109-017 for wind energy. Although the titles differ, the core duties everywhere are the same: plan the lift, supervise it with competent people, use certified gear, and keep traceable inspection records.

Figure 54: Tower construction. Paul Anderson, CC BY-SA 2.0, via Wikimedia Commons.

All cranes, hoists, slings, shackles, and spreader bars must carry a valid inspection certificate—typically renewed every six or twelve months—and their load ratings (WLL/SWL) must be clearly marked. Any custom lifting brackets for blades or gearboxes must be engineer-approved.

Before any major lift, a written rigging plan is produced showing load weight, centre of gravity, sling angles, wind limits, crane position, and exclusion zones. On complex lifts, such as multi-crane or tandem lifts, an engineer may prepare a full "lift study." Everyone involved should review the plan during a pre-lift briefing.

A competent supervisor oversees the lift, ensuring rigging is correct, weather limits are respected, and the plan is followed. Signallers or "slingers" must be trained and authorised (such as through GWO Slinger/Signaller or a national equivalent), and only the appointed signaller gives instructions to the crane operator.

Figure 55: Construction of a Wind Turbine. Greece-China News, Public Domain, via Pexels.

Wind speed, visibility, and ground bearing capacity are checked before each lift. Most lift plans specify maximum allowable wind speeds for each component, and ground mats or temporary roads may be needed for heavy mobile cranes.

Before lifting inside or near a turbine, SCADA and lock-out/tag-out procedures must place the machine in a safe maintenance state. This prevents automatic yaw or pitch movement while personnel and cranes are in the drop zone.

Exclusion zones, tag lines, and barriers control access under suspended loads. A rescue plan must be in place for dropped tools or injuries at height, especially when hoists are used inside towers.

Day-to-Day Checklist for Technicians

Verify that lifting equipment certificates are current, inspect slings, shackles, and tag lines before use, and review the rigging plan so you understand your role. Check wind speed and weather forecasts and pause the lift if conditions exceed limits. Stay out of exclusion zones unless your task requires entry, and maintain clear communication with the lift supervisor and crane operator.

Whether you're working in the UK under LOLER/PUWER, in the US under OSHA crane rules, or elsewhere under equivalent laws, the essentials are the same: every lift must be planned, supervised by competent people, and executed with certified equipment and strict weather limits. On wind farms, where components are massive and heights extreme, these controls are not just paperwork—they're what keep technicians safe and protect the turbine from catastrophic damage.

Lightning and Weather

Wind turbines are tall, conductive structures placed in exposed environments. This makes them prime targets for lightning strikes and vulnerable to high winds, icing, or storms. If lightning or weather controls are neglected, it can lead to catastrophic equipment damage, electrical faults, or severe injury to technicians. IEC 61400-24 sets the international baseline for lightning protection design and maintenance, and most operators align their procedures to it.

After any detected or suspected lightning strike, turbines must be inspected before re-entry. Technicians check blade lightning receptors, down-conductors, and bonding points for visible damage or continuity loss. Blades may need non-destructive testing (NDT) to locate internal delamination caused by strikes. The nacelle and tower earthing system is also visually checked and tested.

IEC 61400-24 requires a verified, low-impedance path from blade receptors through the hub, nacelle, and tower to the grounding grid. During maintenance, you'll see technicians use earth-resistance meters or continuity testers to confirm that all joints, bonding straps, and surge connections are intact. Offshore turbines require special attention to corrosion at grounding points.

Turbine electrical systems incorporate surge arresters and transient-voltage suppressors on the generator, converters, and low-voltage control circuits. Part of routine maintenance is inspecting these devices, replacing expired arresters, and confirming bonding of cable shields to the main earthing system. This is what protects sensitive SCADA and pitch systems from the indirect effects of a strike.

Each turbine has defined cut-in, rated, and cut-out wind speeds in its control software (often around 3–4 m/s cut-in, 12–15 m/s rated, and 20–25 m/s cut-out). When actual wind speeds exceed the cut-out value or lightning is detected within a set radius, the SCADA system automatically feathers the blades and stops the turbine. Site procedures also prohibit climbing or working aloft during high winds, lightning activity, or ice-shedding conditions. This is why weather monitoring and anemometer calibration are part of your safety checks.

Technician practices in the field:

- Always review the site's weather and lightning policy before planning work.

- Use SCADA or the control room to confirm that turbines are locked out and cannot auto-restart.
- Stop climbs if wind gusts exceed the tower's safe-climb threshold.
- Wear arc- and flame-resistant PPE when performing post-strike inspections on electrical gear.
- On offshore sites, coordinate with vessel masters on lightning and storm-avoidance protocols.

Marine/Offshore Transfers

Accessing an offshore wind turbine involves moving from a vessel to a fixed ladder, boat landing, or suspended basket—or sometimes arriving by helicopter. Unlike onshore work, sea state, tide, visibility, and the movement of vessels and structures create constantly changing hazards. Slips, falls, crush injuries, or immersion in cold water can occur in seconds if procedures aren't followed. For this reason, training, correct equipment, and strict permit interfaces are essential.

Crew Transfer Vessels (CTVs): These are fast catamarans or similar boats that push up against the turbine's boat landing. Technicians only step across when the bow is firmly engaged and conditions such as wind, wave height, and current are within agreed limits.

Service Operation Vessels (SOVs): Larger ships equipped with motion-compensated gangways ("walk-to-work" systems). They provide a safer transfer in rougher seas but still require detailed safety briefings, appropriate PPE, and vessel permits.

Helicopter Transfers: Used where distances are long or sea conditions are too rough for boats. These transfers may involve winching directly onto nacelles or platforms and expose crews to rotor downwash and sling or hoist risks.

All offshore technicians must complete specialist training before transfer. GWO Sea Survival provides baseline skills in personal flotation, immersion techniques, sea evacuation, life-raft entry, hypothermia awareness, and offshore hazards. In the UK and North Sea regions, OPITO BOSIET/FOET adds modules such as helicopter ditching, compressed-air emergency breathing systems, and Helicopter Underwater Escape Training (HUET). Many operators also require GWO Advanced Rescue Training (ART) and offshore first aid, particularly for team leads.

Figure 56: Wind turbine construction offshore. Department of Energy and Climate Change, CC BY-ND 2.0, via Flickr.

Key protective equipment includes correctly sized and inspected immersion suits and lifejackets, often fitted with personal locator beacons (PLBs) and AIS man-overboard transmitters. Fall-arrest harnesses are needed for climbing ladders from the boat landing up the tower. Offshore-rated helmets, gloves, and eye protection are mandatory, along with anti-slip footwear for deck work. Radios and other communication devices must be intrinsically safe where required, with coordinated channels between the vessel and turbine.

Before any transfer, the turbine owner's control room, vessel master, and team lead confirm the weather window, access point, lock-out/tag-out status of the turbine, and emergency arrangements through a formal permit-to-work process. Crews regularly conduct man-overboard (MOB) drills using recovery equipment such as Jason's cradles, throw lines, and rescue boats. Each transfer mode has defined weather limits for significant wave height and wind speed, and operations stop if conditions exceed those limits. Clear exclusion zones are marked around boat landings or hoist areas, allowing only authorized personnel inside.

Figure 57: Marine transport of wind turbines. William Starkey, CC BY-SA 2.0, via Wikimedia Commons.

Technicians should always attend the vessel safety briefing, know muster points and alarms, and correctly don immersion suits and lifejackets while checking PLB functionality. During transfer, maintain three points of contact and wait for the "go" signal from the deckhand or gangway operator. Report any slips, near misses, or equipment defects immediately.

Offshore wind transfers combine maritime and high-voltage/height hazards. Operators therefore layer GWO Sea Survival and, in some regions, OPITO BOSIET/FOET on top of national maritime and work-at-height rules. Immersion suits, MOB drills, and close permit-to-work coordination with vessel masters are not just formalities—they are vital measures that keep technicians safe in one of the most demanding environments in renewable energy.

Management Systems

A safety or occupational health & safety (OH&S) management system provides the framework for controlling risk on a wind farm. Instead of treating each job as an isolated activity, the system sets consistent procedures for identifying hazards, communicating controls, training workers, and improving over time. Most operators adopt ISO 45001, the international standard for OH&S management systems, to organise these elements. Although ISO 45001 isn't a law, it is widely required by turbine owners and contractors and aligns with regulatory duties worldwide.

Before any task—climbing a tower, electrical switching, or a major lift—technicians complete a Job Safety Analysis (JSA) or Job Hazard Analysis (JHA). This is a written risk assessment identifying each step of the task, the hazards present (fall, electrical, pinch points, weather), and the controls to be applied (LOTO, harnesses, exclusion zones). Each person involved reviews and signs it so there's a shared understanding of the risks and mitigations.

Once hazards and controls are documented, the team holds a short toolbox talk at the work site. The supervisor or competent person walks through the JSA/JHA, confirms roles, checks PPE, and highlights any new site conditions such as high winds or simultaneous operations. These talks are vital for catching last-minute changes and ensuring everyone understands the plan.

If something goes wrong—or almost does—workers must report it. ISO 45001 systems require a process for capturing incidents, near misses, unsafe conditions, and observations. These are logged into the operator's system (paper form, app, or SCADA-linked database), investigated to find root causes, and used to adjust procedures or training. This is how lessons are learned and shared across sites.

A key part of ISO 45001 is the "Plan–Do–Check–Act" cycle. Management regularly reviews hazard data, training records, audit findings, and incident reports to spot trends. They then update risk assessments, retrain staff, or invest in new equipment to eliminate or reduce hazards. Workers are consulted during these reviews so improvements reflect real conditions at the turbine.

On a well-run wind farm, you'll see clear JSA/JHA templates, sign-in sheets for toolbox talks, lock-out/tag-out checklists, and a simple way to report hazards. You'll also see supervisors closing the loop—sharing investigation results, updating procedures, and explaining why changes are made. All of this is the practical expression of ISO 45001: not paperwork for its own sake but a structured way to keep people safe and continually improve safety performance.

Risk Assessments and Safe Work Processes

Wind turbines bring together several high-risk activities: working at height, exposure to high-voltage equipment, heavy lifts, rotating machinery, and challenging weather conditions. A risk assessment is the formal process of identifying these hazards, evaluating their likelihood and severity, and selecting controls before any work begins. It forms the foundation for safe work processes. Without it, technicians may be exposed to hidden hazards such as arc flash, oxygen-deficient atmospheres, or unstable loads.

Most operators use a Job Safety Analysis (JSA) or Job Hazard Analysis (JHA) to document hazards and controls. These assessments typically include:

- **Task breakdown** – listing each step of the job, such as climbing the tower, opening a converter cabinet, or handling a blade bolt.

- **Hazard identification** – highlighting potential risks like falls, electric shock, manual handling injuries, dropped objects, or weather impacts.
- **Risk evaluation** – rating likelihood and consequence using a colour-coded matrix.
- **Control measures** – applying elimination, substitution, engineering controls, administrative controls, and personal protective equipment (PPE).
- **Sign-off** – workers and supervisors review and sign the document, confirming understanding of the plan.

Risk assessments must be reviewed and updated whenever conditions change, such as when new contractors arrive, new equipment is used, or the weather shifts.

Safe work processes are the practical outcome of risk assessments. They translate identified hazards into clear, step-by-step procedures that technicians follow at the work site. These processes act as the blueprint for carrying out tasks safely, ensuring that all foreseeable risks are controlled before, during, and after work.

A key element is the permit-to-work system, which covers high-risk activities such as confined-space entry, electrical switching, lifting operations, and hot work. Permits specify hazards, control measures, and authorized personnel, ensuring that work is coordinated and monitored.

Lock-out/Tag-out (LOTO) procedures are another cornerstone. They require technicians to isolate and verify that equipment is de-energized before entering or working on it. This protects workers from unexpected start-up or residual energy.

For working at height, safe work processes mandate certified anchor systems, twin-lanyard climbing techniques, and a pre-planned rescue strategy in case of a fall. Similarly, electrical and high-voltage controls include test-before-touch practices, arc-flash personal protective equipment (PPE), and the use of energized work permits when de-energizing equipment is not feasible.

Heavy lifts are controlled through rigging plans and lifting protocols. This includes using certified slings, preparing detailed load plans, and maintaining exclusion zones to keep people clear of suspended loads. Weather conditions are also critical; weather and lightning protocols define shutdown limits and evacuation procedures to prevent work during unsafe conditions.

Finally, emergency preparedness underpins all safe work processes. Rescue kits, first-aid readiness, and drills for fire, man-overboard, or medical emergencies ensure that crews can respond quickly if something goes wrong.

These safe work processes are not generic. They are tailored to the specific turbine model, site conditions, and crew competency. This customization ensures that controls remain effective and relevant to the actual hazards technicians face on a particular job.

Case Study Scenario: Gearbox Exchange on a Coastal Wind Farm

A team of four technicians is assigned to replace a failed gearbox on a 3 MW offshore wind turbine located 15 nautical miles from shore. The operation involves climbing 80 metres up the tower, isolating high-voltage systems, disconnecting the old gearbox, and rigging it out through the nacelle for transfer to a service vessel. Weather conditions are forecast to be marginal, with moderate swell and gusts up to 14 m/s.

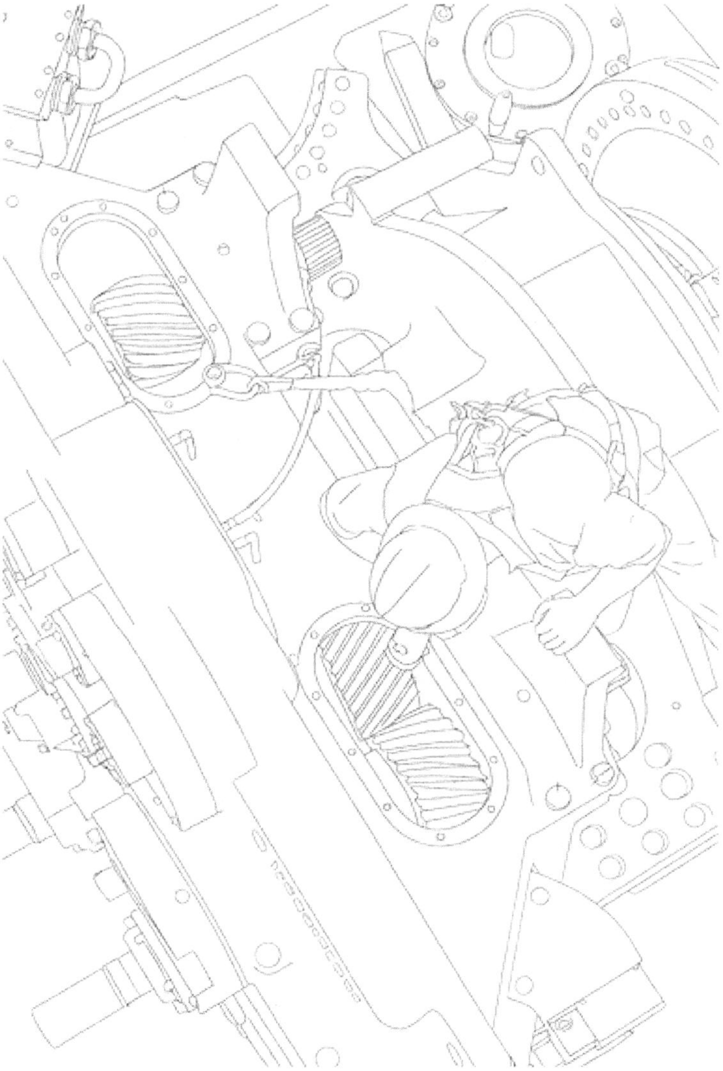

Figure 58: Wind Turbine Gearbox Inspection.

Step 1 – Job Safety Analysis (JSA)

Before departure, the team leader conducts a Job Safety Analysis (JSA) with all crew members and the vessel master. They break the job into tasks: transfer to the turbine by Crew Transfer

Vessel (CTV), climbing the tower, isolating and locking out the electrical system, removing the gearbox, rigging and lifting it to the deck below, and installing the new unit.

For each task, they identify hazards:

- Slips and crush injuries during CTV-to-ladder transfer.
- Falls while climbing or working at height.
- Electric shock or arc flash when isolating the generator and switchgear.
- Dropped tools or components during rigging.
- Load swing and collision during lifting in gusty conditions.
- Heat stress and limited egress inside the nacelle.

They use a colour-coded matrix to rate likelihood and consequence, flagging high-risk steps for extra control measures.

Step 2 – Control Measures

For marine transfer, the team agrees on weather limits and uses immersion suits, lifejackets with personal locator beacons, and a man-overboard (MOB) plan with the vessel crew.

For working at height, each technician is issued a twin-lanyard fall-arrest harness and climbs using the fixed ladder and service lift. A rescue kit and GWO Advanced Rescue-trained personnel are staged at the nacelle.

Electrical hazards are addressed with a formal Lock-out/Tag-out (LOTO): the turbine is put into local control, all feeders opened and locked, capacitors discharged, and a "test-before-touch" verification done on all phases. Arc-rated PPE and insulated tools are used.

Heavy lifting hazards are controlled through a certified rigging plan prepared by an engineer, specifying sling angles, centre of gravity, crane positioning, and exclusion zones. Only the appointed signaller directs the lift.

Weather is monitored in real time; operations stop immediately if wind or sea states exceed limits.

Step 3 – Permit-to-Work and Sign-off

All high-risk activities—confined-space entry into the nacelle, HV isolation, and lifting—are covered by permit-to-work documents specifying hazards, controls, and authorized personnel. Each technician and the vessel master review and sign off, confirming they understand the plan.

Step 4 – Execution and Monitoring

During the job, the team maintains communication with the vessel and the control room via radio. The attendant on the vessel monitors conditions and can call for evacuation if weather deteriorates. The rescue kit and first-aid equipment are kept ready at all times.

Step 5 – Outcome

Thanks to the risk assessment and tailored safe work processes, the gearbox exchange is completed without incident. When wind gusts rise above the pre-agreed limit, the team pauses lifting and shelters inside the nacelle until conditions stabilize. The JSA is updated with lessons learned and submitted for continuous improvement.

This scenario shows how risk assessments and safe work processes—JSA/JHA, permit-to-work, LOTO, working-at-height controls, lifting plans, weather protocols, and emergency preparedness—form a living system rather than paperwork. By breaking down the task, identifying hazards, applying controls, and training technicians to follow the process, operators dramatically reduce the chance of hidden hazards like arc flash, oxygen deficiency, or unstable loads causing injury or equipment damage.

Personal Protective Equipment (PPE) for Turbine Work

Wind turbines combine working at height, high voltage, heavy moving parts, and harsh environmental conditions. Even with engineering and administrative controls in place, technicians still rely on PPE as the last line of defence against injury. PPE requirements are not optional; they're mandated by law (EU PPE Regulation 2016/425, OSHA 29 CFR 1910/1926 Subpart I, CSA Z94/Z195/Z462, AS/NZS 1891/2210, etc.) and by operator procedures.

Core PPE Categories

a) Fall Protection

- **Full-body harness** rated to EN 361 / ANSI Z359 / AS/NZS 1891.
- **Twin-lanyard or Y-lanyard systems** with shock absorbers for 100% tie-off during ladder climbs.
- **Guided type fall arresters** or SRLs (self-retracting lifelines) on fixed ladders.
- **Certified anchor points** and connectors (EN 795 / ANSI Z359.18).
- **Rescue equipment** staged on site: inertia reels, descenders, casualty-haul kits.

b) Head, Eye, and Face Protection

- **Industrial helmet** with chinstrap certified for work at height (EN 12492 / ANSI Z89.1 Type I).

- **Safety glasses** (EN 166 / ANSI Z87.1) or goggles; optional anti-fog and wraparound.
- **Face shields** or visors for grinding, hydraulic work, or arc-flash tasks (arc-rated per IEC 61482-2 or NFPA 70E).

c) Hand Protection

- **General-purpose gloves** for handling components.
- **Cut-resistant gloves** for blade edges or sheet metal.
- **Insulating rubber gloves** (ASTM D120 / EN 60903) with leather protectors for HV work.
- **Arc-rated gloves** for tasks within arc-flash boundary.

d) Foot Protection

- **EH-rated safety boots** (ASTM F2413 / EN ISO 20345) with slip-resistant soles for ladders and decks.
- Offshore: non-metallic midsoles for corrosion and improved insulation.

e) Body Protection / Clothing

- **Arc-rated coveralls** (IEC 61482-2 / NFPA 70E) for electrical work.
- **Weather-resistant outerwear**: windproof, waterproof, breathable layers.
- **Hi-visibility vests** or markings for marine transfers.

f) Respiratory and Hearing Protection

- **Half-mask respirators** with appropriate filters for fibreglass dust, solvents, or welding fumes.
- **Disposable masks** for short-term tasks.
- **Earplugs or earmuffs** for >85 dB noise levels (gearboxes, power tools, helicopters).

g) Marine / Offshore PPE

- **Immersion suits** and **150–275N lifejackets** with sprayhood, whistle, and light.
- **Personal locator beacons (PLB)** or AIS MOB transmitters.
- **Helicopter transfer PPE**: survival suits, HUET modules as required by OPITO/GWO.

Figure 59: PPE for working at height.

On a wind-turbine site, PPE management begins with risk assessment. Each Job Safety Analysis (JSA) or Job Hazard Analysis (JHA) identifies specific hazards for the task at hand and specifies the exact PPE required. This ensures that workers are protected against the particular risks they face—such as falls, electrical hazards, or exposure to harsh weather—rather than relying on generic assumptions.

Once selected, PPE must be inspected and traceable. Harnesses, lanyards, gloves, helmets, and other gear are checked before each use and at prescribed intervals by a competent person. Inspection records are maintained so that every item has a documented safety history, reducing the risk of equipment failure.

Fit and compatibility are equally important. PPE must be properly sized and worn correctly to work effectively. Items like helmets, face shields, and hearing protection must be compatible with one another so there are no gaps or interference that compromise safety. Ill-fitting or mismatched gear can be as dangerous as having no PPE at all.

Training is another essential element. Workers must be trained to correctly don, adjust, use, and maintain their PPE. Refresher training—mandated by Global Wind Organisation (GWO) modules and national safety rules—keeps these skills current and reinforces proper practices.

PPE must be stored under the right conditions. Gear should be kept clean, dry, and protected from ultraviolet light, chemicals, and mechanical damage to preserve its integrity. Proper storage extends the life of PPE and ensures it will perform as expected when needed.

Technicians working on wind turbines need to be familiar with the main PPE standards in their region and internationally. In the European Union, the PPE Regulation (2016/425) requires that harnesses, helmets, gloves, and boots carry the CE mark. This shows that the equipment meets minimum safety, health, and environmental requirements for its intended use.

In the United States, OSHA Subpart I places a duty on employers to provide appropriate PPE and to train workers in its correct use, care, and maintenance. This includes fall protection, eye and face protection, and other specialized gear needed for wind-turbine work.

Canada applies CSA standards, with CSA Z259 covering fall-protection systems and CSA Z462 addressing electrical safety, including arc-flash protection. These standards outline performance criteria for equipment as well as safe work practices.

Australia uses AS/NZS standards, such as AS/NZS 1891 for industrial fall-arrest systems and devices and AS/NZS 2210 for protective footwear. These set out design, testing, and use requirements for equipment used at height and in demanding environments.

Arc-flash hazards are addressed globally through IEC 61482-2 and NFPA 70E. These standards specify requirements for arc-rated clothing, face shields, gloves, and other PPE used during electrical work, ensuring workers are protected from thermal energy released in an arc-flash event.

Before each job, check the JSA/JHA or permit for the specified PPE. Inspect your harness, lanyards, gloves, helmet, and PLB. Verify arc-flash rating for electrical tasks and immersion suit readiness for marine transfers. Using the right PPE—clean, intact, and correctly worn—is just as critical as following LOTO or rescue procedures. In wind energy, PPE is not a backup plan; it's an integral part of the safe work system that keeps you alive at height, with high voltage, and in harsh offshore conditions.

Fall Protection

Before climbing a tower or working on a nacelle roof, you must plan the task and verify anchor points. Review the Job Safety Analysis (JSA) or permit-to-work to see where fall protection is required. Identify rescue procedures and check that a trained rescue team and kit (descender, casualty haul system) are ready.

Inspect and fit your gear:

- **Harness:** Use a full-body fall-arrest harness certified to the relevant standard (CE, ANSI, CSA, AS/NZS). Check webbing for cuts, burns, or UV damage; verify labels and inspection dates.
- **Lanyards:** Inspect shock absorbers, stitching, and connectors. Twin-tail lanyards allow 100% tie-off during climbs.
- **Connectors:** Carabiners, snap hooks, or scaffold hooks must be compatible with anchor points and auto-locking.
- **Fit:** Adjust straps snugly at shoulders, chest, legs, and sub-pelvic area. The dorsal D-ring should sit between your shoulder blades.

Anchoring and 100% Tie-Off:

- **Certified Anchors:** Only clip to anchor points rated and marked for fall arrest (usually 15–22 kN / 3,600–5,000 lb). On towers these are fixed rails, cable systems, or ladder-integrated fall arresters.
- **Climbing:** When moving up or down ladders, stay attached at all times. Use twin lanyards to unclip and reclip one leg at a time or use a guided-type fall arrester on a fixed rail or cable.
- **Working Position:** Where possible, use work-positioning lanyards to keep your hands free, but always back them up with fall arrest.

Maintain good housekeeping: keep tools tethered and avoid slack in lanyards to limit fall distance. Stay within the manufacturer's recommended angle and length for your positioning devices. Always know your fall-clearance distance—the space needed below your anchor to arrest a fall without hitting a lower level or obstacle.

Fall protection isn't just about preventing the fall; it's also about planning for suspension rescue. Rescue kits must be staged at the work area. All crew members must be trained (e.g., GWO Advanced Rescue Training) to retrieve a suspended worker quickly to avoid suspension trauma.

Inspect your harness, lanyards, and connectors for damage, clean them as per manufacturer instructions, and store them dry and away from UV, chemicals, or sharp edges. Quarantine and report any gear subjected to a fall or impact load.

Figure 60: Fall protection in use.

Climbing a Wind Turbine

Climbing a wind turbine is a specialist job that combines height, confined spaces, high-voltage equipment, and demanding weather conditions. It can only be performed by trained and certified personnel following strict safety protocols. Every stage—from preparation to post-climb inspection—is designed to reduce risk.

Before anyone climbs, they must hold valid certifications such as Global Wind Organisation (GWO) Working at Heights, First Aid, Fire Awareness, and Manual Handling. Technicians also undergo periodic health checks to ensure they are physically and medically fit to work at height. A pre-job safety briefing and risk assessment are carried out to review hazards, weather conditions, and rescue arrangements.

Each climber wears a full-body fall-arrest harness connected to a vertical safety system. This may be a lanyard, self-retracting lifeline, or guided fall arrester attached to an approved anchor point inside the tower. A safety helmet protects against head injuries, gloves improve grip, and sturdy non-slip boots provide secure footing. Many climbers also wear flame-resistant or weatherproof suits to protect against environmental hazards.

Most turbines have an internal ladder running the full height of the tower. The ladder is equipped with a vertical safety rail or cable system to which the climber connects their fall-arrest device. Climbs are made in stages, with technicians resting on intermediate platforms which also serve as emergency exits if fatigue or other issues arise. Self-retracting lifelines automatically take up slack to minimize fall distance.

Some modern turbines include a small service lift that carries workers part or all of the way up the tower, reducing physical strain. Before use, technicians must inspect the lift, check weight limits, and follow operating procedures and emergency protocols.

At the top of the tower, the final section may require transitioning from the ladder into the nacelle (the housing for the gearbox, generator, and controls). Once inside, technicians immediately secure themselves with additional lanyards or anchor points to remain tethered while working.

The nacelle is often treated as a confined space with limited ventilation and restricted access. Technicians follow confined-space entry procedures and coordinate with the ground team. Blade work may require rope access techniques from the nacelle, or the use of suspended platforms or cranes for maintenance and inspections. Throughout, workers maintain radio contact with the ground to report progress and condition.

After completing tasks, technicians descend the tower using the same ladder or lift, staying attached to the fall-arrest system at all times. In an emergency, an approved emergency descent device allows controlled rappelling to the ground. All personnel are trained in self-rescue and casualty rescue techniques before being authorized to climb.

At the end of the climb, equipment is inspected for wear or damage and cleaned for future use. Any hazards, near misses, or incidents are reported, and a debriefing is conducted to identify improvements in procedures or equipment.

Work at height on wind turbines is strictly limited during high winds, storms, or lightning. Technicians are trained to assess weather conditions and stop work if limits are exceeded. They also undergo regular refresher courses to keep certifications current and to stay proficient in climbing, rescue, and emergency procedures.

Chapter 4

Emergency Preparedness and Rescue

Wind energy sites are among the most complex and isolated workplaces in modern industry. Whether on a windswept hill in rural Australia or hundreds of kilometres offshore in the North Sea, technicians often work at great heights, in confined spaces, and around high-voltage systems and heavy machinery. In such environments, even a minor incident can rapidly escalate into a serious emergency. *Chapter 4: Emergency Preparedness and Rescue* introduces the principles, systems, and practical procedures that keep wind technicians safe when the unexpected happens.

This chapter begins by explaining why emergency preparedness is an essential part of every wind technician's professional training and daily mindset. The unique configuration of wind turbines—narrow towers, enclosed nacelles, and limited escape routes—demands a proactive approach to risk management. Because emergency services are rarely able to reach a turbine immediately, technicians must be capable of initiating an effective first response, performing rescues, and stabilising casualties until help arrives.

Readers are then guided through the major emergency categories encountered on wind farms: fires, medical incidents, falls, electrical accidents, and severe weather events. The chapter examines how each scenario develops, what hazards are present, and how response procedures are tailored to the wind energy environment. It highlights the critical role of fire prevention and control, including how turbines are designed to comply with international safety standards such as IEC 61400-1 and IEC 61400-24, and how early detection and suppression systems integrate with SCADA monitoring.

Equally important are evacuation and rescue procedures. The chapter outlines how site-specific emergency plans are developed and tested through drills, ensuring that every worker knows how to exit a turbine quickly and safely. Step-by-step guidance is provided on evacuation routes, the use of descent devices, coordination with control rooms and offshore vessels, and the responsibilities of each team member during an emergency.

A key focus of this chapter is high-angle rescue—the specialised set of techniques used to recover an injured or incapacitated person from height or confined areas within the tower or nacelle. Using the Global Wind Organisation's Advanced Rescue Training (ART) framework as a baseline, readers learn how technicians prepare, stage equipment, and execute rescues safely under pressure. Detailed examples show how rescue kits, anchors, and mechanical systems are used in real-world scenarios, both onshore and offshore.

Finally, the chapter discusses first aid, hypothermia prevention, and incident reporting. These sections emphasise that a technician's role does not end when the casualty is reached—the ability to provide immediate care and document events accurately is critical for continuous improvement and legal compliance. Reporting near misses and learning from incidents form the foundation of a strong safety culture.

Emergency Preparedness and Rescue ties together every aspect of safety management covered so far in the book—planning, communication, technical competence, and teamwork. It equips future and current wind turbine technicians with the confidence and procedural knowledge needed to respond effectively when things go wrong, ensuring that every worker can return home safely after each shift, no matter how remote or challenging the environment.

Fire Safety and Evacuation Plans

Wind turbines contain high-voltage switchgear, transformers, converters, hydraulic oil, resins in blades, and confined electrical cabinets. Heat from electrical faults, friction, lightning strikes, or maintenance activities can ignite fires inside the tower or nacelle. Because escape routes are narrow and vertical, even a small fire can create a life-threatening situation. That's why every turbine site is required to have written fire-safety and evacuation procedures.

Figure 61: Damage to a wind turbine which was destroyed by fire on 3 August 2022. Paul Harrop , CC BY-SA 2.0, via Wikimedia Commons.

Fire Hazards Technicians Face

Wind turbines combine high-voltage systems, mechanical components, and confined spaces, all of which can create unique fire risks. One of the most common hazards is electrical faults. Arcing, short circuits, or overheating in converters, transformers, or cables can ignite nearby materials and quickly escalate into a serious fire inside the nacelle or tower.

Another major risk is hydraulic and lubricating oils. These fluids are flammable under pressure, and leaks around hot components or electrical equipment can greatly increase the chance of ignition. Combined with the confined nature of many turbine spaces, even small leaks can pose a significant fire hazard.

Technicians also face hazards during hot work activities such as welding, grinding, or cutting during maintenance. Sparks and heat from these tasks can easily ignite oils, insulation, or composite materials if proper precautions and fire watches are not in place.

Lightning strikes are another critical risk factor. They can damage blades or nacelle components, and in some cases cause secondary fires in electrical systems or structural materials if surge protection or earthing is compromised.

Finally, confined spaces inside the tower, nacelle, or hub can allow smoke, toxic gases, and heat to accumulate rapidly if a fire occurs. Limited ventilation and narrow escape routes make early detection, quick evacuation, and adherence to fire-safety procedures essential for technician safety.

Preventive Controls

Wind turbines incorporate several layers of design features to minimise fire risks. They are built to comply with IEC 61400-1 for structural loads and IEC 61400-24 for lightning protection, ensuring that grounding and surge systems are robust from the start. Many nacelles also use fire-resistant materials, contain automatic fire-suppression systems, and provide dedicated escape hatches to improve technician safety in an emergency.

Regular inspections are a critical part of fire prevention. Technicians routinely check earthing and surge-protection continuity to verify that lightning protection remains effective. Fire extinguishers, suppression cylinders, and alarms are also inspected and maintained during scheduled service intervals to ensure they will function correctly if needed.

All hot-work activities—such as welding, grinding, or cutting—require strict control measures. Before starting, technicians must obtain a hot-work permit, set up a fire watch, and remove or shield any combustible materials nearby. This process helps prevent sparks or heat from igniting oils, insulation, or other flammable items.

Good housekeeping is equally important. Keeping the tower base, nacelle, and work platforms free of oily rags, packaging, and debris eliminates fuel sources that could worsen a fire. Clean, organised work areas reduce hazards and support faster evacuation if an incident occurs.

On-Site Fire Safety Procedures

Wind turbine sites are equipped with portable fire extinguishers positioned strategically at the tower base, intermediate platforms, and inside the nacelle. These may be Class ABC extinguishers or specialist types such as CO_2 or clean-agent units for electrical or sensitive equipment fires. Technicians are trained to select and use the correct extinguisher for each fire type to contain an incident quickly and safely.

Fire extinguisher classes vary internationally, but commonly include Class A (ordinary combustibles like wood and paper), Class B (flammable liquids and gases), Class C (electrical equipment in the US or flammable gases in Australia/Europe), Class D (flammable metals), and Class K/F (cooking oils and fats). While a unified global system doesn't exist, Australia uses the A, B, C, D, E, F system, whereas the US uses A, B, C, D, K.

Understanding fire classes:

- **Class A:** Ordinary combustibles like wood, paper, cloth, and textiles.
- **Class B:** Flammable liquids (petrol, oil, grease) and gases (methane, propane).
- **Class C:** Energized electrical equipment.
- **Class D:** Combustible metals, such as magnesium, aluminium, or titanium, typically in industrial settings.
- **Class E:** Electrical fires (used in Australia).
- **Class K/F:** Cooking oils and fats (Class K in the US, Class F in Australia/Europe).

Regional differences:

- **Australia:** Utilizes a system with Classes A, B, C, D, E, and F.
- **United States:** Primarily uses Classes A, B, C, D, and K.
- **Europe:** Uses a different system, distinguishing flammable gases under Class C and cooking oils under Class F.

Extinguisher Types by Fire Class

- **Water/Water Mist:** Effective for Class A fires.
- **Foam:** Suitable for Class A and Class B fires.
- **Dry Powder (ABC):** Can be used on Class A, B, and C fires, as well as electrical fires.
- **Carbon Dioxide (CO2):** Ideal for electrical (E) and Class B (liquid) fires.
- **Wet Chemical:** Specifically designed for Class F (cooking oil) fires.
- **Specialized Dry Powder:** Used for Class D (metal) fires.

Fire detection systems form another key layer of protection. Smoke, heat, and gas detectors are installed in critical areas and are integrated with the turbine's SCADA system. When activated, these detectors automatically transmit alarms to the control room, enabling rapid response and coordination of emergency actions.

Fire suppression in wind turbines is critical due to the unique operational environment and the high risk of fire hazards. Wind turbines house various electrical, mechanical, and hydraulic systems in a compact nacelle, which presents substantial fire risks associated with sources of ignition such as arc faults, overheating components, and lightning strikes. The nacelle's elevation and narrow escape routes mean that even minor fires can escalate rapidly, putting lives at risk and threatening significant property damage [90, 91].

Some turbines also include automatic suppression systems within high-risk areas like converters or transformers. These systems, which may use aerosol or water-mist agents, are designed to knock down or contain a fire before it spreads, giving technicians time to evacuate and reducing damage to the equipment.

The variety of fuels present within wind turbines, including hydraulic oils and composite materials used for blades, exacerbate the fire risk. With the potential for intense and widespread fires, employing effective fire suppression systems is essential. Kim et al. emphasize the necessity of employing fire prevention systems to mitigate the hazards posed by both accidents and mechanical failures during operations [92]. Comprehensive fire suppression strategies not only entail the installation of automated systems but also necessitate rigorous maintenance and periodic checks to ensure operational efficiency [90, 91].

Advanced fire suppression methodologies in wind turbines typically integrate automated detection and suppression technologies rather than relying solely on manual extinguishing methods. The predominant technologies include aerosol generators, clean-agent gas systems, water-mist systems, and hybrid systems that combine various suppression techniques. Aerosol generators release fine particles that interrupt the chemical reactions during combustion, while clean-agent gas systems effectively reduce oxygen levels to extinguish flames without damaging sensitive equipment [90, 91]. Water-mist systems deploy fine droplets that cool and smother fires, representing a strategic choice in minimizing water damage [91].

The operational sequence of fire suppression systems in wind turbines typically involves several critical steps: detection, alarm and SCADA integration, automatic discharge, containment, and potential manual override by technicians. Sensors detecting smoke, heat, or gases play a pivotal role in the early identification of fire situations. This information is crucial for integrating with the turbine's SCADA system, prompting timely actions that might include shutting down operations and repositioning the turbine to mitigate risk [90, 91].

Furthermore, a technician's role in the ongoing maintenance of fire suppression systems cannot be overstated. Regular inspections ensure that all components are functioning optimally, risks are minimized, and that personnel are familiar with emergency protocols, increasing preparedness in the event of a fire [91].

Evacuation Planning

Because escape routes inside wind turbines are vertical and narrow, evacuation plans must be detailed and site-specific. They cannot rely on generic procedures; every turbine model and site layout requires its own plan.

Two-way communication is essential so that alarms and instructions are received instantly. Radios should be linked to the control room onshore and, for offshore turbines, to the vessel supporting the work. This ensures that technicians can coordinate with emergency responders as soon as an incident occurs.

The primary evacuation route is usually the internal tower ladder or the service lift, if it is still safe to use. However, plans must also include a secondary escape method. At nacelle level, dedicated escape hatches or descent devices allow technicians to exit quickly without using the main ladder. All crew members must be trained to deploy and use these devices for rapid evacuation.

Once outside the turbine, workers proceed to designated muster points onshore or on a vessel deck for roll-call. These assembly points must be clearly marked and communicated in the emergency plan to prevent confusion during an evacuation.

Everyone in the team must know their specific role in an emergency. This includes who raises the alarm, who carries the first-aid kit, and who initiates the rescue. Clear role assignments make evacuation faster and more coordinated, reducing the risk to both the affected worker and the rest of the crew.

Wind Turbine Fire Emergency & Evacuation Plan (Sample)

1. Purpose

To provide clear instructions for all personnel working inside or near the turbine on how to respond to a fire, protect life, and minimize equipment damage.

2. Scope

Applies to all technicians, contractors, and visitors entering the tower, nacelle, hub, or offshore platforms associated with this turbine.

3. Detection & Alarm

- **Automatic detection**: Smoke, heat, or gas detectors in nacelle, tower base, and transformer compartments are linked to SCADA.
- **Alarm notification**: Audible/visual alarm locally, and automatic alert to the site control room or vessel (offshore).
- **Manual reporting**: Any worker who discovers smoke, flames, or overheating must immediately activate the nearest manual alarm point and call the control room via radio.

4. Immediate Actions on Alarm

1. **Stop work** immediately.
2. **Communicate** location and nature of the fire to the control room / vessel master.
3. **Shut down**: SCADA or local E-stop triggers rotor feathering and generator trip.
4. **Evacuate** following the designated primary or secondary route.
5. **Do not attempt to fight the fire** unless it is incipient, you are trained, and the correct extinguisher is at hand.

5. Evacuation Routes

- **Primary route**: Use the internal ladder or service lift to descend to the tower base if safe (check for smoke/heat first).
- **Secondary route**: Deploy the dedicated escape hatch or emergency descent device at nacelle level. Technicians must be trained to use the device.
- **Assembly points**:
  - **Onshore**: Muster point at [location] near the substation.
  - **Offshore**: Muster point on deck of the CTV/SOV at [location].

6. Personal Protective Equipment During Evacuation

- Always wear your harness and lanyard when moving on ladders or open areas.
- For offshore, don immersion suit and lifejacket before leaving the nacelle if time allows.
- Carry your personal locator beacon (PLB) or AIS MOB transmitter.

7. Roles & Responsibilities

- **Team Lead/Permit Holder**: Raises alarm, confirms headcount, contacts control room/vessel master, ensures all personnel evacuate.
- **Technicians**: Follow evacuation route, bring portable gas detector and first-aid kit if close at hand, assist injured colleagues.
- **Attendant (if posted)**: Stays outside entry point to monitor conditions and provide rescue support.
- **Control Room / Vessel Master**: Initiates emergency response plan, coordinates external emergency services if needed.

8. Firefighting Measures (Only if Safe)

- Portable extinguishers staged at tower base, intermediate platforms, and nacelle:
  - Class ABC for general use.
  - CO_2 or clean-agent extinguishers for electrical equipment.
- Use only if the fire is small, your escape route is clear, and you've been trained.
- Never use water on energized electrical equipment.

9. Rescue & Medical Response

- **Rescue kit** staged at tower base and nacelle (descender, casualty haul system).
- **First aid**: Trained personnel provide immediate care for burns, smoke inhalation, or injuries.
- **External services**: Call 000/112 (onshore) or coordinate via vessel master (offshore) for medical evacuation.

10. Post-Incident Actions

- Conduct a headcount at the muster point and report to control room or vessel master.
- Do not re-enter the turbine until the all-clear is given by the incident commander or emergency services.
- Complete incident report and participate in debriefing.

11. Training & Drills

- All personnel must hold valid **GWO Basic Safety Training (BST)** Fire Awareness module.
- Offshore staff must complete **GWO Sea Survival** and any operator-specific helicopter/boat transfer training.
- **Emergency descent and rescue drills** performed at least every 12 months (or per site policy).

12. Review & Continuous Improvement

- Plan reviewed annually or after any incident.
- Feedback from drills and actual emergencies incorporated into updated procedures.

Training and Drills

Fire awareness training is an essential part of the Global Wind Organisation (GWO) Basic Safety Training (BST). It covers fire theory, the correct use of fire extinguishers, and evacuation drills. This ensures that technicians understand how fires start, how to respond quickly, and how to evacuate safely under realistic conditions.

Emergency descent and rescue skills are also practiced regularly. Through GWO Advanced Rescue Training (ART) or equivalent programs, technicians learn how to exit the turbine under their own control if lifts or ladders become unavailable. These drills prepare workers for rare but high-consequence events, reducing panic and ensuring safe escape.

First aid and hypothermia management are included to address injuries that can occur during fire or evacuation scenarios. Training covers treatment of burns, smoke inhalation, and shock so technicians can stabilise themselves or their colleagues until professional medical care arrives.

Offshore and Severe Weather Considerations

In offshore environments, evacuation plans must be tightly coordinated with marine operations. Crew Transfer Vessels (CTVs) or Service Operation Vessels (SOVs) are often required to pick up personnel, and vessel masters must be fully integrated into the emergency plan. This coordination ensures that everyone involved understands the procedures, communication channels, and timing needed for a safe evacuation.

Severe weather adds another layer of complexity. High seas, strong winds, or poor visibility can delay rescue operations, meaning technicians must be prepared to remain self-sufficient for extended periods. This includes wearing immersion suits, carrying personal locator beacons (PLBs), and having access to survival equipment so they can stay safe until evacuation is possible.

High-Angle Rescue and Confined Space Procedures

High-angle rescue refers to retrieving an injured or incapacitated person from a position where vertical or near-vertical movement is required. In the wind-turbine context, this typically involves rescuing someone from inside the tower, nacelle, hub, or even from the outside of a blade. Because turbine towers are tall and access is through narrow ladders or hatches, traditional emergency services cannot reach a casualty quickly. For this reason, trained co-workers must perform the initial rescue using dedicated equipment.

High-angle rescue is needed because of several factors. Onshore turbines are frequently located in remote areas far from emergency responders, while offshore turbines may be hours away from a vessel or helicopter. Nacelles and hubs are both high and confined, and an unconscious technician can block ladders or hatches. Electrical equipment, rotating machinery, and adverse weather can all complicate rescue operations. Time is critical: falls, electric shock, or medical emergencies require rapid retrieval to a safe area for treatment.

The global training baseline for high-angle rescue on wind sites is the GWO Advanced Rescue Training (ART) programme. These modules teach technicians how to evacuate an injured person from the nacelle, tower, or hub using descent devices, rescue stretchers, and rope systems. They also cover how to transfer a casualty from a ladder fall-arrest system to a rescue kit and coordinate with external emergency services such as onshore ambulances or offshore vessels and helicopters. Other countries layer national requirements on top of GWO training—for example, the UK Work at Height Regulations require rescue plans, OSHA 29 CFR 1926/1910 applies in the U.S., DGUV Rule 113-004 governs rope access in Germany, and CSA Z259 standards apply in Canada.

Typical rescue equipment includes sealed, pre-packed rescue kits containing ropes, descenders, pulleys, slings, connectors, and edge protection. Casualty-haul systems use mechanical-advantage pulleys or winches to lift an injured worker up to a platform or down to the base. Controlled-descent devices allow single or two-person evacuation from nacelle level. Rescue stretchers or cradles secure the casualty for vertical or horizontal lifts, and certified rescue anchors separate from the system that failed provide safe attachment points.

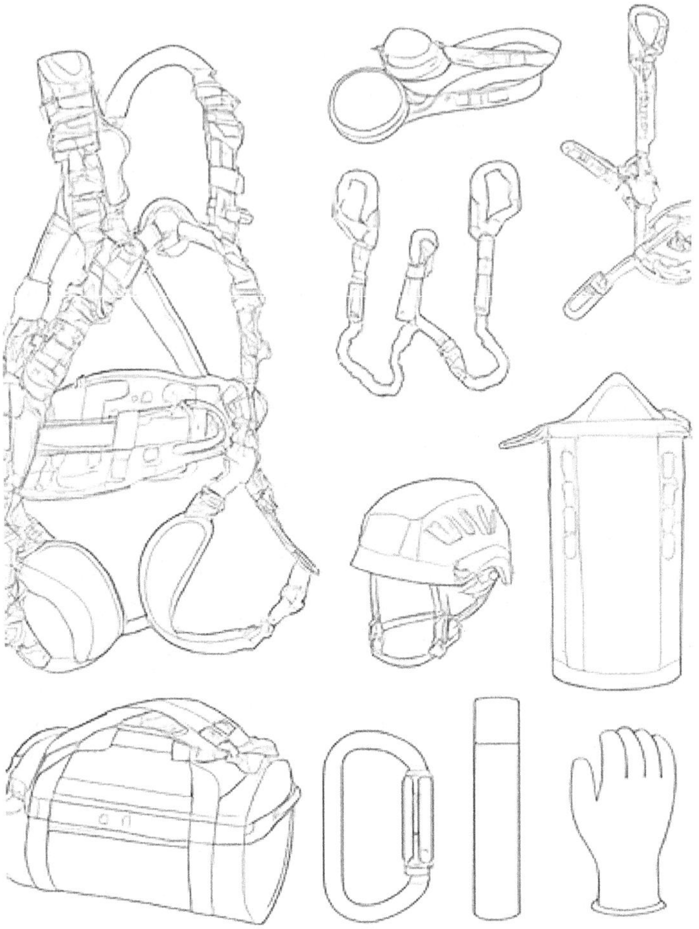

Figure 62: Example wind turbine rescue kit.

A wind-turbine rescue kit (sometimes combined with a climber's kit) is assembled so that trained technicians have everything needed to prevent a fall, arrest a fall, and then carry out a rescue or evacuation in the demanding onshore or offshore environment. It typically contains the following components and features:

Harnesses and Belts: At the core of the kit is a full-body fall-arrest harness designed specifically for climbing wind-turbine ladders. It usually has a raised rear attachment point to reduce chafing when climbing and an integrated waist belt for work positioning. A well-designed harness also makes it easier to attach and handle a casualty in a confined space or at the front of a nacelle during rescue.

Twin Fall-Arrest Lanyards: A pair of energy-absorbing lanyards with locking connectors allows "100% tie-off" when moving between anchor points. Wide-opening, high-strength connectors reduce the chance of accidental side-loading—a common risk when clipping to ladder rails or narrow beams inside towers.

Work-Positioning Lanyard: A separate adjustable lanyard lets the user secure themselves in place to work hands-free once at height. This differs from fall arrest because it holds the technician steady while connected to a main anchor point.

Helmet and Head Protection: A technical safety helmet rated for industrial height work protects against falling objects and head strikes inside towers. Modern models feature configurable chinstraps for climbing versus ground work, adjustable vents, and compatibility with visors, hearing protection, and communication headsets.

Rescue and Evacuation Equipment: Many kits also include a sealed rescue pack or modules—descenders, pulleys, slings, connectors, edge protection, and a casualty-haul system—so a co-worker can retrieve an incapacitated person vertically up or down the tower or from the nacelle. Some sites stage the "climbing kit" and "rescue kit" separately, but both are normally available together.

Storage and Transport Bag: A rugged transport bag keeps all components together, protected from UV, salt, oil, and mechanical damage, and is easy to carry up ladders or between vessels. Bags are usually rated for a working load and have clear labelling for inspection records.

What's Not Typically Included: Fall-arrest travellers (the device that locks onto the fixed ladder rail or cable) are normally site-specific to each turbine model and are issued separately. Rope-access or blade-rescue gear may also be separate modules.

Every task at height must have a documented rescue plan in place before work starts. This plan identifies likely rescue scenarios—such as a fall on a ladder, incapacitation in a nacelle, suspension trauma, or an offshore transfer incident—along with equipment staged at the tower base and nacelle, the number of trained rescuers required, communication links with the control room or vessel master, and the evacuation route to medical care whether onshore ambulance or offshore CTV/SOV/helicopter.

A typical rescue unfolds in a clear sequence. The alarm is raised, the control room or vessel master is notified, and emergency services are alerted. The scene is stabilised by stopping turbine rotation and isolating electrical and hydraulic systems. Rescuers reach the casualty safely using fall protection and backup systems, then transfer the casualty from their fall-arrest lanyard to the rescue system. The casualty is lowered or raised to the nearest safe platform or the ground, and first aid or CPR is provided until external responders take over. After the incident, equipment is debriefed, inspected, and restocked.

Offshore or blade-work rescues have additional considerations. Technicians working on blades may hang outside the nacelle, requiring rope-to-rope transfer or a winch from a platform or vessel. CTVs or SOVs must be ready to receive the casualty, and immersion suits and personal locator beacons (PLBs) support survival until pickup. High winds or sea states may delay external rescue, so crews must be self-sufficient.

Technicians play a central role in making high-angle rescue possible. They must maintain valid GWO ART or equivalent certifications, inspect and stage rescue gear before starting work, know the rescue plan for each task, practise realistic drills such as vertical lifts, horizontal evacuations, and confined-space extractions, and immediately report any hazards or equipment defects.

High-angle rescue in wind turbines is not an optional extra but a core part of working at height. With trained technicians, pre-staged equipment, and clear coordination with control rooms and vessels, wind farms ensure that if something goes wrong at 100 metres up, help can reach the casualty within minutes rather than hours.

Rescue Procedure

The first priority is to secure the scene. Rescuers immediately raise the alarm by notifying the control room, SCADA operator, or, for offshore sites, the vessel master. This includes confirming that the turbine is placed in "safe maintenance" mode, which stops rotation, disables yaw and pitch movement, and isolates electrical and hydraulic systems. Emergency services are then called with the turbine's location, GPS coordinates, nearest access point, the nature of the injury, and any hazards such as electrical faults, fire, or hydraulic leaks. At the same time, non-essential personnel are cleared from the area, and tools or loads are secured to prevent secondary hazards.

Roles are clearly assigned: one person leads the rescue, another manages equipment, and a third stays with the casualty. Offshore rescues also require coordination with vessel crews for eventual transfer. Rescuers check that everyone involved is wearing full fall protection—including harness, twin lanyards, helmet, and gloves—and carrying radios for two-way communication. The sealed rescue kit containing a descender, pulleys, slings, connectors, edge protection, and a rescue stretcher is brought to the casualty's location or to the nearest safe platform above or below.

Before any movement, rescuers perform a quick assessment of the casualty's condition—checking consciousness, breathing, and visible injuries—and give first aid if safe to do so. The environment is also scanned for live electrical parts, moving machinery, or falling-object risks. In confined spaces, the atmosphere is tested for oxygen levels and hazardous gases. Based on the casualty's injuries, available space, and access points, the team decides whether to lift the person up to a platform or lower them down the tower.

Certified rescue anchors independent of the system that failed are used to set up the rescue rig. Turbine ladders and nacelle frames typically have designated rescue anchors. Edge protection such as pads or rollers is placed at hatch edges to protect ropes from abrasion. Wherever possible, a second safety line is rigged for redundancy.

If the casualty is suspended from their fall-arrest lanyard, a rescue lanyard or rope is attached first so they are always secure. Using a pick-off strap or rescue connector, the rescuer takes the casualty's weight and disconnects them from the failed or obstructed system. Unconscious or seriously injured persons are placed in a rescue stretcher or cradle to maintain their airway and spinal alignment during movement.

If conditions allow, a controlled-descent device is used to lower the casualty vertically down the tower or externally from nacelle level to the ground or vessel deck. Where lowering isn't possible—for example, if a ladder is blocked or the base is flooded—a mechanical advantage haul system such as pulleys or a winch is used to lift the casualty to the nearest safe platform. Throughout the operation, constant radio contact is maintained with the ground crew or vessel to signal when the casualty is moving, when obstacles are reached, and when the handover is ready.

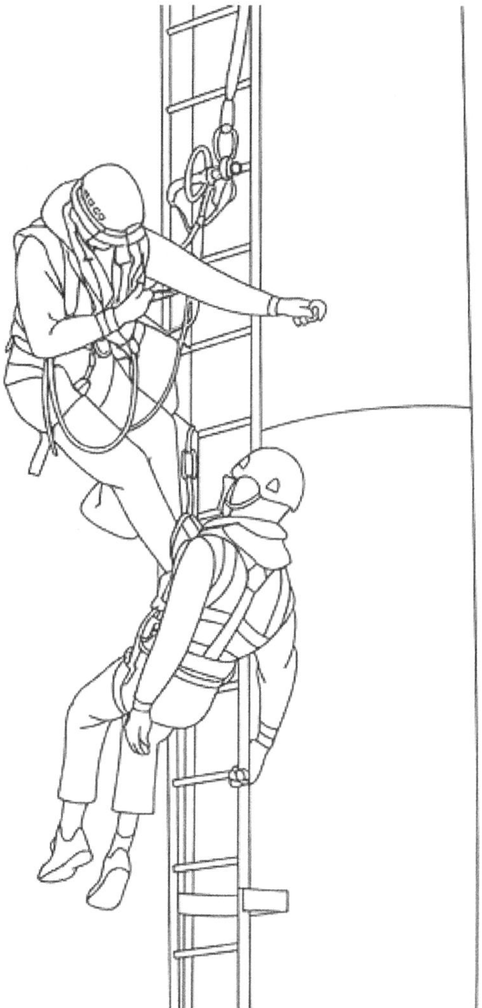

Figure 63: Using a descent device.

Once the casualty is safely at ground or vessel level, they are moved to the designated muster or medical area. First aid or CPR is continued until external responders take over. The rescue team records the time of rescue, injuries observed, and any hazards encountered for incident reporting and lessons learned.

After the rescue, the team debriefs to discuss what went well and identify improvements. Rescue equipment is inspected and replenished immediately so it's ready for the next job. Any new hazards or lessons are used to update the rescue plan and risk assessment.

For blade work, rescuers may need to transfer the casualty from one rope system to another or winch them from a vessel or platform. Marine coordination is critical to ensure a crew transfer vessel (CTV) or service operation vessel (SOV) is in position to receive the casualty, with immersion suits and personal locator beacons (PLBs) used if there's a risk of falling into the water. High winds or heavy seas may delay evacuation, so rescuers must be prepared to stabilize the casualty and provide extended care until transport is possible.

First Aid and Incident Reporting

Wind turbines are remote, tall, and filled with high-risk equipment. Even with strong controls, injuries can happen—falls, electrical shocks, burns, crush injuries from heavy components, or hypothermia offshore. Because external emergency responders may take a long time to arrive (especially offshore), wind technicians must be prepared to provide immediate first aid until professional help takes over.

Most operators require technicians to hold at least GWO First Aid certification (or a nationally recognized equivalent). This training focuses on conditions most likely on a wind farm: trauma from falls, electrical burns, CPR/AED, hypothermia, bleeding control, and casualty management in confined or vertical spaces. Offshore workers also learn immersion and hypothermia care, stretcher handling in narrow ladders, and helicopter winch coordination.

On site, the first step in any emergency is a quick initial assessment. Technicians should first ensure the area is safe, then check the casualty's responsiveness, breathing, and circulation before approaching. This protects both the rescuer and the injured person from further harm.

If there is bleeding, it must be controlled immediately. Apply direct pressure to the wound, use dressings to cover large injuries, and, if trained, apply a tourniquet to stop severe bleeding. This rapid intervention can be life-saving in remote wind-turbine locations.

Cardiac arrest and electrical shock are major risks. Every turbine should have an accessible AED (Automated External Defibrillator) at the tower base or nacelle level, and all crew members must know its location and how to use it. Combining CPR with an AED gives the best chance of survival.

For burns and electrical injuries, the burn area should be cooled as soon as possible, the casualty monitored for shock, and entry and exit wounds treated carefully. Technicians must also be ready for sudden cardiac arrest following electrical contact.

When an injured person needs to be moved, proper immobilization is critical. Use splints or rescue stretchers to support broken bones or spinal injuries and to safely lower casualties down ladders or through narrow hatches without worsening their condition.

Offshore or in cold weather, hypothermia is a serious concern. Wet clothing should be removed, the casualty insulated with dry layers, and emergency blankets or immersion-suit liners used to retain body heat until evacuation.

Finally, all first-aid equipment—kits, oxygen if supplied, AED batteries—must be inspected at regular intervals and records kept. This ensures everything is functional and ready to use at a moment's notice. Together, these practices form the backbone of effective first-aid response on wind sites.

Once the immediate emergency has been stabilized and the casualty cared for, the next critical step is incident reporting. This is not simply a bureaucratic exercise; it is the mechanism that drives safety improvements and protects both crews and employers legally. Around the world, regulators such as OSHA in the U.S., RIDDOR in the UK, the EU Framework Directive 89/391/EEC, Canada Labour Code Part II, the Model WHS Act in Australia, and Germany's DGUV rules all require employers to document—and, in serious cases, notify authorities of—injuries, near misses, or dangerous occurrences.

For technicians, the process begins with immediate notification. As soon as it is safe to do so, inform the site supervisor or control room about the incident. This allows management to start emergency or regulatory procedures without delay.

Next, secure the scene. Preserve evidence wherever possible—photograph hazards, failed equipment, or environmental conditions before anything is moved. This information is crucial for accurate investigations.

Then complete the incident report form. Record the date and time, location, task being performed, weather conditions, a clear description of the injury or near miss, and any factors that may have contributed to it. Accuracy here helps identify root causes later.

Technicians may also be asked to participate in the investigation by giving factual statements and suggesting practical improvements to prevent recurrence. This is an opportunity to share front-line insights that managers or engineers might miss.

Finally, follow up on the lessons learned. These should be reviewed at toolbox talks or safety meetings and used to update job safety analyses, rescue plans, or work procedures. Importantly, near misses must be reported with the same diligence as actual injuries because they often reveal hidden hazards before someone is seriously hurt. This closes the loop and ensures that each incident leads to stronger controls and a safer workplace.

Chapter 5

Industry Standards and Certifications

The global wind energy industry operates in a highly technical, safety-critical, and internationally connected environment. Every turbine, no matter where it stands—from the plains of Texas to the coasts of Denmark or the deserts of South Australia—must meet a consistent standard of quality, safety, and environmental compliance. *Chapter 5: Industry Standards and Certifications* introduces the frameworks, qualifications, and regulatory systems that make this global consistency possible.

This chapter explains how international and national standards ensure that technicians, engineers, and operators across the world work to the same high level of competence and safety. It begins with the Global Wind Organisation (GWO)—the central body responsible for harmonising training across the wind sector. GWO standards, though not legal requirements, have become the global benchmark for technician certification. They ensure that every person working on a turbine, whether onshore or offshore, possesses the same foundational skills in safety, technical operations, and emergency response. Readers will learn how GWO's modular system—including Basic Safety Training (BST), Basic Technical Training (BTT), and Advanced Rescue Training (ART)—forms the professional entry point into the wind industry, with certifications recorded in the international WINDA database.

The chapter then explores the complex world of electrical licensing and high-risk work permits, where regional regulations play a critical role. Wind turbines contain both low- and high-voltage systems, making it essential that only authorised, qualified personnel perform installation, testing, and energisation tasks. Because licensing laws vary between countries, the chapter compares requirements across major wind-producing regions, including Australia, the United Kingdom, the United States, Canada, Germany, and France. It also explains how permit-to-work systems manage activities such as confined-space entry, hot work, lifting operations, and electrical isolation, ensuring every task is controlled, documented, and safely executed.

A further focus is on environmental and wildlife protection compliance, which is increasingly central to modern wind operations. Wind power may be a clean energy source, but its construction and maintenance still affect land, water, and ecosystems. This section describes how international conventions, such as the Environmental Impact Assessment (EIA) Directive, the U.S. National Environmental Policy Act (NEPA), and Australia's Environment Protection and Biodiversity Conservation (EPBC) Act, guide sustainable site development and operation. It also covers how technicians contribute to environmental compliance through spill prevention, waste management, and wildlife protection practices. For offshore operations, readers will learn about MARPOL (the International Convention for the Prevention of Pollution from Ships) and how its pollution control rules extend to vessels supporting offshore wind farms.

Throughout this chapter, the emphasis is on professional accountability. Adhering to recognised standards and certifications is not only a matter of compliance—it is the foundation of safety, environmental integrity, and public trust. Wind technicians, supervisors, and managers all rely on these frameworks to demonstrate competence, maintain site access, and uphold the reputation of the renewable energy industry as a whole.

By the end of *Industry Standards and Certifications*, readers will understand how international standards, national laws, and company-specific requirements intersect to form the backbone of safe and ethical wind energy operations. This knowledge prepares technicians to work confidently in a regulated global industry, ensuring that every turbine—no matter where it spins—operates safely, sustainably, and in full compliance with the world's most rigorous standards.

Global Wind Organisation (GWO) Standards

The Global Wind Organisation (GWO) is a non-profit body founded by major wind-turbine manufacturers and operators to create a common set of safety and technical training standards for the industry. Its purpose is to ensure that technicians working on wind turbines—onshore or offshore—receive consistent, high-quality training that is recognised worldwide.

GWO standards aren't laws; they're voluntary benchmarks. But because almost every major owner and operator requires them as a condition of site access, they effectively form the industry baseline for technician safety and competence.

GWO maintains a set of training standards, each divided into modules that address specific skills and hazards.

The Basic Safety Training (BST) module covers the fundamental safety skills required for all turbine work. It includes Working at Heights, Manual Handling, Fire Awareness, First Aid, and Sea Survival for offshore environments. This forms the entry-level foundation for anyone working on or inside turbines.

The Basic Technical Training (BTT) module builds on this by providing core technical knowledge. It introduces the principles of mechanical, electrical, hydraulic, and bolt-tightening systems so that technicians can carry out supervised tasks confidently and safely.

The Advanced Rescue Training (ART) module focuses on high-angle rescue, teaching technicians how to retrieve injured or incapacitated colleagues from towers, nacelles, hubs, or blades. This ensures that immediate, effective rescue can be performed even before external responders arrive.

Specialised modules such as Slinger/Signaller & Lift User address rigging, signalling, and lifting operations for major component exchanges, while Enhanced First Aid (EFA) equips technicians with more advanced trauma and medical care skills for remote or offshore sites.

The Control of Hazardous Energies (CoHE) module trains workers on lock-out/tag-out, test-before-touch, and energy isolation principles tailored to wind turbines. In addition, Blade Repair and Inspection modules provide practical training for working with composite blade materials.

Each module specifies its duration, learning objectives, and assessment criteria, ensuring a consistent and measurable standard of competence across the global wind industry.

When you complete a GWO course at an accredited training provider, your certificate is uploaded to the WINDA database—GWO's global online registry. This allows site operators anywhere in the world to verify your credentials quickly and securely before granting site access.

Basic Safety Training (BST) certificates are valid for two years. After this period, you must complete a refresher course to maintain your qualification and ensure your skills remain current.

Basic Technical Training (BTT) is considered an "enduring" qualification with no fixed expiry date. However, many operators still expect technicians to show proof of recent practice or complete refresher training to demonstrate currency of skills.

Other GWO modules also specify validity periods in their standards, typically two years. Technicians are responsible for monitoring expiry dates and arranging refresher courses as required to stay compliant and site-ready.

GWO standards don't replace legal training like OSHA in the U.S., Work at Height Regulations in the UK, or DGUV rules in Germany. They are designed to complement and harmonise these national requirements so that a technician trained in one country can work safely in another without re-training from scratch. Operators usually layer any local mandatory training (for example, UK "harness inspection" or French "habilitation électrique") on top of GWO modules.

For a technician, holding valid GWO certificates is essentially your passport to work in the wind industry. It shows you've learned the minimum global safety and technical skills, from climbing safely to performing high-angle rescues. It also standardises the language and procedures you'll use on site, making teamwork smoother when crews come from multiple countries.

Electrical Licensing and High-Risk Work Permits

Wind turbines combine low-voltage (400–690 V) generator systems with medium- to high-voltage (5–66 kV) switchgear, transformers, and export cables. Work at these levels isn't just a safety risk—it's tightly regulated. Most countries separate duties into:

- Licensed electricians / HV operators – legally authorised to install, test, or energize equipment.
- Technicians with limited authorisation – trained to perform inspections, measurements, or assist under supervision.

In addition, many activities on a turbine are classed as "high-risk work" and require permits even if you're qualified.

Electrical Licensing by Region

Australia

- Electrical work above extra-low voltage (>50 V AC / >120 V DC) normally requires a state or territory electrical worker's licence.
- High-voltage switching permits are issued separately (HV operator authorisation).
- The Model WHS Regulations and Safe Work Australia Codes of Practice make it an offence to allow unlicensed persons to do prescribed work.
- Wind-farm operators typically require both a licence and GWO CoHE training for technicians who isolate or test equipment.

United Kingdom & European Union

- There's no single "electrician licence" like Australia or the U.S.; instead, competence is demonstrated through qualifications (NVQs, apprenticeships) plus employer authorisation.
- The Electricity at Work Regulations 1989 (UK) and national transpositions of the EU "Use of Work Equipment" Directive require that only "competent persons" work on or near electrical installations.
- Many operators use a "safety rules / permit system" model (similar to power utilities) with written authorisations, switching programmes, and key interlocks.

United States

- Each state licenses electricians separately, but OSHA also enforces 29 CFR 1910 Subpart S and §1910.269 for power-generation work.

- "Qualified persons" must be trained to identify hazards, use PPE, and understand clearances.
- High-voltage switching, grounding, or work on export cables usually requires a utility-style authorisation from the employer plus NFPA 70E arc-flash training.

Canada

- Provincial OHS Acts require work to be done by licensed or certified electricians under the Canadian Electrical Code.
- High-voltage access is governed by provincial utilities or site operators; technicians must hold CSA Z462 (arc-flash/electrical safety) training and be on an authorised-persons list.

Germany

- The Arbeitsschutzgesetz and DGUV Vorschrift 3 require electrical work to be performed by an Elektrofachkraft (qualified electrical specialist) or under their supervision.
- Wind operators issue "Schaltberechtigung" (switching authorisations) to named individuals for HV operations.

France

- The Code du Travail requires "habilitation électrique" (H0/B0, BS/BE, B2V, etc.)—specific electrical authorisations granted by the employer after training.
- Without the right "habilitation," a worker cannot legally open or test electrical panels.

Offshore / Marine Context

- In addition to electrical licences, offshore work may require STCW / OPITO or GWO Sea Survival for transfers, plus permits for hot work, confined spaces, or energised operations on platforms.

Across all these regions, high-risk tasks are controlled by permit-to-work systems:

- Electrical permits: issued after isolation and testing ("LOTO"), showing the boundaries of the safe work zone.
- Hot-work permits: welding, grinding, or cutting in nacelles or towers.
- Confined-space permits: entering hubs, tower bases, or converter rooms.
- Lifting permits: major component exchanges under LOLER/OSHA crane rules.
- Work-at-height rescue plans: required under UK Work at Height Regs, EU Directives, OSHA Subpart D, DGUV rope-access rules.

Permits specify: the work to be done, hazards, isolations applied, authorised personnel, PPE, and emergency arrangements. They are signed by both the issuer (site supervisor or HV operator) and the recipient (technician/team lead).

Environmental and Wildlife Protection Compliance

Wind power is marketed as “clean” energy, but building and running turbines affects land, water, and wildlife. Governments and regulators therefore impose strict environmental conditions on everything from site selection to daily maintenance. Breaching these rules can lead to heavy fines, shutdowns, or the loss of operating licences. For technicians, this means environmental protection is not just the developer’s responsibility—your work practices directly influence compliance.

Environmental Impact Assessments (EIAs)

Before any wind turbine construction begins, almost every country requires an Environmental Impact Assessment (EIA). An EIA is a formal study that identifies sensitive habitats, species, and cultural or heritage sites that could be affected by the project. It also prescribes mitigation measures to reduce or eliminate those impacts. In the European Union, EIAs are mandated under the Environmental Impact Assessment Directive (2011/92/EU, amended 2014). In the United States they are required under the National Environmental Policy Act (NEPA). Australia applies the Environment Protection and Biodiversity Conservation (EPBC) Act, and Canada uses the Impact Assessment Act.

Wildlife Protection Laws

Wind turbine developers must also comply with laws specifically designed to protect wildlife. For birds and bats, these include the EU Birds and Habitats Directives, the U.S. Migratory Bird Treaty Act and Endangered Species Act, the UK Wildlife and Countryside Act, and Australia’s EPBC Act threatened species lists, as well as provincial or state endangered species laws in other countries. For marine mammals and fisheries in offshore areas, key regulations include the EU Marine Strategy Framework Directive, the U.S. Marine Mammal Protection Act, and various regional fisheries management agreements.

ISO 14001 Environmental Management Systems

Many wind farm operators integrate the ISO 14001 environmental management system standard with ISO 45001 for health and safety. This combined approach provides a structured process for identifying environmental aspects and impacts, setting measurable objectives, monitoring compliance, and continually improving performance. It ensures that environmental protection is embedded into everyday operations rather than treated as a separate activity.

Typical Conditions

Technicians often work under restrictions designed to protect sensitive wildlife. This includes maintaining no-go zones around nesting birds, bat roosts, or marine mammal haul-outs. In many locations, work is restricted to certain seasons or times of day to avoid disturbing animals during critical breeding or migration periods.

Maintenance activities may be limited during breeding seasons or at night to comply with noise regulations. These limits help reduce stress or displacement of wildlife living near wind farms.

Turbine sites frequently use down-shielded or specially coloured lights at night to reduce the attraction of birds or bats. Technicians must ensure lights remain in compliance with these specifications during inspections or replacements.

All oils and chemicals, such as hydraulic or gearbox fluids, are stored in bunded (secondary containment) areas at the base of towers. Spill kits are staged on site, and any leaks must be reported and cleaned up immediately to prevent soil or water contamination.

Technicians are expected to segregate waste into dedicated bins for oily rags, metals, composites, and general waste. Hazardous materials like resins, solvents, or used filters must be disposed of under controlled conditions following environmental regulations.

Controls are implemented to prevent sediment runoff, concrete washout, or paint overspray near waterways. Offshore turbines must follow MARPOL discharge rules to protect marine ecosystems.

The International Convention for the Prevention of Pollution from Ships (MARPOL) was established in 1973 by the International Maritime Organization (IMO) as a comprehensive global legal framework aimed at mitigating pollution from ships. Its primary goal is to address both operational and accidental discharges into the marine environment. MARPOL is applicable not only to traditional maritime shipping but also to various offshore industries, particularly those operating vessels used in offshore wind activities, such as crew transfer vessels (CTVs) and service operation vessels (SOVs). These vessels are treated as "ships" under the convention and are thus required to comply with its regulations aimed at protecting the marine environment [93, 94].

MARPOL is structured through six annexes, each addressing a specific category of pollution. Annex I focuses on the prevention of oil pollution, setting forth protocols for the management and discharge of oily mixtures, with a primary concern for bilge water and oily residues. Annex II pertains to noxious liquid substances carried in bulk, providing guidelines for their handling and disposal. Annex IV addresses sewage, imposing treatment requirements and discharge limitations depending on proximity to land. Annex V regulates the disposal of garbage, particularly plastics, mandating segregation and proper disposal practices. Lastly, Annex VI addresses air pollution from ships, restricting emissions and targeting greenhouse gases [95-97].

For offshore wind operations, the most relevant annexes are I, IV, and V, which have a significant impact on daily operations, safety protocols, and environmental compliance for involved vessels. For example, the discharge of oily mixtures is prohibited within 12 nautical miles of land, requiring the use of oily-water separators and comprehensive record-keeping in the Oil Record Book. Similarly, untreated sewage discharge is restricted to beyond this distance, with specific treatment requirements imposed closer to shore. The strict prohibition of plastic waste under Annex V reflects the increasing emphasis on environmental sustainability in marine operations, necessitating responsible waste management practices [94, 98, 99].

To meet MARPOL compliance, offshore wind vessels must be equipped with pollution prevention technologies like oily-water separators and sewage treatment plants. Record-keeping systems, including Oil, Sewage, and Garbage Record Books, are essential for tracking waste disposal and prevention. Crew training on MARPOL regulations and emergency response, alongside regular inspections by flag state or port state authorities, are critical for effective compliance strategies. Therefore, a thorough understanding of MARPOL's provisions is essential for technicians working on CTVs and SOVs, as adherence to these environmental standards and procedures is vital during their operational activities [100, 101].

Composite blade disposal and recycling is becoming increasingly regulated. In some jurisdictions, operators must prepare end-of-life recycling or reuse plans for blades, and technicians may be involved in segregating and preparing components for compliant disposal or transport.

Wildlife Monitoring and Adaptive Management

Wind projects typically begin with extensive pre- and post-construction surveys to understand local wildlife activity. This includes monitoring bird and bat populations through visual counts, acoustic detectors, and even radar tracking systems to identify flight paths and migration patterns. These surveys provide the baseline data needed to design mitigation measures.

One of the most common mitigation measures is curtailment protocols. These systems automatically slow or stop turbines during periods of high risk, such as at night or during seasonal migration, to reduce bird and bat collisions. The control software links real-time wildlife data or weather conditions to turbine operation so that blades are only curtailed when necessary, minimizing both risk and lost energy production.

Developers also deploy deterrent systems to prevent wildlife strikes before they happen. Examples include ultrasonic devices to discourage bats from approaching turbines, ultraviolet or contrasting paint stripes on blades to make them more visible to birds, and lighting adjustments to reduce attraction at night.

Finally, collision and carcass reporting forms part of the adaptive management plan. Technicians on site may be required to log and photograph any wildlife fatalities or injuries they encounter. These records are then submitted to regulators to evaluate the effectiveness of mitigation measures and, if necessary, refine the operating protocols to better protect local species.

Technician Responsibilities on Site

Technicians must be familiar with the permit conditions that govern their work area. This means carefully reading the site's Environmental Management Plan (EMP) and completing the environmental section of the site induction. These documents outline seasonal restrictions, spill-reporting numbers, and procedures for protecting wildlife, so understanding them is essential before starting any task.

Preventing spills and leaks is a key part of day-to-day work. Technicians should use drip trays, absorbent pads, and approved transfer methods when handling oils, fuels, or chemicals. Regular inspections of nacelle drip pans and tower base bunding help catch small leaks before they become major incidents.

Respecting exclusion zones is equally important. Fenced or flagged habitats around sensitive areas such as nesting birds or bat roosts must not be entered without explicit clearance. These zones are there to protect both the environment and the crew from penalties or hazards.

Waste and materials handling must follow the EMP. Chemicals should be stored, labelled, and disposed of according to site procedures. Segregating waste streams and using the correct containers prevents contamination and makes audits easier.

Any environmental incident—whether it's a spill, a wildlife strike, or a suspected non-compliance—must be reported immediately to the site supervisor or control room. Environmental events are treated just like safety incidents and require the same urgency in notification.

Technicians should actively participate in toolbox talks, where environmental topics are often discussed alongside safety briefings. This ensures everyone on site stays up to date on current restrictions, new procedures, and lessons learned from previous events.

Offshore-Specific Controls

Marine coordination is critical when working offshore. Vessels must comply with designated speed limits and maintain safe approach distances to protect marine mammals and other wildlife. For activities like pile driving or other noisy maintenance work, "soft-start" procedures are often required—gradually increasing noise levels to give animals time to leave the area before full operations begin.

Fuel and oil transfers at sea are governed by strict MARPOL and flag-state regulations. Every vessel involved in offshore wind work must carry spill-response gear, such as absorbent booms and pads, so any accidental release can be contained and cleaned up immediately. Transfers are closely planned and supervised to minimize the risk of spills.

Waste management is equally important offshore. Except for specific discharges allowed under international rules—such as macerated food waste or limited greywater—waste must be brought back to shore for proper disposal. Crews must segregate and store all waste securely on board to prevent accidental release into the marine environment.

Part 3

Wind Turbine Components and Systems

Chapter 6

Mechanical Systems

At the heart of every wind turbine lies a finely engineered combination of mechanical systems that convert the kinetic energy of the wind into rotational and, ultimately, electrical power. *Chapter 6: Mechanical Systems* explores the intricate machinery within a wind turbine—from the massive nacelle assemblies and gearboxes to the precision-engineered bearings, blades, and yaw mechanisms that enable turbines to operate safely, efficiently, and reliably in some of the harshest environments on Earth.

This chapter opens by examining the nacelle and drivetrain, which together form the mechanical core of the turbine. Readers are introduced to each major component within the nacelle—the main shaft, bearings, gearbox, generator, brake system, yaw system, cooling and lubrication circuits, and control cabinets. Each component plays a critical role in translating the aerodynamic torque from the rotor into electricity. The chapter provides detailed explanations of how these systems interact, including the flow of mechanical energy through the drivetrain and how modern engineering design minimises vibration, wear, and fatigue under fluctuating wind loads.

From there, the discussion expands into gearbox and bearing technologies, which are among the most critical and maintenance-intensive parts of any wind turbine. The section explains how planetary and cylindrical gear trains increase rotational speed for the generator, and how high-precision spherical roller bearings manage immense axial and radial loads. Readers learn how alignment, lubrication, and thermal management are essential to long-term reliability and how modern turbines employ advanced condition-monitoring systems to detect wear before it leads to costly downtime.

The chapter then introduces direct-drive systems, an increasingly popular alternative to conventional geared designs. Direct-drive turbines eliminate the gearbox entirely by coupling the rotor directly to a large permanent-magnet synchronous generator. While these systems simplify the drivetrain and improve reliability, they introduce new challenges in terms of weight, size, and magnetic design. Both inner-rotor and outer-rotor direct-drive configurations are explored in detail, illustrating the latest innovations in generator construction and materials science that enable multi-megawatt machines to operate at very low speeds.

Subsequent sections explore the yaw and pitch systems—the mechanical subsystems responsible for orienting the turbine into the wind and controlling the aerodynamic performance of the blades. Readers will learn how active and passive yaw mechanisms work, how hydraulic and electric pitch systems adjust blade angles for optimal energy capture, and how control systems integrate sensor data to maintain stability and efficiency in changing wind conditions. The explanation of yaw gearboxes, hydraulic brakes, and control sensors provides a clear understanding of how precision mechanics and automation work together to maintain alignment and protect the turbine from overspeed or structural overload.

The chapter also delves into blade and hub design, exploring how aerodynamic theory, composite materials, and structural engineering converge in modern rotor construction. Topics such as blade geometry, airfoil optimization, tip-speed ratios, noise reduction, and lightning protection are discussed alongside the materials science that makes them possible. The section covers fibre-reinforced polymer composites, carbon fibre reinforcements, nano-enhanced coatings, and the challenges of blade recycling—illustrating the intersection of mechanical performance and sustainability.

Finally, *Mechanical Systems* examines the maintenance and reliability aspects that underpin the operation of wind turbines worldwide. From routine gearbox oil sampling and bearing inspections to advanced vibration analysis and blade repair techniques, the chapter provides insights into how mechanical integrity is monitored, preserved, and restored over the turbine's 20–25-year service life.

By the end of this chapter, readers will have a complete understanding of the mechanical architecture of wind turbines—how the physical components work in harmony to convert wind into energy, how design decisions influence performance and maintenance, and how emerging innovations are shaping the next generation of wind technology. This knowledge forms the technical foundation for anyone pursuing a career as a wind turbine technician, engineer, or maintenance specialist in the global renewable energy industry.

Nacelle and Drivetrain Overview

Nacelle

The nacelle is the large housing mounted on top of the tower, behind the rotor. It contains the main mechanical and electrical components of the turbine, protecting them from weather and allowing maintenance access. Typical nacelles are equipped with service platforms, ladders, lighting, fire suppression, and ventilation systems so technicians can work safely inside.

Although layouts vary by manufacturer, most nacelles include:

- **Main Shaft (or Rotor Shaft):** Connects the hub and blades to the gearbox, transmitting torque from the rotor. Often hollow to carry pitch-control cables and hydraulic lines to the hub.
- **Main Bearings:** Support the rotor shaft and absorb axial (thrust) and radial loads from the wind. Usually large spherical roller bearings.
- **Gearbox (if not direct-drive):** Steps up the slow rotor speed (e.g. 10–25 rpm) to the higher speed required by the generator (1,000–1,800 rpm). Many use a combination of planetary and spur gears for a high ratio in a compact space.
- **High-Speed Shaft and Coupling:** Exits the gearbox and connects to the generator; flexible couplings dampen vibrations and misalignment.
- **Generator:** Converts mechanical rotation into electrical energy. May be an induction generator (doubly fed) or a permanent-magnet synchronous generator (direct drive).
- **Brake System:** Disc brake on the high-speed shaft or generator side plus aerodynamic braking via blade pitch. Used for emergency stops and maintenance.
- **Yaw System:** A ring of electric or hydraulic drives beneath the nacelle rotates it to face the wind. Brakes lock it in position.
- **Cooling and Lubrication Systems:** Gearbox oil pumps, filters, and heat exchangers; generator air or liquid coolers. Critical for reliability.
- **Control and Power Cabinets:** SCADA controllers, converters, transformers, and switchgear mounted inside the nacelle.
- **Slip Ring Assembly:** Transfers power, signals, and hydraulic fluid between the rotating hub and the stationary nacelle.

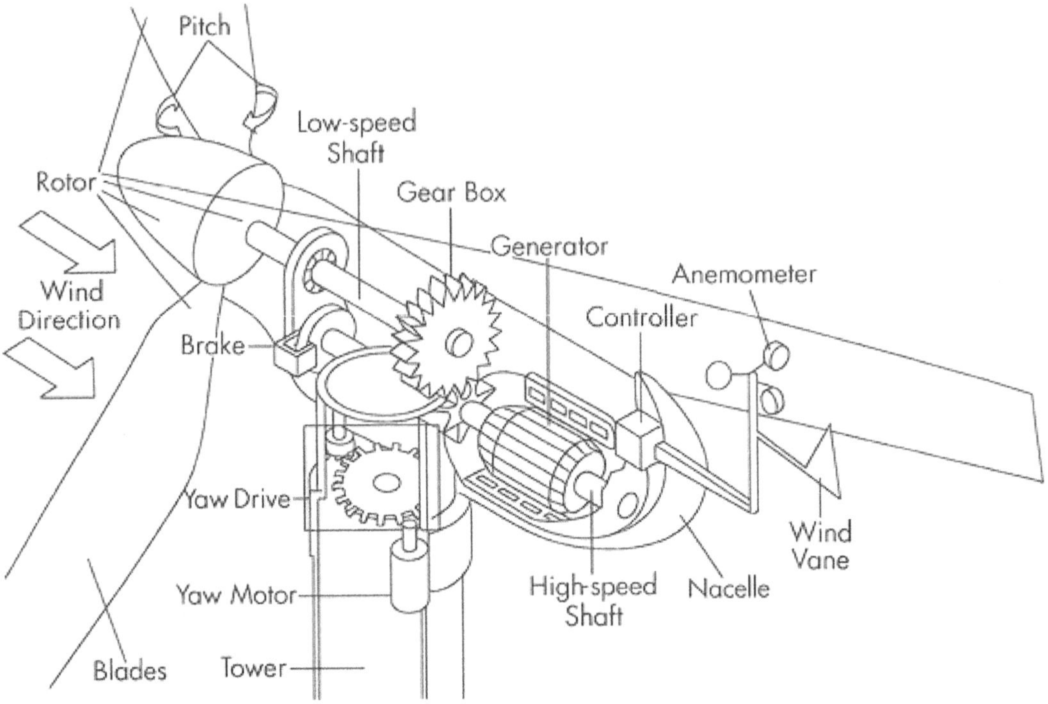

Figure 64: Major components in the nacelle.

Main Shaft (Rotor Shaft)

The main shaft is the first element of the drivetrain. It connects the hub (where blades are bolted) to the gearbox input. It carries low-speed, high-torque rotation — for a 3 MW turbine, torque can exceed 2 MN·m at 10–15 rpm.

Most shafts are forged alloy steel for strength and fatigue resistance. Many are hollow so hydraulic hoses, power cables, and signal lines for blade pitch actuators can run through to the hub. Hollow shafts reduce weight while still carrying torque. The front end is flanged to the hub; the rear end has a coupling to the gearbox. Shafts are often slightly tilted upward a few degrees to increase blade clearance from the tower and reduce overhang moments.

Main Bearings

Wind turbines have many rotating parts—rotor blades, shafts, gears, yaw drives, and pitch mechanisms. All of these depend on bearings to support loads and allow smooth movement. The three most critical bearings are:

- **Main Shaft Bearings**: Support and guide the main rotor shaft.
- **Yaw Bearings**: Allow the entire nacelle to rotate and face the wind.

- **Pitch Bearings**: Allow each blade to rotate about its axis for pitch control.

Unlike standard industrial bearings, these are extremely large (inner diameters over 1 m for main bearings) and must tolerate variable loads, misalignment, and harsh environmental conditions.

The main shaft bearings sit between the rotor hub and the gearbox inside the nacelle. Typically there are two bearings—a front bearing near the hub and a rear bearing toward the gearbox. Together, they carry:

- **Radial Loads**: The combined weight of the hub and blades pressing downward.
- **Axial (Thrust) Loads**: Wind pushing directly on the rotor.
- **Bending and Oscillation**: Wind turbulence and gusts cause the shaft to flex and oscillate.

Because the shaft bends slightly under load, the bearings must allow for small misalignments without seizing or generating high stress. This is called self-aligning capability.

Spherical Roller Bearings – Structure

The most common main-shaft bearings in geared wind turbines are spherical roller bearings. Their key feature is a spherical raceway in the outer ring, which allows the rollers to follow small angles of misalignment between the inner and outer rings.

They are built as follows:

- **Outer Ring**: Its inner surface is a continuous spherical raceway, with the centre of curvature aligned to the bearing's centre.
- **Inner Ring**: Has two separate raceways, each angled to the bearing axis.
- **Rollers**: Two rows of large, drum-shaped (barrel) rollers that run in the inner ring raceways and share the outer spherical raceway. This shape spreads load along a line of contact rather than a point, increasing capacity for heavy and shock loads.
- **Cage**: Separates and guides the rollers to prevent friction and distribute load evenly.

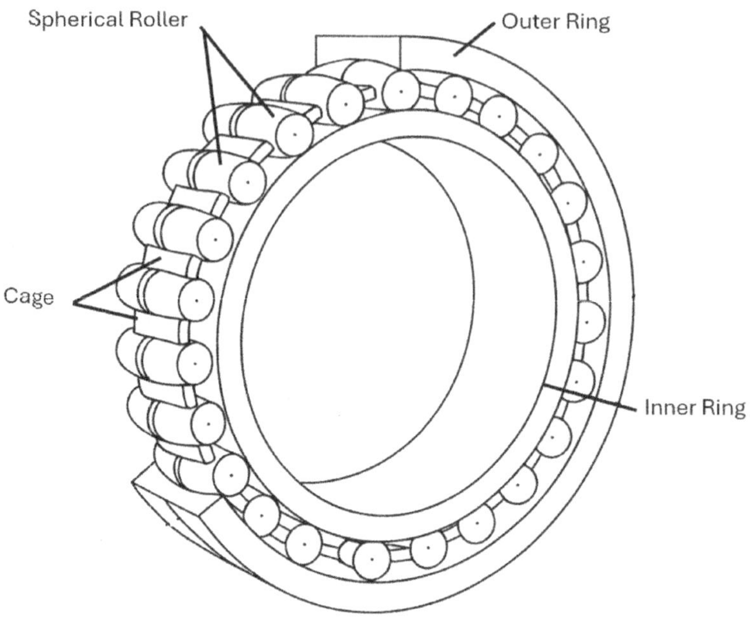

Figure 65: Spherical Roller Bearing.

The spherical roller bearing is mounted in a bearing pedestal—a large housing bolted to the nacelle frame. Bearing end shields on both sides hold it in position and seal it against contamination. A slight clearance or “end-shield space” is left to allow for the small tilt of the inner ring relative to the outer ring during operation. This clearance, often only about 0.5°, ensures the rollers stay correctly aligned as loads shift.

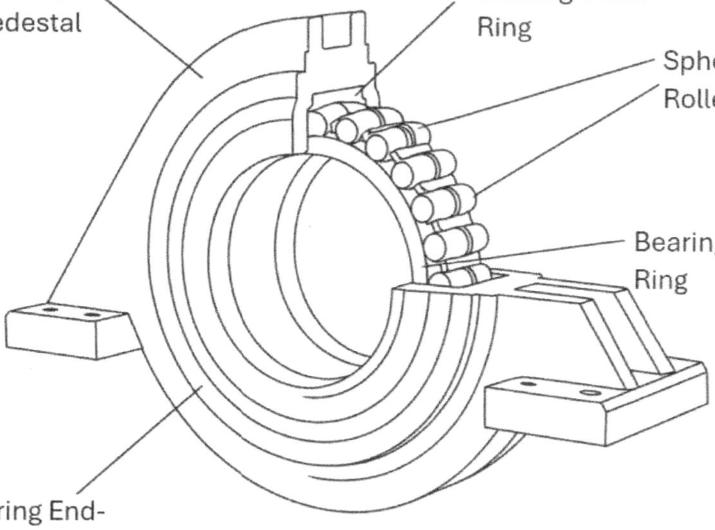

Figure 66: Wind turbine spindle bearing.

Gearbox (Speed-Increase Stage)

Because a generator needs thousands of rpm to produce 50/60 Hz AC economically, the gearbox multiplies the rotor's slow rpm to the generator's speed.

Typical ratios: 1:60–1:100.

Most turbines use a multi-stage gearbox:

- **Planetary stage(s):** compact, coaxial input/output, shares load among multiple planet gears for high torque and small size.
- **Parallel (spur/helical) stage(s):** fine-tunes speed, aligns to generator shaft. Planetary gear stages handle the huge input torque, while spur stages refine speed and direction. Gearboxes are fitted with integrated oil pumps, filters, and coolers. Bearings inside are high-speed, precision types with splash or forced lubrication.

High-Speed Shaft and Coupling

The high-speed shaft emerges from the gearbox at 1,000–1,800 rpm carrying much lower torque. It connects to the generator via a flexible coupling.

Flexible couplings (elastomeric or metallic disc types) allow small angular or axial misalignments between gearbox and generator shafts, dampen torsional vibrations, and reduce bearing loads. Many couplings also incorporate a brake disc for the mechanical brake.

Generator

This converts mechanical rotation to electrical energy. Two common types:

- Doubly Fed Induction Generator (DFIG): Most common in geared turbines. Rotor windings are connected to a power converter that controls slip, allowing variable speed while outputting grid-frequency AC.
- Permanent Magnet Synchronous Generator (PMSG): Used in direct-drive and some medium-speed turbines. Large-diameter, multi-pole design eliminates excitation losses and can produce variable-frequency AC for conversion to grid power. Generators are cooled (air-to-air or liquid) and have insulation rated for high voltage (400–690 V LV or up to MV directly). They're mounted rigidly on the nacelle frame for alignment.

Brake System

Two layers:

- **Mechanical Brake:** Usually a multi-caliper disc brake on the high-speed shaft or generator. Used mainly for holding the shaft still during maintenance or final stopping after aerodynamic braking.
- **Aerodynamic (Pitch) Brake:** Blades rotate to a feathered position (around 90°) to drastically reduce lift and slow the rotor. This is the *primary* emergency brake, as it uses the wind's own aerodynamics rather than mechanical friction. Together they ensure a controlled stop and hold.

Yaw System

Located under the nacelle where it sits on the tower. It includes:

- **Yaw Bearing/Raceway:** A large slewing ring bearing, 3–4 m in diameter, carrying the nacelle's weight.
- **Yaw Drives:** Several electric or hydraulic planetary gear motors engaging a large gear ring to rotate the nacelle.
- **Yaw Brakes:** Hydraulic or spring-applied brakes clamp the nacelle in place against wind forces.

Sensors from the wind vane tell the controller which direction to face; the yaw system turns the nacelle until aligned.

Cooling and Lubrication Systems

Because gearboxes and generators dissipate tens of kilowatts of heat:

- **Gearbox:** Pumped oil through filters and heat exchangers (air or liquid cooled). Monitored for temperature, pressure, and metal particles.
- **Generator:** Air-to-air heat exchangers or liquid cooling loops to maintain winding temperature.

These systems ensure oil film integrity, prevent overheating, and extend component life.

Control and Power Cabinets

Inside the nacelle are cabinets containing:

- **SCADA Controllers:** Microprocessor units that handle yaw, pitch, and power output.
- **Converters:** Rectify and invert variable-frequency AC from the generator to grid-stable AC.
- **Switchgear/Transformers:** Switch and step up voltage before sending down the tower cables.
- **Fire Detection & Safety Systems:** Alarms, suppression controls, and emergency-stop circuits.
  Technicians interface with these cabinets for diagnostics and maintenance.

Slip Ring Assembly

Because the hub rotates with the blades but the nacelle stays stationary, power and signals for blade pitch actuators must cross a rotating interface.

A slip ring assembly contains conductive rings and brushes (for electricity/signals) and may include rotary unions for hydraulic fluid. It's mounted on the main shaft/hub interface. Slip rings allow continuous rotation without twisting cables.

Drivetrain

The drivetrain is the path from the rotor hub to the generator:

1. Rotor Blades capture kinetic energy from the wind and rotate the hub.
2. Main Shaft transmits this low-speed, high-torque rotation to the gearbox.
3. Gearbox multiplies the speed so the high-speed shaft turns fast enough for the generator.
4. Generator produces electrical power, which is conditioned by converters and sent down the tower cables.
5. Brake System and pitch control work together to slow or stop the drivetrain when required.

In direct-drive designs, the rotor connects straight to a large, multi-pole generator with no gearbox. This reduces mechanical complexity but requires a heavier nacelle and larger generator.

Direct Drive System

Wind turbine rotors turn relatively slowly compared to traditional generators. Even small turbines only reach a few hundred revolutions per minute (rpm), while medium and large turbines often rotate at just tens of rpm—or as low as 10–15 rpm for very large models. However, conventional electrical generators are designed to run at much higher speeds (e.g. 1,000–1,800 rpm for a four-pole machine or up to 3,000 rpm for a two-pole alternator at 50 Hz).

In most geared turbines, a speed-increasing gearbox sits between the rotor and the generator. This multiplies the slow rotor rpm up to generator rpm. The downside is that gearboxes are complex, heavy, and subject to wear, especially the high-speed stages. They also require careful maintenance in a cramped nacelle environment, and replacing a failed gearbox at height is one of the most expensive operations in the wind industry.

To eliminate the gearbox entirely, manufacturers developed direct-drive generators that operate at the same slow speed as the rotor. These machines have:

- Many magnetic poles (sometimes several hundred poles instead of two or four).
- Very large diameter stators and rotors to produce sufficient torque at low speed.

Because they rotate at rotor speed, they can be coupled directly to the hub without mechanical gearing. Most are permanent-magnet synchronous generators (PMSGs), although some use electrically excited designs.

In a direct-drive turbine, the nacelle contains fewer major components than a geared design:

- **Generator**: The central piece of equipment, often several meters in diameter. The stator (stationary part) is bolted to the nacelle frame or base, while the rotor (rotating part with permanent magnets) is fixed to the same shaft as the hub and blades.

- **Main Shaft**: Often integrated with the rotor of the generator. It transmits torque from the blades directly to the generator without any gearing.
- **Yaw Bearing and Drive**: The entire nacelle sits on a large yaw bearing at the top of the tower. Electric or hydraulic yaw drives rotate it to face the wind.
- **Frame/Base**: A heavy-duty structure mounted on the tower, supporting the stator, rotor, and other ancillary systems.

This layout simplifies the drivetrain but requires a much larger generator.

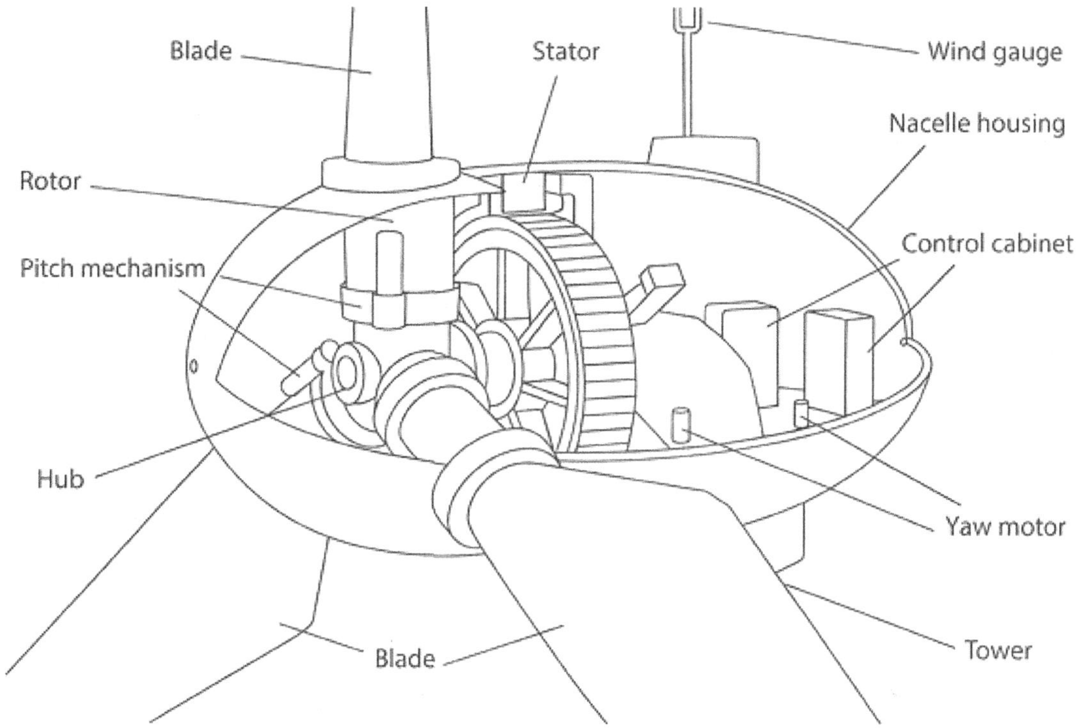

Figure 67: Direct drive system nacelle with stator and permanent magnet rotor.

A typical direct-drive permanent-magnet synchronous generator (PMSG) uses an inner-rotor design. The rotor, fitted with rows of permanent magnetic poles, is mounted directly on the inner shaft that connects to the turbine blades. Surrounding this rotor is a stator core with a three-phase winding embedded in it. The stator is fixed securely to the nacelle frame so it remains stationary as the rotor turns. Between the rotor and stator is a carefully controlled radial-flux air gap; the magnetic field crosses this gap from rotor to stator, inducing electricity in the stator windings as the blades rotate. Because this generator is both large and heavy, the entire nacelle yaw system—including the yaw bearing and tower top structure—must be extremely robust to support its mass.

Direct-drive generators produce variable-frequency AC because their speed changes with wind conditions. This cannot be connected directly to a 50 Hz or 60 Hz grid. Instead, the output passes through a full-scale AC–DC–AC power converter. First, the AC from the generator is rectified to DC. The DC is then processed by power electronics, and finally an inverter converts it back into a stable, grid-synchronized 50 Hz (or 60 Hz) AC supply. This full-power converter must handle the entire turbine output—often several megawatts—so it is a major cost item, but it enables variable-speed operation, which maximizes energy capture and reduces mechanical stress.

Removing the gearbox means less mechanical complexity and fewer wear parts. Direct-drive turbines run more quietly, can offer higher reliability, and capture energy more efficiently at variable speeds because they aren't limited to a fixed rotational rate.

The main drawback is that these generators are very large and heavy, increasing nacelle weight and structural demands on the tower. Full-scale converters are expensive and must be extremely robust to handle all output power. While maintenance access is generally easier than with gearboxes, the components themselves remain massive and require careful handling.

Despite these challenges, advances in high-voltage, high-power semiconductors and permanent-magnet technology are steadily reducing costs. As a result, direct-drive turbines are becoming more common, and some of the world's largest offshore machines—rated at 9 MW and above—now use direct-drive designs.

Inner Rotor Permanent Magnet Direct-Drive Wind Turbine

An inner-rotor permanent-magnet direct-drive generator has the same fundamental principles as any multi-pole alternator but is adapted to very low rotational speeds. It uses a multi-salient-pole rotor design, with the magnetic flux in the air gap running radially (perpendicular to the shaft). This arrangement allows the stator windings to cut the magnetic field efficiently at low rpm. The structure must also be lighter than a conventional generator of similar capacity because it sits high in the nacelle and is supported entirely by the tower and yaw system.

Because large wind turbines rotate slowly—often only tens of revolutions per minute—the generator needs a very high pole count to produce useful frequency and voltage. Modern direct-drive machines commonly have 40 or more pole pairs. In teaching models or diagrams, the pole count and coil slots are often reduced so that the magnetic layout can be seen clearly.

The stator core is built from thin laminated silicon-steel sheets to reduce eddy-current losses and provide high magnetic permeability. On the inner circumference of the core, a series of evenly spaced slots hold the stator windings. These windings form a three-phase coil system, with each phase made up of many coils arranged in a defined sequence to produce balanced three-phase power. Once wound, the stator is assembled into a robust casing that also provides mounting points for the main shaft, rotor, and generator flange.

The rotor is a multi-pole structure with rows of permanent magnets fixed around the outer surface of a rotor yoke. Adjacent magnets have opposite polarities so that a series of north and south poles face the stator. As the rotor turns, the alternating poles create a changing magnetic field across the narrow air gap. The rotor yoke is fixed to the main shaft by a rotor bracket. Bearings in the stator casing support the rotor so that the air gap can be kept extremely small—enough for efficient flux transfer but still allowing free rotation.

The air gap between rotor and stator is one of the most critical tolerances. It must be as small as practical to maximize flux density but large enough to prevent contact under deflection or vibration. When the wind turns the blades and rotor, the stator windings cut through the magnetic field, inducing an alternating potential. Because the generator speed varies with wind conditions, this AC output is variable frequency and must later be converted for grid use.

The nacelle at the top of the tower contains a heavy-duty base that supports the entire generator and hub. The lower chassis carries yaw motors to rotate the nacelle, while the upper flange holds the stator. Also housed in the nacelle are the main control cabinet, cooling system, and sensors for wind speed and direction. This layout keeps the stator fixed to the nacelle while the rotor spins inside it.

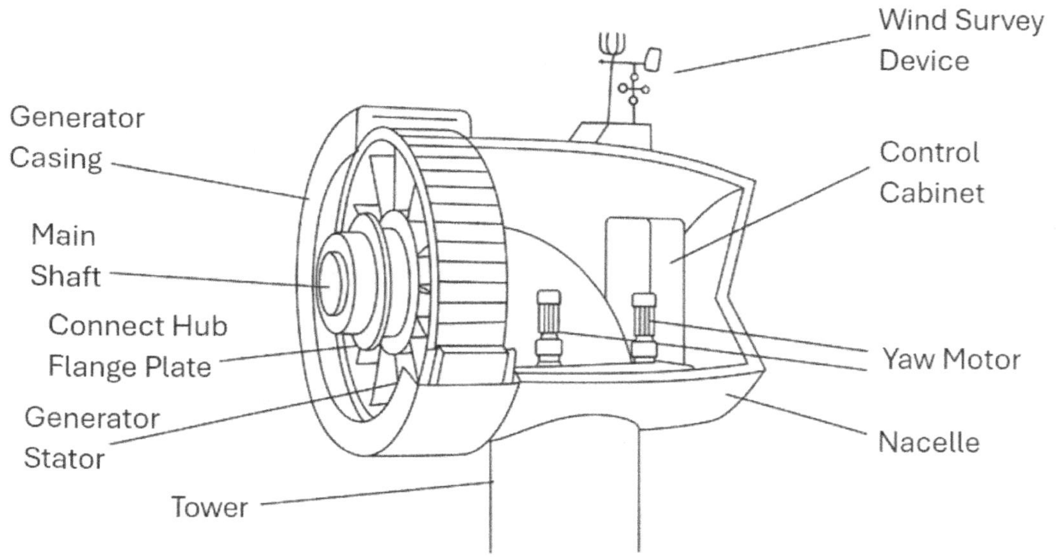

Figure 68: Inner Rotor Permanent Magnet Direct-Drive Wind Turbine.

During erection, the nacelle is mounted on the tower first. The inner-rotor permanent-magnet generator is then hoisted into place and bolted to the nacelle flange with high-strength bolts. Once the generator is secured, the hub spinner and blades are assembled on the ground or in stages, then hoisted and attached to the generator shaft. The hub and rotor are fixed to the generator flange with additional high-strength bolts to form the complete rotor assembly.

After the mechanical assembly is complete, technicians connect all signal, control, and power cables in the nacelle, link the generator to the converter system, and test auxiliary systems such as cooling and yaw drives. Once commissioning checks are finished, the turbine can enter service.

This design—an inner rotor with permanent magnets inside a fixed stator—eliminates the gearbox and allows direct drive at very low speed. The result is high efficiency, fewer moving parts, and a compact drivetrain, but it requires a large, heavy generator and precise installation to maintain the small air gap and alignment needed for efficient operation.

Outer Rotor Permanent Magnet Direct Drive Wind Turbine

An outer-rotor permanent-magnet generator is essentially the mirror image of the more common inner-rotor design. Instead of the rotor sitting inside the stator, the stator is fixed at the centre of the machine and the rotor turns around it. This is still a radial-flux arrangement: the magnetic field crosses the air gap radially between rotor and stator. The stator is often called the "inner stator" because it sits in the middle, while the rotating part is called the "outer rotor."

The inner stator core is laminated from thin silicon-steel sheets with high magnetic permeability to reduce eddy-current losses. On the outer circumference of this core are evenly spaced slots that hold the generator windings. These windings are laid out according to a three-phase distribution pattern so that balanced three-phase electricity is produced. In large machines the stator may have 180–240 slots.

The stator is mounted on a rigid stator bracket. One end of this bracket has a flange that bolts to the nacelle frame, while the other end carries the main shaft that supports the entire rotor assembly. This shaft and flange must be extremely strong because they bear the weight of the rotating mass and transmit the wind-driven torque.

Surrounding the stator is the outer rotor—a cylindrical "barrel" of ferromagnetic material that acts as the rotor yoke. On the inside of this barrel, permanent-magnet poles are fixed in alternating north–south order. Because the magnets are on the inside surface of the rotating drum, centrifugal forces press them against the yoke rather than pulling them off, which makes fixation easier and more secure at high speed.

The outer rotor sleeve is mounted on the generator's main shaft through two large bearings. This sleeve not only carries the magnetic poles but also connects directly to the wind-rotor hub, so the entire blade set rotates with the outer rotor. The bearings have to handle enormous radial and axial loads as well as bending moments from wind gusts.

As the outer rotor spins, the permanent magnets sweep past the stator windings across a very small radial air gap. The magnetic flux alternates direction from pole to pole, and the stator conductors cut this changing field to induce a three-phase voltage. The number of pole pairs is typically 30–40 in large direct-drive turbines to allow power generation at the very low rotational speeds of the wind rotor.

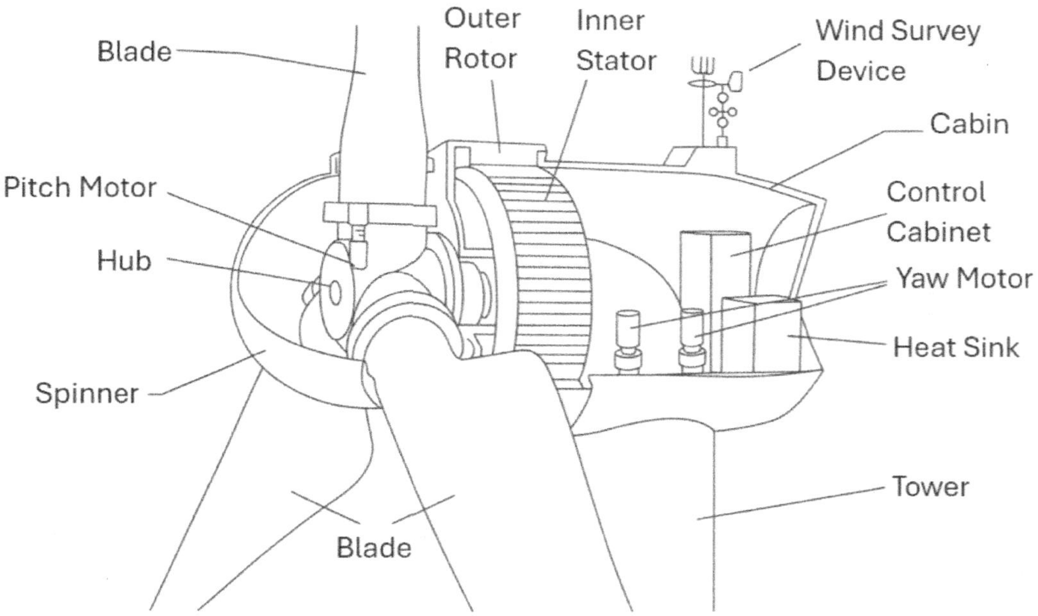

Figure 69: Outer Rotor Permanent Magnet Direct Drive Wind Turbine.

During installation, the nacelle carcass (its main frame) is hoisted to the top of the tower first and fixed in place. The direct-drive generator is then hoisted into the nacelle and bolted to the frame flange with high-strength bolts. Inside the nacelle you also find the electrical cabinets, control systems, cooling systems, and wind-measuring devices needed to run and monitor the turbine.

The blades are installed onto the hub to form the complete wind rotor. This hub is then hoisted and bolted to the outer-rotor flange so that the blade set and the generator's outer rotor turn together as one unit. A spinner or nose cone is usually fitted over the hub for aerodynamic protection.

Because the outer rotor encloses the stator, it can be larger in diameter than an inner-rotor machine for the same nacelle envelope. This allows more permanent magnets and higher torque at lower speed. The structure is also mechanically advantageous because the magnets are retained by centrifugal force rather than opposed by it. However, the design still requires extremely strong main shafts, bearings, and nacelle frames to support the mass of the rotating drum and blades.

An outer-rotor permanent-magnet direct-drive wind turbine reverses the positions of stator and rotor compared with the inner-rotor type. The stator windings are fixed at the centre; the rotor—a large permanent-magnet drum—rotates around them and is directly connected to the blade hub. This configuration allows the turbine to generate electricity at very low rpm without a gearbox, combining high torque capability with simplified drivetrain mechanics but demanding very robust support structures and precision assembly in the nacelle

Gearboxes, Brakes, and Yaw Systems

Gearboxes

Wind turbines rotate relatively slowly compared to the speed that most generators need. Small turbines might spin at a few hundred revolutions per minute (rpm), but large onshore and offshore turbines rotate at only a few tens of rpm, sometimes as low as 10–20 rpm. By contrast, standard generators are designed to run much faster—typically 3,000 rpm for a two-pole alternator, 1,500 rpm for a four-pole, or 1,000 rpm for a six-pole alternator [102]. To bridge this enormous gap, a speed-increasing gearbox sits between the rotor shaft and the generator. Its job is to multiply the slow rotor speed up to the high generator speed. The ratio can be tens to hundreds of times, depending on the turbine design.

This gearbox is the primary drivetrain component and is usually mounted in the centre of the nacelle. It transmits torque from the low-speed main shaft to the high-speed shaft connected to the generator. Because of the high forces and fluctuating loads from gusting winds, the gearbox must be very robust and precisely aligned.

Besides the main speed-increasing gearbox, there are two other important gearboxes:

- Yaw-drive gearboxes: small but powerful reduction gearboxes on each yaw motor. They turn the nacelle slowly to face the wind and hold it steady against side loads.
- Pitch-drive gearboxes: mounted in the hub to rotate each blade about its axis, adjusting the pitch angle to control aerodynamic torque and limit loads.

These auxiliary gearboxes operate at low speeds but must handle very high torque and be extremely reliable because they are critical for safe control of the turbine.

A common way to achieve speed change is by using cylindrical gears. The teeth of these gears are cut on the surface of a cylinder and mesh with another cylindrical gear. When two gears mesh, the smaller "pinion" drives the larger gear or vice versa:

- If the pinion (fewer teeth) drives the large gear (more teeth), the large gear turns slower and in the opposite direction: this is gearing down.
- If the large gear drives the pinion, the pinion spins faster than the large gear: this is gearing up, which is what the main turbine gearbox does.

The transmission ratio equals the number of teeth of the driven gear divided by the number of teeth of the driver. For example, a 12-tooth pinion driving a 36-tooth gear produces a 1:3 ratio (three times slower). Reversing the drive produces a 3:1 ratio (three times faster).

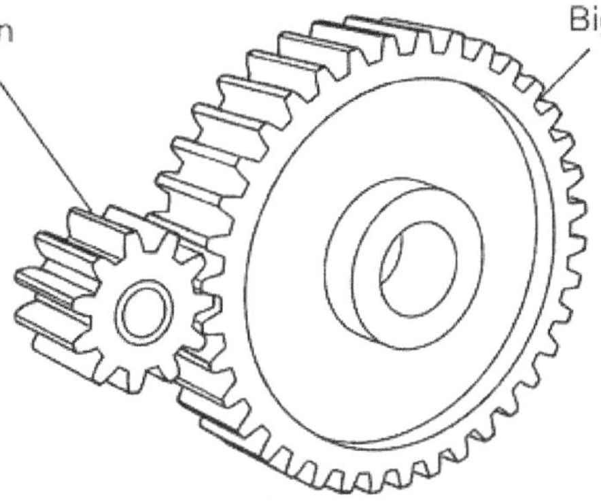

Figure 70: External gearing spur gear.

Each gear is fixed to a rotating shaft supported by rolling bearings at both ends. These shafts are mounted in a robust housing or pedestal. The lower half of the housing holds the gears and bearings, while the upper half serves as a cover. When assembled, the unit forms a gearbox. In a wind turbine, the low-speed shaft from the rotor enters the gearbox and meshes with a series of gears to progressively increase speed. The output high-speed shaft exits to the generator.

A single-stage cylindrical gearbox can achieve a ratio of about 4:1 (maximum 10:1), but to reach the very high ratios needed for turbines (50:1 or more), manufacturers combine multiple stages. The first stage often uses a planetary gear set for compactness, followed by spur or helical stages for further speed increase.

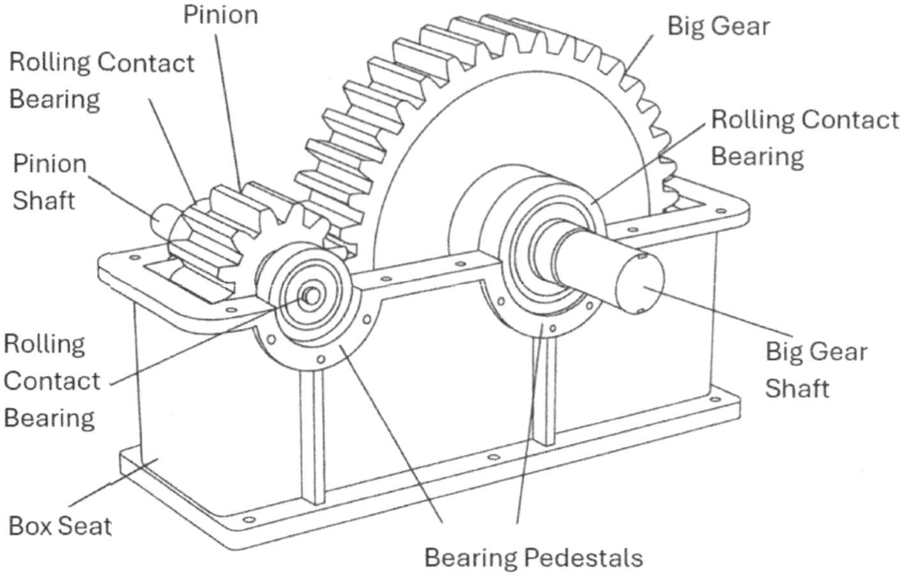

Figure 71: Primary cylindrical speed reducing gear box.

Within the "cylindrical gear" family are several tooth profiles:

- **Spur gears**: teeth cut straight and parallel to the shaft; simplest and most common but noisier.
- **Helical gears**: teeth cut at an angle, giving smoother and quieter operation and higher load capacity.
- **Herringbone gears**: two mirrored helical gears joined together to eliminate axial thrust, used in heavy-duty, high-torque applications.

Planetary gear systems, which are also made from cylindrical gears but arranged differently (sun, planet, and ring gears), are widely used in wind turbines for their high torque density and coaxial input/output shafts. These will typically form the first stage of the gearbox, followed by spur or helical stages to reach the final generator speed.

In a wind turbine, the main speed-increasing gearbox is essential for matching the slow rotor speed to the fast generator requirements. Cylindrical (spur, helical, herringbone) and planetary gear systems are combined to achieve very high ratios in a compact, reliable package. Auxiliary yaw and pitch gearboxes handle the precise positioning of the nacelle and blades. All of these gear systems must be robust, carefully maintained, and regularly inspected because failures are costly and difficult to repair in the nacelle environment.

<u>Gearbox Schematic Symbols</u>

In engineering drawings for wind turbines (or any mechanical system), it's impractical to show every gear, shaft, and bearing in full 3-D detail on every diagram. Instead, designers use schematic symbols—simple shapes and lines—to represent the function of components without cluttering the drawing. This is especially useful when you need to illustrate how power flows from one stage of a gearbox to another [102].

For a cylindrical gear drive mechanism, the schematic symbol shows only the essential parts: the input shaft, output shaft, and meshing gear teeth. This lets engineers and technicians quickly understand the direction of rotation, speed ratio, and which shafts are connected without worrying about every bearing or housing detail [102].

The first drawing (Figure 72) represents a generic cylindrical gear pair. Two circles or discs show the meshing gears; the relative size difference reflects the gear ratio. Arrows or small tick marks may indicate the direction of rotation. The parallel arrangement of the gear symbols corresponds to parallel gear shafts. It's a shorthand for saying, "this stage uses spur or helical gears with parallel axes to transmit power" [102].

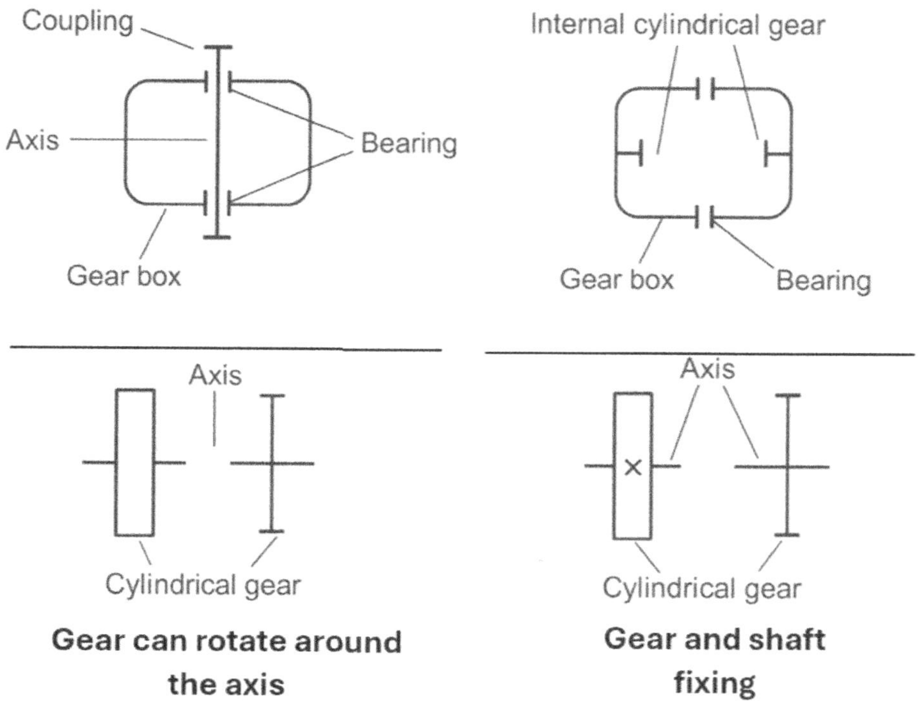

Figure 72: Basic schematic symbol of cylindrical gears drive mechanism.

The second drawing (Figure 73) takes the real gearbox shown earlier—the primary cylindrical speed-reducing gear box—and boils it down to its schematic essentials. Instead of showing casings, bearings, and seals, it shows only the input pinion gear on its shaft and the output gear on its shaft. This simplified diagram clearly shows [102]:

- Which shaft is the input (connected to the rotor or motor).
- Which shaft is the output (connected to the next stage or generator).
- The speed ratio (depicted by gear size difference).
- The direction of rotation (arrows).

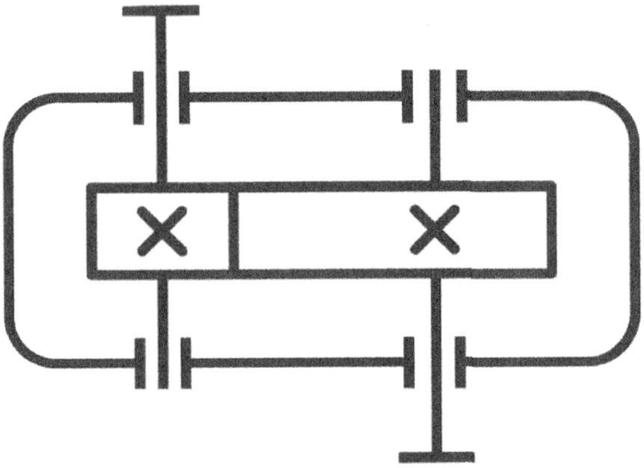

Figure 73: Simplified diagram of the primary cylindrical speed reducing gear box.

This is enough for technicians to trace power flow and plan maintenance or inspections without misinterpreting how the gears engage.

Being able to read these schematic symbols is part of understanding the wind-turbine drivetrain. During troubleshooting or maintenance planning, you may only have a schematic rather than a cutaway view of the gearbox. Knowing that a particular stage is a single-stage cylindrical gear with a certain ratio, for example, helps you anticipate the torque and speed on that shaft, the kind of lubrication required, and the correct procedure for lock-out/tag-out before opening the casing.

<u>Main Form of the Cylindrical Gears Wind Turbine Speed-Increasing Gearbox</u>

A single-stage cylindrical gear drive has a limited transmission ratio—its gearing "step up" capacity is typically only up to about 10:1 and more commonly around 4:1. However, wind turbines must increase rotor speeds from just tens of revolutions per minute to several hundred or even a thousand revolutions per minute to drive a generator. This high ratio is achieved by using multi-stage gear trains, where two or more gear pairs are connected in sequence to progressively raise speed. In wind turbines, the resulting arrangement is called a gear train.

Two-Stage Cylindrical Gear Drive (Unfolding Type): This is the simplest multi-stage layout. The pinion on the first shaft drives a large gear on an intermediate shaft, and then a second pinion on that intermediate shaft drives a second large gear on the output shaft. Its structure is straightforward, but the gears sit asymmetrically relative to the bearings. This uneven arrangement puts bending loads on the shafts, which means the shafts must be stiff and carefully aligned. It's generally used where loads are fairly stable. The high-speed stage is often cut with helical teeth for smoother running, while the low-speed stage may use straight (spur) teeth. Recommended total transmission ratios range from about 8:1 to 40:1, with a practical maximum of 60:1.

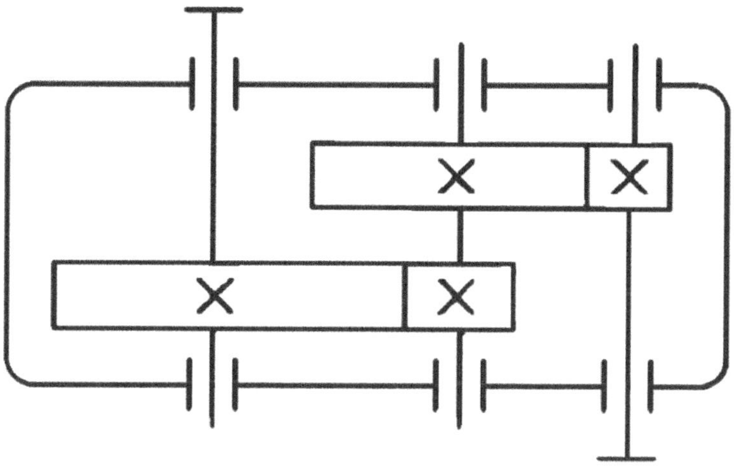

Figure 74: Two-stage cylindrical gear drive.

Two-Stage Cylindrical Gear Drive (Distributary or "Symmetrical" Type): Here the gears are arranged symmetrically with respect to the bearings. This spreads the load more evenly along the gear face width and reduces uneven bearing loads compared with the unfolding type. The arrangement is more complex mechanically but is better suited for variable or fluctuating loads—common in wind turbines. The high-speed stage usually uses helical gears, and the low-speed stage can use straight or herringbone gears. Again, recommended ratios are 8–40 with a maximum of 60.

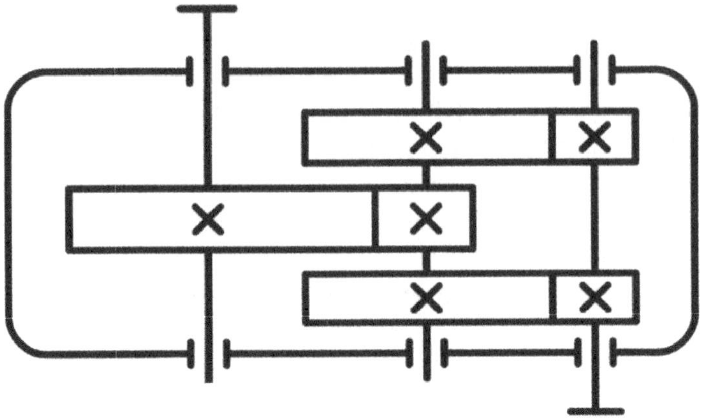

Figure 75: Two-stage cylindrical gear transmission (distributary type).

Two-Stage Cylindrical Gear Drive (Coaxial Type): In this layout the input and output shafts are on the same axis (coaxial), so the overall transverse dimension (width) of the gearbox is small. Both gear pairs are immersed to similar depths in the oil bath, which helps with cooling and lubrication. However, the axial dimension (length) and overall weight are greater. This arrangement can be advantageous where nacelle space is limited but coaxial alignment is required. The ratio range is the same: 8–40 recommended, 60 max.

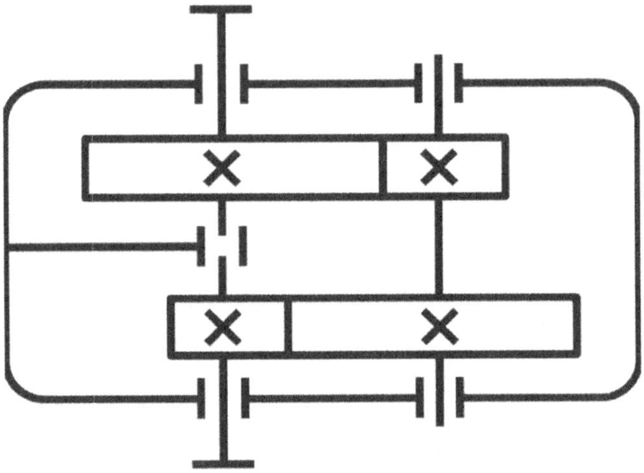

Figure 76: Two-stage cylindrical gear transmission (coaxial type)

Two-Stage Cylindrical Gear Drive (Coaxial Split Type): This design splits the torque between two parallel gear paths. Each pair of meshing gears carries only half the total load, which reduces the load on the intermediate shaft and allows smaller journal diameters. The input and output shafts see only torque (not bending), improving reliability and longevity. This makes it suitable for very high-power wind turbines. Ratio ranges are again 8–40 recommended, 60 max.

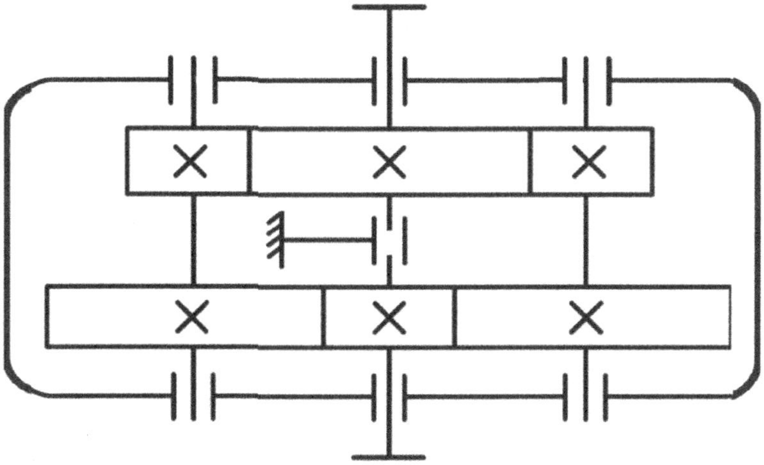

Figure 77: Two-stage cylindrical gear transmission (coaxial split type).

Three-Stage and Combined Gear Trains: For large wind turbines, where the speed increase must exceed 100:1, a three-stage gearbox is common. Each stage may have a ratio of 3–5, and combining them yields the high total ratio needed to match the generator's rated speed. In modern designs, a hybrid arrangement is often used: a planetary gear stage on the low-speed side (to handle large torque compactly) combined with one or two cylindrical gear stages on the high-speed side for final speed-up.

Each gearbox layout balances size, weight, cost, efficiency, and durability. By choosing the right combination of cylindrical gear stages—and in many cases combining them with planetary stages—wind-turbine designers achieve the very high speed ratios required while still keeping gearboxes compact, maintainable, and robust enough to handle fluctuating wind loads.

<u>Planetary Gear Speed Increasers</u>

Planetary gear speed increasers are the primary transmission system in most wind turbine gearboxes. They are particularly suited to wind applications because they combine a large speed-increase ratio with high load-carrying capacity, yet remain compact and lightweight. This is critical in a nacelle where space, weight, and maintenance access are limited. Another major advantage is that the input and output shafts share the same axis, simplifying alignment and making the gearbox more efficient. These characteristics make planetary gears ideal for increasing the low rotational speeds of a wind turbine rotor to the high speeds needed by a generator.

A planetary gearbox consists of several gears working together in a compact "planetary" arrangement:

- **Ring gear (internal gear):** A large outer gear with inward-facing teeth.
- **Sun gear (external gear):** A smaller central gear that meshes with the planet gears.
- **Planet gears (external gears):** Multiple gears (commonly three) positioned between the sun and ring gears.
- **Planet carrier:** A rotating disk or frame that holds the planet gears' shafts and supports their rotation.

This setup allows each planet gear to do two things simultaneously:

1. Rotate on its own axis (like a planet spinning).
2. Revolve around the sun gear (like a planet orbiting the sun) as the carrier turns.

Because the planet gears mesh with both the sun gear and the ring gear, torque can be transmitted smoothly and evenly, allowing high-speed increases in a compact form factor.

The planetary carrier is typically a disk with three or more shafts set at 120° intervals. Each shaft supports a planet gear. The carrier itself is mounted on bearings and connected to the low-speed shaft flange, which links to the main shaft of the wind turbine hub. When the rotor spins, it drives the carrier, which in turn causes the planet gears to move relative to the fixed ring gear and sun gear.

Each planet gear is mounted on its own shaft and can rotate freely around that shaft. These planet gears simultaneously mesh with the ring gear on the outside and the sun gear in the centre. This dual engagement allows power to be split across multiple gear meshes, which spreads out the load and increases the gearbox's torque capacity without increasing size.

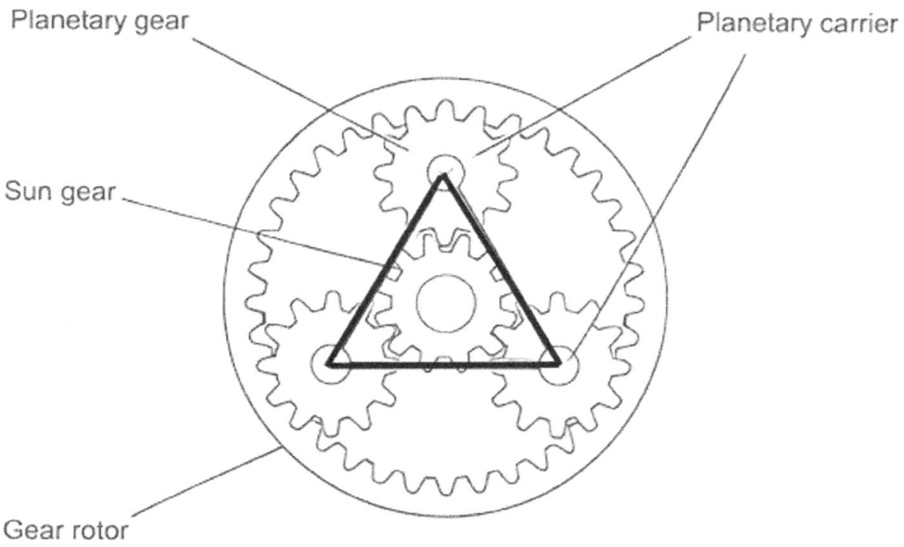

Figure 78: Planetary gears.

The ring gear (internal gear) is mounted inside the front terminal cover of the gearbox. Its internal teeth mesh with the outer sides of the planet gears. As the carrier rotates, the planet gears roll along the inner circumference of the ring gear, transmitting torque to or from the sun gear depending on the configuration.

The sun gear sits at the centre of the planetary system and is usually connected to the high-speed output shaft of the gearbox. As the carrier rotates (driven by the rotor), the planet gears orbit around the sun gear, causing the sun gear to spin much faster than the carrier. This rotation is what drives the generator at high speed.

A single-stage planetary gearbox is built up in steps:

1. **Carrier assembly:** The planet gears are mounted to their shafts on the carrier.
2. **Carrier installation:** The carrier is inserted into the front terminal cover with bearings.
3. **Ring gear installation:** The ring gear with its internal teeth is fitted inside the cover to mesh with the planet gears.
4. **Sun gear placement:** The sun gear is installed centrally, meshing with all planet gears and connected to the high-speed shaft.
5. **Rear end cover:** This closes the assembly and provides bearings for the sun gear shaft.

Once complete, the planetary gearbox allows the low-speed shaft (rotor-driven) to turn once while the high-speed shaft (sun gear) can rotate several times—effectively increasing speed.

The speed-increase ratio for a single-stage planetary gear system is given by:

$$I = 1 + \frac{Z_2}{Z_1}$$

Where:

- Z_1= Number of teeth on the sun gear
- Z_2= Number of teeth on the ring gear

For example, if $Z_2$is much larger than Z_1, the sun gear rotates significantly faster than the carrier. In practice, single-stage planetary gearboxes in wind turbines usually achieve ratios up to about 8:1. If higher ratios are needed, multiple planetary stages are combined or mixed with cylindrical gear stages.

Speed Increasing Based on Planetary Gear Transmission

A single-stage planetary gear transmission is the simplest form of planetary speed increaser. It consists of one planetary stage made up of a sun gear, planet gears, a ring gear, and a carrier. In this arrangement, the rotor drives the carrier, which turns slowly, while the sun gear spins much faster to drive the high-speed shaft connected to the generator. This type of transmission is compact and lightweight compared to cylindrical gearboxes of the same ratio, with coaxial input and output shafts. However, it requires very high manufacturing precision and careful assembly, and the ratio is limited to about *i = 2.8–12.5*, meaning the sun gear can rotate up to approximately 12 times for each rotation of the carrier. For this reason, single-stage planetary gearboxes are often combined with other gear stages to achieve higher speed ratios.

A first-class planetary plus two-stage cylindrical gear transmission starts with a planetary stage on the low-speed shaft, directly connected to the turbine rotor, followed by two cylindrical gear stages—spur or helical—that increase speed further at the generator shaft. In this arrangement, the planetary stage splits the input power and moderately increases speed, while the cylindrical gears provide additional speed increase. This setup allows designers to distribute the speed-increase ratio more evenly across stages, improving efficiency and reducing stress. It is both compact and efficient for medium-to-high ratios and retains coaxial input and output at the planetary stage, with a total gear ratio range of *i = 20–80*.

A secondary planetary gear transmission uses two consecutive planetary stages rather than a planetary stage followed by cylindrical gears. The first planetary stage moderately increases speed, and the second stage increases it further. This design provides very high ratios while maintaining a compact footprint, still benefiting from coaxial shafts and load-sharing among multiple planet gears. With a gear ratio range of *i = 24–160*, it is particularly suited for very large turbines needing significant speed increases for the generator. However, this configuration is more complex and demands greater precision in manufacturing than single-stage designs.

A secondary planetary plus primary spur gear transmission combines two planetary stages with a final spur gear stage. The planetary sections handle the lower-speed, high-torque input, while the spur gear manages the highest-speed stage. This setup achieves extremely high ratios with a relatively small gearbox and efficiently distributes torque across the planetary gears, reducing wear and weight. It is commonly used in large-scale wind turbines with high power output and very slow rotor speeds.

Wind turbine gearboxes also include special construction features to suit the nacelle environment. The low-speed shaft connected to the rotor hub is often hollow, allowing routing of pitch mechanism drives, power cables, signal lines, and oil pipes through the centre of the shaft into the hub. These integrated channels enable hydraulic or electrical systems controlling blade pitch to reach the hub without extra external conduits. Additionally, high-strength bearings and housings are built to ensure long service life under fluctuating loads, turbulence, and misalignment, which are typical conditions for wind turbine operation.

Brakes

Slowing and stopping an 80-metre-diameter turbine rotor involves converting its large amount of kinetic energy into heat through the braking system. This is why the design of rotor brakes is so important. Rotor brakes serve three main purposes: controlling overspeed, holding the rotor stationary ("parking"), and providing emergency braking when other systems fail. The brakes can be mounted on the rotor or low-speed shaft, on the generator (high-speed shaft), or, in some cases, on both shafts. Each location has advantages and trade-offs in terms of torque requirements, heat dissipation, and mechanical stress [103].

Mounting the brake on the low-speed shaft is relatively straightforward. Because this shaft turns slowly, a large disc brake with an extensive friction-lining area can be installed, making it easier to dissipate heat. However, this arrangement must generate very high braking torque to stop the rotor because the shaft transmits full rotor torque directly. Although easier to package physically, the brake and its mounting structure must be very robust to handle these loads [103].

For most modern turbines, the most cost-effective position is on the high-speed shaft between the gearbox and generator. The gearbox multiplies the shaft speed while reducing torque, so a brake on the high-speed shaft can achieve the same energy dissipation with much lower torque than at the rotor shaft. The total energy to dissipate remains the same regardless of brake location, so the friction-lining area must still be sufficient to absorb the heat of an emergency stop. Because space and disc diameter are limited at high rotational speeds, designing a brake with adequate pad area and heat dissipation capacity is more challenging. Nevertheless, this arrangement has been common on turbines up to about 750 kW. As turbines grow larger, the trend is moving back toward rotor-shaft braking because of the higher energy levels involved [103].

One drawback of braking on the gearbox-output shaft is the potential for gear tooth damage. When a stationary turbine is braked at the high-speed shaft, wind gusts can cause the rotor to rock within the backlash of the input and output gears. Without forced lubrication between the mating teeth, this rocking can lead to fretting or micro-damage, which eventually becomes expensive gear wear. This is one reason designers weigh brake location carefully in large turbines [103].

Large turbines also use high-torque static holding brakes on their pitch drives—the systems that rotate the blades. These brakes are typically electrically released, spring-engaged units designed to withstand harsh conditions. They are compact compared to the motor they serve, adding minimal length to the pitch drive assembly. For example, a brake rated at 135 Nm may be just 6.5 inches in diameter and slightly over 2 inches long, yet achieve a design life of 500–1,000 stops (or more). Electric brakes also react quickly, often in 0.20 seconds or less, which is crucial for rapid blade pitch adjustments [103].

Determining the correct braking torque level for rotor brakes is critical. On the rotor shaft, the maximum permissible braking torque is limited by the structural capacity of the blades and their anchorage to the gearbox input shaft. On the high-speed shaft, it is limited by the maximum permissible gear-tooth loading. There is also a minimum torque requirement below which the brake's variable frictional forces under different conditions could leave the rotor insufficiently controlled. Designers therefore build in a safety margin or "service factor" to ensure reliable performance in all climates and operating scenarios [103].

For example, consider a 1-MW turbine with an aerodynamic (load) torque of 100,000 Nm. Applying a brake during an emergency at 20 % overspeed, the rise in disc temperature and stopping time will depend on the chosen service factor. A commonly applied factor of 2.0 is expressed as [103]:

$$T_b/T_L = 2$$

where T_b is brake torque and T_L is load torque. Using this formula, the required braking torque would be:

$$T_b = 2 \times 100{,}000 = 200{,}000 \text{ Nm.}$$

This illustrates how engineers size brakes not just for average loads but for worst-case scenarios, ensuring the rotor can be brought safely to a stop without damaging components.

Yaw System

In small wind turbines, the simplest and most common way to keep the rotor facing the wind is by using a tail vane. The tail vane is a large fin mounted behind the rotor that automatically aligns the turbine with the prevailing wind direction. When the wind shifts, the vane acts like the rudder on a boat, swinging the nacelle so the blades remain correctly oriented. This passive system works without any power supply, making it ideal for rural or off-grid installations. Many agricultural multi-blade turbines also use tail vanes, sometimes with two-leaf designs to generate greater stabilizing force against the resistance of the heavy multi-blade rotor. These larger or dual tail vanes ensure that the rotor remains steady even in turbulent winds.

Some traditional windmills and a few special turbines use a more complex system called side wind rotors to face into the wind. This arrangement places two small steering rotors on either side of the nacelle at the back. These are mounted on a horizontal shaft perpendicular to the main turbine shaft. When the main rotor is aligned with the wind, the side rotors are parallel to the wind flow and remain stationary. If the nacelle begins to drift out of alignment, the side rotors rotate, generating a torque that turns the nacelle back into the wind through gears or worm gear drives. This mechanical system allows a turbine to adjust its yaw even when no electricity is available. Dutch windmills famously use this method to automatically orient themselves toward the wind.

To maximize power output and protect equipment, a wind turbine must measure wind direction and speed very accurately. Knowing when the wind reaches cut-in speed (typically 3–4 m/s) to start, rated speed (around 12–14 m/s) to deliver full power, and cut-out speed (about 25 m/s) to shut down safely all depends on precise wind data. For this reason, modern turbines mount wind vanes and anemometers on the top rear of the nacelle. The most common devices are mechanical (three-cup anemometers and wind vanes) because of their low cost and easy maintenance, although ultrasonic sensors are also used. Typically, two redundant sensor sets are installed side by side to provide backup and cross-checking of readings.

Passive yaw systems for wind turbines are increasingly recognized for their simplicity and cost-effectiveness, relying on natural wind forces to maintain rotor alignment rather than complex motor or hydraulic systems. These systems are typically limited to smaller turbines or specific downwind designs due to the inherent constraints in managing larger yawing moments [104, 105]. The fundamental design incorporates a roller bearing connection that facilitates free rotation of the nacelle, and in some configurations, the rotor itself can generate the necessary yaw torque, obviating the need for a tail fin [105, 106].

The effectiveness of passive yaw systems largely depends on their ability to maintain proper alignment with the wind direction while minimizing mechanical stresses. For smaller turbines, conventional designs featuring tail fins can efficiently achieve this by acting as rudders, while in downwind arrangements, the turbine rotors can orient themselves to the wind autonomously [105, 106]. Research has shown that passive systems, while cheaper and easier to maintain, are not without their challenges. Rapid yawing can impose substantial gyroscopic loads on blades and drivetrain systems, particularly if the system experiences low friction, leading to dynamic oscillations that can compromise performance and reliability [104, 106].

To enhance the operational reliability and prolong turbine life, specific passive configurations have been developed. These include systems that utilize a roller bearing for complete freedom of motion, sometimes enhanced by a brake that locks the nacelle once optimal alignment is achieved, reducing potential vibrations [105, 106]. In contrast, designs employing gliding bearings create intrinsic friction, thus artificially slowing the yaw motion, which further aids in stabilizing the system without necessitating additional active components [105].

Notably, the passive design principle is also utilized effectively by manufacturers like Vergnet, which produce mid- and large-scale self-orienting downwind turbines, demonstrating that larger applications can benefit from this technology under specific design conditions [107]. This capability underscores the importance of careful tuning and system design in optimizing yaw performance, as misalignment can result in significant energy losses and mechanical stress [105, 107]. While passive systems offer distinct advantages in lower maintenance costs and simplified operations, the selection of materials and designs must be optimized for the specific environmental and operational conditions faced by each turbine configuration.

Figure 79: Vergnet wind turbine at Gommerville in Eure-et-Loir on 9 April 2015. Lionel Allorge, CC BY-SA 3.0, via Wikimedia Commons.

Active yaw systems are critical components in modern wind turbines, as they enable the precise orientation of the nacelle to ensure optimal rotor alignment with the prevailing wind direction. These systems are especially prevalent in medium and large-scale turbines where significant forces must be managed to adjust the heavy nacelle and rotor assemblies effectively. The complexity of these systems involves several essential elements, including the yaw bearing, yaw drive, yaw brake, and control system, each working in concert to facilitate accurate and responsive operation.

The yaw bearing serves as the rotatable connection between the top of the tower and the nacelle, allowing for smooth rotation. Commonly implemented yaw drives can be either electric or hydraulic, providing the necessary torque to rotate the nacelle efficiently. In comparison, the yaw brake ensures that once the nacelle is correctly oriented, it is securely locked in place to avoid unintended movement, thereby reducing fatigue on the mechanical components [105]. The control system, utilizing automatic signals from wind direction sensors, processes information to dynamically align the nacelle with changing wind directions, thereby maximizing energy capture [108, 109].

The design variations of these active yaw systems include the roller yaw bearing paired with either electric or hydraulic drives, both of which exhibit unique benefits and challenges. For instance, electric drives are prominent for their reliability, while hydraulic drives offer higher power-to-weight ratios albeit with potential issues such as fluid leaks [110]. Furthermore, innovations in yaw systems, such as the gliding yaw bearing approaches which forgo a separate yaw brake by leveraging friction for nacelle restraint, illustrate the ongoing evolution in yaw system designs aimed at cost-efficiency and performance enhancement [111].

Wind direction fluctuations significantly impact yaw system performance. As wind conditions change, the yaw mechanisms must respond to correct the nacelle's orientation actively. Research indicates that misalignment due to inadequate instrumentation or control failures can result in reduced aerodynamic efficiency and increased mechanical strain on turbine components [107]. To improve the responsiveness of yaw controls, advanced control strategies such as model predictive control have gained traction, allowing for real-time adaptations to varying wind conditions and enhancing overall turbine performance [112].

Successful yaw control not only enhances energy extraction but also contributes to minimizing the mechanical loads, which is crucial for the longevity of turbine systems. Methods such as LiDAR-assisted measurement and adaptive control systems have been proposed to further refine the accuracy of yaw adjustments. These approaches address delays in response time that traditional systems face, ensuring that turbines maintain optimal alignment even amidst turbulent conditions [113].

Larger horizontal-axis wind turbines cannot rely on passive tail vanes. Instead, they use an active yaw system to rotate the nacelle on top of the tower so that the rotor faces the incoming wind. The yaw system includes a massive yaw bearing, drive motors, gear sets, control electronics, and the wind measurement devices described above.

The yaw bearing sits at the tower's top flange. Its outer ring is fixed to the tower, while the inner ring supports the nacelle chassis (the frame holding the main shaft, gearbox, generator, and other equipment). The yaw bearing must carry high axial loads, radial impacts, and overturning moments. A yaw gear is integrated into the outer edge of the bearing. The nacelle chassis is bolted to the bearing's inner ring and can rotate around the tower axis when driven.

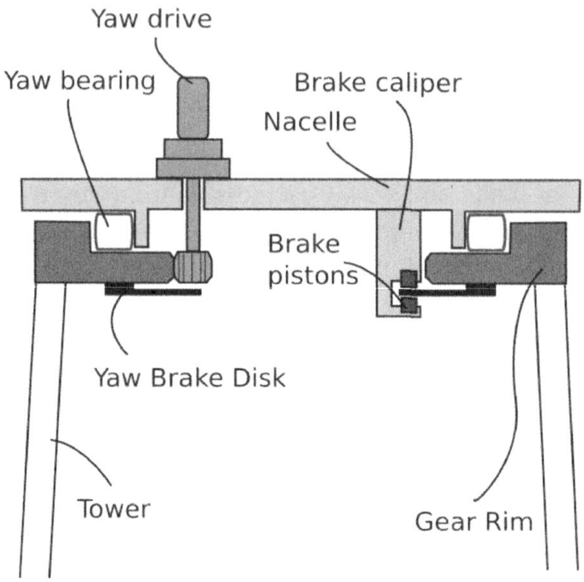

Figure 80: Wind turbine yaw system. Hanuman Wind at en.wikipedia, CC BY-SA 3.0, via Wikimedia Commons.

Several electric yaw motors—typically two to four on small MW turbines and up to six or eight on very large turbines—are mounted on the nacelle chassis. Each motor drives a pinion gear through its own gearbox. The pinions mesh with the yaw gear on the bearing's outer ring. When the control system senses that the turbine is off-wind, it energizes the yaw motors to rotate the nacelle until the rotor is properly aligned.

Because power and signal cables run from the nacelle down through the tower, designers must prevent them from twisting as the nacelle yaws. Small turbines solve this with slip rings at the yaw axis so the cables can rotate freely. On larger turbines, where voltage and current are too high for slip rings, cables are fixed, and the nacelle must yaw back and forth rather than continuously in one direction. This requires a yaw encoder to track rotation angle. The encoder gear meshes with the yaw gear and outputs precise position pulses to the control system, which then reverses or stops yawing to untwist the cables. Some small and medium turbines instead use electromagnetic sensors to count gear teeth for the same purpose.

Finally, a yaw brake locks the nacelle in place when the wind direction is stable or when the motors are not actively yawing. This prevents gusts from causing oscillations and protects the yaw system from wear. Yaw brakes are normally closed and can be implemented either as mechanical brake discs and callipers under the yaw bearing or by electrically braking the yaw motors themselves.

Yaw Brake Hydraulic System

The main parts of the yaw brake hydraulic system are:

Accumulator: The accumulator is essentially a pressurised fluid reservoir. It stores hydraulic fluid under high pressure so that, if the pump or main supply loses pressure, there's still enough hydraulic energy available to keep the yaw brake clamped. This prevents the nacelle from rotating freely in gusts and protects the yaw bearing from damage. It also smooths out pressure fluctuations during normal operation.

Bleed Valve: The bleed valve is a service component. It allows technicians to safely drain fluid from the accumulator when maintenance or repairs are needed. By relieving stored pressure, it ensures that hydraulic lines and brake components can be opened or replaced without danger.

Brake Callipers: The brake callipers sit around the yaw bearing and, when pressurised, they squeeze against brake discs or pads to hold the nacelle firmly in position. The clamping force of the callipers resists wind-induced torque and prevents unwanted yaw movement during normal operation.

Check Valve: This valve acts as a non-return or "trap" valve. If system pump pressure drops or a line breaks, the check valve closes automatically, sealing the pressurised fluid inside the brake circuit. Combined with the accumulator, it keeps the brake engaged long enough for the turbine to remain stable until pressure is restored.

Control Valve: The control valve is spring-loaded and normally directs full system pressure—around 180–200 bar—to the brake callipers. When the yaw control system calls for a position change, the solenoid in the valve shifts it to a different position, reducing the pressure in the callipers so they release slightly. This lets the yaw motors rotate the nacelle but still applies some drag for stability. When reorientation is complete, the valve shifts back to reapply full pressure.

Pressure Reducer (8 bar / 116 psi): This component supplies the low, controlled pressure needed for partial release during yawing movements. Excess hydraulic fluid from this stage is returned to the tank to avoid over-pressurising the circuit.

Pressure Transducer: The transducer continuously monitors the hydraulic pressure in the yaw brake system and sends this data to the turbine's control system. This allows automated checks of brake status, alarm triggers if pressure drops, and closed-loop control of yaw braking.

Relief Valve (230 bar / 3335 psi): This safety valve protects the system against accidental over-pressure. If pressure rises above the set point—say due to a stuck pump or temperature changes—the valve opens to divert fluid back to the tank, preventing damage to hoses and seals.

Test Port: Test ports are service access points where technicians can attach pressure gauges or diagnostic equipment to measure real-time pressure in different parts of the circuit. They're essential for troubleshooting and periodic maintenance.

System Operation: Brake Applied: In normal "locked" mode, pump pressure lifts the check valve, sending fluid through the control valve to the brake callipers. Fluid also charges the accumulator and feeds the pressure transducer and service ports. System pressure is available at the pressure reducer and the relief valve, while excess fluid returns to the tank.

System Operation: Brake Released: When the yaw system needs to rotate the nacelle, the control valve shifts, closing off full-pressure supply and routing reduced pressure from the pressure reducer to the callipers. This releases the brake just enough to let the yaw motors turn, while still applying a damping force to keep motion controlled.

Fail-Safe Function: If system pressure is lost entirely (pump failure, line rupture), the check valve automatically seats to trap fluid in the brake. The accumulator maintains enough pressure to keep the brake locked until pressure can be restored. This fail-safe design ensures the nacelle doesn't freewheel in the wind and prevents damage to yaw components.

Yaw Gearbox

The yaw motors (electric or hydraulic) turn the nacelle so that the rotor faces the incoming wind. These motors run at relatively high rotational speeds but produce low torque. In contrast, the nacelle is extremely heavy—often hundreds of tonnes—and must be rotated slowly but with very high torque to overcome wind-induced loads. The yaw gearbox is the device that converts the high-speed, low-torque output of the motor into the low-speed, high-torque motion required to rotate the nacelle.

Modern yaw drives almost always use a planetary gearbox because this design delivers very high reduction ratios in a compact, lightweight, and robust package. The yaw motor is mounted vertically on top of the gearbox and drives into a stack of reduction stages. Each stage multiplies torque and slows the rotational speed further. By combining multiple stages, the gearbox can be tailored to match the exact torque and speed requirements of a specific turbine and nacelle size.

If you look at a cross-section or drawing of a yaw gearbox, you'll see that the gears in each successive stage get larger as you move downward. This is because as torque increases through the stages, the gears and bearings must be physically larger to withstand the rising loads without premature wear or deformation. At the very bottom of the gearbox sits the output shaft, which carries a large output gear or pinion. This output gear meshes with the yaw ring gear that is built into the nacelle or the top of the tower, providing the direct mechanical interface to rotate the nacelle. Large bearings support the output shaft to handle the significant radial and axial loads at this point.

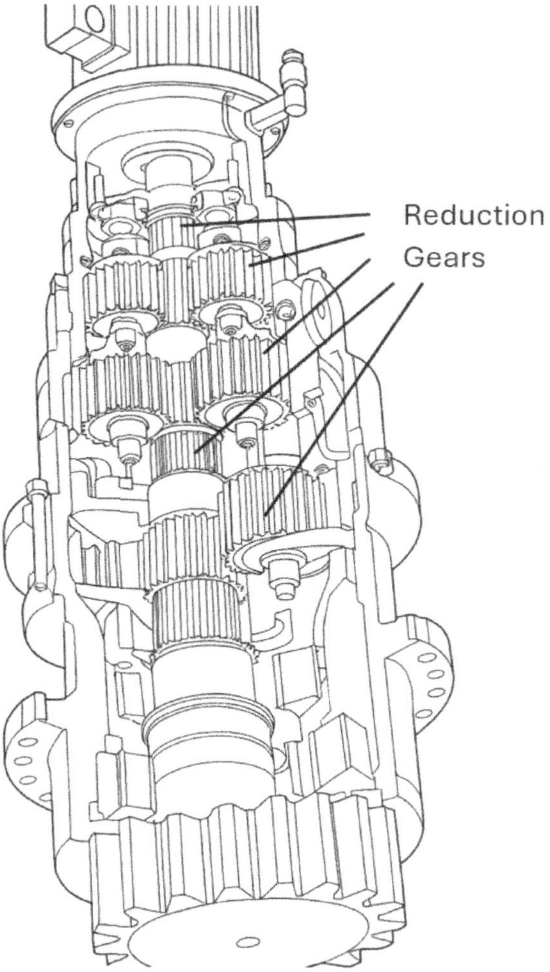

Figure 81: Yaw Gearbox (Cross-section).

A planetary gearbox is named after its geometry. In each stage:

- The sun gear sits at the centre and is driven by the motor or the preceding stage.
- Around it are multiple planet gears, which mesh externally with the sun gear.
- The planet gears also mesh internally with a fixed ring gear.
- The planet carrier holds the planet gears and transfers torque to the next stage or to the output shaft.

This arrangement allows the planet gears to rotate on their own axes (like planets spinning) while also revolving around the sun gear (like planets orbiting the sun). Because several planet gears share the load simultaneously, the design is efficient, compact, and capable of transmitting very high torque. The coaxial layout of the shafts also makes the system mechanically simpler and better balanced.

In the example given, the sun gear rotates about six times for each rotation of the planet gear carrier, which means the reduction ratio is 6:1. If the motor runs at 600 RPM, then after a single 6:1 stage, the carrier and output shaft will rotate at only 100 RPM, while the torque will be multiplied by the same factor (six times).

In practice, yaw gearboxes often use multiple stages stacked together. Two 6:1 stages in series, for instance, produce a total reduction ratio of 36:1. Large wind turbines may use several planetary stages combined with spur or helical gear stages to achieve ratios as high as 2,000:1. This allows relatively small yaw motors to smoothly and precisely rotate massive nacelles under high wind loads.

Yaw Controls

A wind turbine is not a single machine but a complex system of interconnected subsystems, each with its own role. Some sections are designed to capture wind energy—primarily the blades and rotor, which convert the wind's kinetic energy into rotational mechanical energy. Other sections, such as the gearbox and generator, convert that mechanical energy into electrical power. These components cannot work independently; they must be carefully coordinated to ensure optimal performance and safety. This coordination is handled by the turbine's control system.

A wind turbine uses a closed-loop control system to operate efficiently. In closed-loop control, the system continuously measures real-world variables (feedback), compares them to a desired state (the set point), and makes automatic adjustments to maintain optimal conditions. For example, the system measures wind speed, wind direction, nacelle position, and rotor speed, then compares these to target values to ensure the turbine is properly aligned and operating within safe limits.

This control method allows for continuous fine-tuning rather than relying on fixed settings. When a variable deviates from the set point—such as a change in wind direction—an error signal is generated. The control system then issues a correction command to return the turbine to the desired state.

Sensors are essential for collecting the data needed by the control system. They translate physical quantities (like wind speed, wind direction, shaft speed, vibration, oil temperature, or pressure) into electrical signals (voltage, current, or frequency) that can be processed by the controller. For the yaw system specifically, sensors measure the nacelle's orientation relative to the wind and detect how many rotations the nacelle makes, allowing the controller to monitor cable twists and the yaw angle.

This sensor data is known as feedback—it tells the controller what is actually happening so it can decide what to do next.

The controller is the brain of the wind turbine. It receives feedback from the sensors and compares it to programmed instructions or set points. For example, if the nacelle's direction differs from the wind direction by more than 5 degrees, the controller activates the yaw motors to rotate the nacelle until it is aligned with the wind. Once alignment is achieved, the controller shuts the motors off to prevent unnecessary wear.

Actuators, such as electric motors or hydraulic cylinders, carry out the controller's commands. They convert the controller's electrical signals into mechanical motion to adjust nacelle position, pitch angle, or brakes as needed.

The yaw system rotates the nacelle so the rotor faces the wind. There are three primary reasons for yawing:

1. **Aligning with the Wind for Power Generation:** The system continually monitors wind direction and makes small corrections to keep the rotor facing the wind for maximum energy capture.

2. **Managing Cable Twists:** As the nacelle rotates with the wind over time, the flexible power cables inside the tower can become twisted. Sensors count the nacelle's rotations, and if excessive twisting is detected, the controller will pause operation and rotate the nacelle in the opposite direction to untwist the cables.

3. **Furling or Turning Away from the Wind:** In high winds or during maintenance/shutdown, the nacelle can be turned out of the wind to protect the turbine and stop rotor rotation.

During normal operation, when wind speeds are above 3–4 m/s, the controller actively keeps the nacelle aligned with the wind. However, during low or no wind conditions, it is unnecessary—and wasteful—to keep tracking minor wind shifts. Doing so would cause unnecessary wear and tear on the yaw motors and brakes. Therefore, the control system minimizes yawing activity in such conditions.

Because the nacelle rotates back and forth over time, its power and signal cables can twist inside the tower. Sensors monitor how many degrees of rotation have accumulated. If the twist exceeds a safe threshold, the controller will automatically rotate the nacelle to unwind the cables before resuming normal operation. This prevents damage to the cables and minimizes downtime.

Yaw Control Sensors

Potentiometer-Based Wind Direction Sensors: A potentiometer is one of the simplest devices for measuring wind direction. It is a variable resistor with three terminals: two connected to a resistive strip and one connected to a movable "wiper." In a wind direction sensor, a wind vane (the fin that points into the wind) is mechanically linked to the wiper. As the wind changes direction, the vane rotates the wiper along the resistive strip. If a voltage is applied across the two end terminals of the potentiometer, the output at the wiper becomes a variable voltage. This voltage is proportional to the vane's position and can be read by the turbine's controller.

Potentiometers are inexpensive, simple, and provide a continuous output with good linearity. However, they have some limitations: the wiper can wear out over time, they have a "dead band" where the wiper crosses from one end of the strip to the other, and they can be susceptible to icing. Heaters and protective designs are used to reduce icing problems.

Encoder-Based Wind Direction Sensors: Some wind direction sensors use encoders instead of potentiometers. An encoder consists of a coded, perforated disk attached to the wind vane, several light-emitting diodes (LEDs) as light sources, and corresponding detectors. As the disk rotates, light passes through or is blocked by the disk, creating binary bits that represent the vane's angular position. When these bits are combined from multiple concentric tracks on the disk, the encoder can determine the vane's absolute position within 360 degrees of rotation.

Most encoders use gray code, a binary system where only one bit changes per step. This reduces errors and makes data transmission more reliable. For example, a 4-bit gray code disk divides the full 360° into 16 sections (22.5° each), while an 8-bit disk divides it into 256 parts (1.41° resolution). More tracks and bits give finer resolution.

Ultrasonic Anemometers for Wind Speed and Direction: Unlike potentiometers and encoders, ultrasonic sensors have no moving parts. They measure wind speed and direction by sending ultrasonic pulses between pairs of transducers. The time taken for the pulse to travel from one transducer to another depends on wind direction and speed—wind blowing with the pulse speeds it up, while wind against it slows it down. By comparing transit times in multiple directions, the controller calculates both wind speed and direction.

These devices avoid the wear, dead zones, and icing problems of mechanical sensors, making them highly reliable, although more expensive.

Yaw Position Sensors: To keep the turbine correctly oriented, the controller must also know the nacelle's position relative to the tower. Yaw position sensors provide this information and can use several technologies, including encoders, limit switches, and proximity (prox) sensors.

Incremental Encoders for Yaw Position: Incremental encoders are similar to absolute encoders but simpler. They have one or more light source/detector pairs arranged around a disk. A single-track encoder outputs a series of on/off pulses, one per slot on the disk. By counting pulses, the controller calculates how far the nacelle has rotated. For example, 360 pulses per revolution gives a resolution of 1 degree.

However, a single track cannot tell direction. Adding a second track offset from the first creates a quadrature encoder, which indicates whether rotation is clockwise or counterclockwise. Some encoders also include a third "Z track" that produces one pulse per full revolution to reset the controller's counters.

In yaw systems, the encoder disk is usually driven by a gear meshed with the yaw ring teeth. Because the disk turns many times for each nacelle rotation, the controller can measure yaw angle with very high accuracy.

Limit Switches: Limit switches are simple mechanical devices actuated by physical contact with a moving part such as a cam. In yaw sensors, the same gear that drives the encoder shaft can also rotate cams that trip limit switches. These switches provide additional signals to the controller—for example, indicating when the nacelle has rotated far enough or warning of excessive cable twist.

Limit switches are versatile: they can be wired to AC or DC, consume no power, and are easy to install. Most have both normally open and normally closed contacts. Typical yaw limit switch gear ratios are around 150:1.

Proximity (Prox) Sensors: Prox sensors are electronic devices that detect metallic targets without contact. Inductive prox sensors emit an oscillating magnetic field. When a metal target enters this field, it absorbs energy and stops the oscillations, triggering a signal to the controller. Prox sensors switch faster than limit switches, don't wear out, and don't cause sparking. They're ideal for detecting gear teeth, cams, or hydraulic cylinders in motion.

However, prox sensors must be installed very close to their target (within a few millimetres), and their voltage and polarity requirements must match the sensor's design—unlike limit switches, they are not universally compatible.

Integration of Multiple Sensors: In practice, yaw position systems often combine several sensor types. For instance, an encoder might provide high-resolution position data, while limit switches or prox sensors act as safety or backup devices to alert the controller of cable twist or extreme nacelle positions. Together, these sensors ensure the turbine can yaw safely, efficiently, and with minimal wear.

Blades and Pitch Control Mechanisms

The design of wind turbine blades is essential for optimizing energy efficiency in wind power generation. A pivotal measure in this context is the tip-speed ratio (TSR), defined as the ratio of the blade rotational speed to the wind speed. This ratio typically ranges from 6 to 8 for high-efficiency three-blade turbines, making it critical for maximizing energy capture, especially during gusts of wind when maintaining an optimal TSR is crucial [114].

Efforts to enhance blade performance have showcased the advantages of using composite materials, which significantly reduce the weight and rotational inertia of newer turbines. This reduction allows turbines to adjust their speeds quickly in response to fluctuating wind conditions, effectively maintaining a consistent TSR [115, 116]. Furthermore, the optimization of blade design has been significantly influenced by aerodynamic considerations. Innovations like specialized airfoil shapes and the strategic placement of vortex generators serve to smooth airflow around the blade, minimizing turbulence and flow separation. These design enhancements contribute to an impressive lift-to-drag ratio that can reach up to 60–70 for advanced designs, which is considerably higher than that of conventional aircraft [117].

It is important to note that noise production is a byproduct of increasing tip speeds, which poses both operational challenges and potential structural loads on the turbine [118, 119]. Therefore, studies on the aerodynamics of marine life, such as the humpback whale's flippers, have informed blade designs by promoting features that reduce stall and improve noise characteristics. Certain turbine models have incorporated designs inspired by these marine features, leading to improved performance and quieter operations [120].

Additionally, the customization of airfoil cross-sections along the blade's length plays a crucial role in maximizing efficiency across varying wind speeds [121]. Research into developing computational design tools has advanced the state of wind turbine blade engineering, allowing designers to simulate performance under various operational conditions and optimize for structural integrity and aerodynamic efficiency [116, 122]. This methodological approach not only streamlines the design process but also enhances the predictive capacity regarding a blade's performance and longevity under operational stress [123].

Applications of Inertial Measurement Units

The application of Inertial Measurement Units (IMUs) in wind power generation, particularly for monitoring blade dynamic deformation and load, is a burgeoning field of research. IMUs are crucial for measuring the dynamic characteristics of wind turbine components, including blades, hubs, and tower tops, by capturing real-time three-axis acceleration and angular velocity data.

IMUs play an essential role in tracking the motion states of wind turbine components using inertial navigation algorithms. They calculate parameters like position, velocity, and attitude by integrating data through algorithms such as Kalman filters and GNSS (Global Navigation Satellite System) data enhancement. This integration mitigates cumulative errors and enhances the precision of blade deflection and load estimations, thus offering vital insights needed for operational load monitoring of wind turbines [124-126].

Real-time data acquisition from IMUs allows for detailed monitoring of angular velocity and acceleration changes, which are integral in assessing the flexural behaviour during turbine operation. The combination of high-precision and low-precision IMUs enhances this capability. High-precision IMUs are strategically located at blade root bases, providing reference data on the overall dynamics of the blade system, while low-precision IMUs distributed along the blade track measure local dynamic deformations. This hierarchical approach allows for robust data fusion, leading to enhanced reliability and fault tolerance of the monitoring system [126].

In terms of structural health monitoring and fault prediction, the fusion of IMU data with information from other sensors (such as vibration and stress sensors) significantly improves fault detection sensitivity. Studies indicate that equipping turbine components, particularly the main shaft, with multiple IMUs allows effective extraction of tower acceleration signals, enabling specific fault identification through interpretative data analysis [125, 126]. Advanced data fusion algorithms have been proposed to achieve reliable multi-node data fusion and effective monitoring of blade deformations [126].

Moreover, during dynamic testing of wind turbine blades, leveraging both high-precision and low-precision IMUs is crucial for capturing the effects of external forces on blade motion. Employing a federated Kalman filter to amalgamate data supports a superior understanding of the blade's three-dimensional flexural behaviour. Simulation results indicate that this fusion algorithm minimizes measurement errors, providing accurate depictions of local dynamics, particularly at critical points along the blade [126].

In summary, IMUs are instrumental in enhancing the precision and reliability of dynamic deformation monitoring and load assessments in wind turbines. Advancements in sensor technology and data fusion strategies concerning IMUs highlight a promising direction in wind energy research, considerably contributing to operational efficiency and safety [124-126].

Basic Hub Designs

The hub is the central component of the rotor that connects the turbine blades to the main shaft. It transfers the aerodynamic forces from the blades into the drivetrain, allowing the rotor to drive the generator. In its simplest form, the hub is a cast or fabricated steel component with mounting points for each blade.

In basic fixed-pitch designs, each blade is bolted directly to the hub without any means of rotation around its axis. This type of design is common in small wind turbines where simplicity and cost are priorities. Because the blades cannot change their angle of attack, the turbine relies on aerodynamic stall (the natural loss of lift above a certain wind speed) to limit power. However, this leads to reduced control and higher loads during strong winds, which can increase fatigue on the blades and hub.

In modern, larger turbines, each blade is bolted to a pitch bearing instead of directly to the hub. This pitch bearing allows the blade to rotate about its longitudinal axis, changing its angle of attack relative to the wind. By adjusting blade pitch, the turbine can maximize energy capture at low to medium winds and limit power output or stop the rotor in high winds.

The pitch bearings themselves are large, precision-engineered rolling-element bearings capable of supporting the enormous radial and axial loads from the blades while still allowing smooth rotation for pitch adjustments. These bearings are bolted securely to the hub, forming a critical interface between the blade and the rotor structure.

The rotation of the blades around the pitch bearings is performed by pitch control systems, which can be hydraulic or electric:

- Hydraulic Systems: These use hydraulic cylinders or motors powered by pressurized fluid to rotate the blades. They can deliver very high torque, which is necessary for large blades, and respond quickly to control signals. However, hydraulic systems require maintenance to prevent leaks and keep valves clean.

- Electric Systems: These use electric servomotors powered by onboard batteries or ultracapacitors inside each blade root or hub. Electric systems are simpler to maintain, eliminate hydraulic leaks, and are increasingly common in newer designs.

Both types of systems include their own control units and emergency backup power sources (such as batteries or ultracapacitors) so the blades can be pitched to a safe position even if grid power is lost.

The hub itself is bolted or keyed to the rotor shaft, which transmits the torque from the rotating blades to the drivetrain. In direct-drive turbines, the rotor shaft connects directly to the generator. In gearbox-based designs, the rotor shaft connects to a low-speed shaft feeding into the gearbox, which then steps up the speed for the generator. In either case, the hub must be strong enough to handle the combined bending moments and torque from all the blades under varying wind loads.

The hub is not just a passive connection point; it's a highly engineered structure that must:

- Transfer massive aerodynamic and gravitational loads from the blades to the shaft.
- Provide precise alignment and support for the pitch bearings.
- House pitch control systems, sensors, and wiring or hydraulic lines routed through the hollow shaft into the hub.

A well-designed hub reduces fatigue loading on the blades and drivetrain, extends turbine life, and ensures the pitch system can function reliably under extreme conditions.

Blade Design Parameters

Choosing the Number of Blades

The number of blades on a wind turbine is a careful compromise between aerodynamic efficiency, cost, and system reliability. Adding blades generally improves the ability of the rotor to extract energy from the wind, but the improvement is not linear. Going from one blade to two blades yields about a 6 % increase in aerodynamic efficiency, and adding a third blade adds roughly another 3 %. Beyond three blades, the gains are very small and the engineering drawbacks grow. With each added blade, stiffness requirements increase while the blades themselves become thinner to keep weight reasonable. Thin blades are more flexible and can deflect excessively, reducing tower clearance and increasing fatigue loads.

Another factor is noise. Blade noise — especially from the trailing edge and the tips — rises sharply with tip speed, following approximately the fifth power of the blade's speed. This means that a small increase in tip speed can dramatically increase acoustic emissions. Designers must consider both blade count and rotor speed to meet noise regulations, especially near populated areas. Rotor placement also matters: upwind rotors (the most common) tend to be quieter because they avoid tower wake effects, while downwind rotors can produce distinctive "whooshing" or "thumping" as blades pass the tower.

Cost and Component Considerations

Fewer blades mean lower material and manufacturing costs for the rotor itself, and allow the turbine to run at a higher rotational speed. Higher speeds reduce peak torque loads on the drive train and generator, which can translate into smaller, lighter, and less expensive gearboxes and generators. This is one reason two-blade and even single-blade prototypes have been built — they're cheaper and lighter to manufacture.

However, lower blade counts also change the dynamic loading on the hub and drive train. During yaw (when the nacelle turns to follow the wind), each blade experiences a cyclic load depending on its position in the rotation. With three blades, these loads are evenly balanced around the drive shaft, leading to smoother operation. One- or two-blade turbines can experience greater asymmetric loading unless they use a teetering hub or other mechanisms to "absorb" the imbalance. Modern two-blade designs often use pivoting or teetered hubs to nearly eliminate these cyclic loads and improve reliability. A well-known example was a Chinese 3.6 MW two-blade turbine tested in Denmark in 2012.

Blade Size and Power Capture

Increasing blade length has been the single biggest driver of rising wind turbine capacities — moving from 1 MW-class machines to offshore giants exceeding 10 MW. A larger rotor sweeps more area and effectively raises the tip-speed ratio at a given wind speed, which means the rotor can extract more energy. However, larger blades also add mass, which in turn increases gravitational and aerodynamic loads. Advanced engineering software, such as HyperSizer (originally developed for aerospace), is now used to optimize blade designs and materials.

As of the mid-2010s, onshore turbines had rotor diameters reaching around 130 m, while offshore turbines reached 170 m. Those numbers have only grown since then, and the material requirements are staggering — in 2001 alone, an estimated 50 million kilograms of fiberglass laminate went into turbine blades globally.

Blade Weight and Structural Loads

Controlling blade weight is critical because blade mass scales roughly with the cube of the radius. As blades get longer, gravity loads — axial, tensile/compressive, and bending — grow rapidly. Each rotation subjects the blade to fluctuating stresses: when a blade is at the top of its rotation it experiences different gravitational and aerodynamic loads than at the bottom. These cyclic loads reverse edgewise bending moments every 180 degrees of rotation.

Typical large turbines rotate at about 10 rpm and are designed for a service life of roughly 20 years. Over that lifetime, each blade may endure on the order of 10^9 loading cycles from wind variability, gusts, and shear. Aerodynamic lift causes flatwise bending (out of plane), while airflow around the blade induces edgewise bending (in plane). Both types of bending involve alternating tension and compression along different surfaces of the blade (pressure vs suction sides, leading vs trailing edge). This cyclic stress is a major driver of fatigue and eventual failure if not properly managed.

Failure Modes and Safety

Blade failures — such as partial or total loss of a blade — are one of the most significant safety risks associated with modern wind turbines. Because of the natural variability of wind and the stress responses of materials, both loads and strength have statistical distributions. Designers must consider the "ultimate loading" scenario (worst-case loads) and design safety factors accordingly. As turbines get larger and blades longer, the public safety implications of blade failures grow, especially for turbines near roads or buildings.

Materials and Future Trends

Given the escalating size and stresses, researchers are focused on new materials with higher strength-to-mass ratios, such as advanced composites, carbon fibre reinforcements, or hybrid laminates. These materials aim to keep blades light while maintaining stiffness and strength, extending service life and reducing fatigue risk. They also help manufacturers meet noise and cost constraints by enabling thinner, stronger blades.

Blade Materials

Modern wind turbine blades are no longer made from simple materials. They must be light enough to reduce gravitational forces yet strong enough to resist bending, fatigue, and environmental damage. For this reason, designers increasingly use carbon-fibre girders inside the blades to stiffen and lighten their internal spar structures. Regardless of the exact material, all blade materials are chosen to meet key criteria: they must be widely available and easily processed to keep costs down; they must have low weight (density) to limit gravitational loads; they must be strong and stiff to resist wind and gravity while holding their aerodynamic shape; they must endure billions of fatigue cycles over decades; and they must be tough enough to handle environmental hazards like moisture, UV light, salt, and lightning strikes.

Early windmills used wood and canvas sails because these were cheap, readily available, and easy to work with. But they required constant maintenance and limited the blade to a flat-plate profile, which is much less aerodynamically efficient than a modern airfoil. As turbines grew, builders moved to metals such as steel and aluminium, which were robust and could form solid blades. However, their weight severely restricted blade length and forced higher energy consumption to keep them rotating.

The shift to fiberglass composites marked a breakthrough. Fiberglass blades were much lighter than metal blades and enabled larger diameters. But fiberglass had its own problems — it can degrade under UV light and moisture, causing delamination, and it is notoriously hard to recycle at end of life. The next step was carbon fibre, which offers an exceptional stiffness-to-weight ratio. Carbon-fibre spar caps or girders inside the blade allow much longer blades without excessive weight, increasing energy capture per turbine. Today, researchers are experimenting with bio-based composites and nano-enhanced resins to further reduce weight and environmental impact.

Most commercial blades today are made from fibre-reinforced polymers (FRPs). These are composite materials where long fibres (glass, carbon, basalt, or natural fibres) carry the tensile and bending loads, while a polymer matrix binds them together and provides toughness, impact resistance, and resistance to delamination. Glass-fibre reinforced plastic (GFRP) and carbon-fibre reinforced plastic (CFRP) have the highest indices for power efficiency, fatigue life, and thermal stability.

Within FRPs, two main matrix types are used: thermosets (epoxies, polyesters) and thermoplastics. Thermosets dominate because they have lower viscosity (easy to infuse through thick laminates), cure at lower temperatures, and are well understood in large-scale blade manufacturing. Thermoplastics offer recyclability and higher fracture toughness but are harder to process because they need higher temperatures and have higher viscosity, which makes infusing large blade moulds difficult.

Blades in the 40–50 m range are commonly produced with fiberglass composite infusion techniques. Resin infusion, vacuum-assisted resin transfer moulding (VARTM), and pre-impregnated ("prepreg") fabrics are used to minimize volatile organic compound (VOC) emissions and improve laminate quality. Some designs mix wood, fiberglass, and carbon fibre in an epoxy matrix for hybrid performance.

One challenge in thick laminates is resin distribution: resin must flow evenly before it gels to avoid voids. Partially impregnated fabrics or special flow media can help by creating channels for resin flow during evacuation, resulting in a uniform laminate. As blades have surpassed 60 m, these infusion techniques became standard because traditional injection methods were too slow relative to resin gel times.

Epoxy resins are now the workhorse of large turbine blades because they cure faster, improve durability, and yield better surface finish than older polyester resins. Specialized epoxies have been developed to match viscosity, pot life, and toughness to large blade production. Using carbon-fibre spar caps within epoxy laminates can cut total blade mass by up to 38% and reduce cost by around 14% compared to all-glass designs, while also reducing laminate thickness so resin infusion is easier.

Beyond traditional glass and carbon fibres, researchers are incorporating carbon nanotubes (CNTs) and carbon nanofibers (CNFs) into blade resins or onto fibres. CNT/polymer nanocomposites can dramatically change material properties even at very low filler contents (<5% by weight), improving modulus, strength, and fracture toughness and reducing crack propagation between layers. CNFs on blade surfaces improve erosion resistance (critical in desert or offshore environments where sand, salt, and rain erode the leading edge), enhance electrical conductivity for lightning strike protection, and increase damping and impact resistance.

A growing priority is to reduce the environmental impact and improve recyclability of blades. Bio-composites use natural fibres like flax, hemp, or basalt instead of synthetic fibres. These have much lower embodied energy but weaker mechanical performance due to high cellulosic content and moisture sensitivity. Research is ongoing to improve their structural and thermal properties. Basalt fibres, which are naturally moisture-resistant, are promising candidates for partially replacing glass fibres.

Recycling remains a major challenge: fiberglass is almost impossible to recycle, and carbon fibre requires complex processes to recover usable fibres. Developing recyclable thermoplastic matrices or bio-based composites with sufficient mechanical properties could help close this loop.

Even with strong internal structures, blade surface coatings matter for performance and longevity. Coatings must resist UV, salt spray, rain erosion, sand abrasion, and ice accretion. Super-hydrophobic surfaces (SHS) cause water to bead and roll off, reducing ice formation down to −25 °C. Combined with heating elements, SHS coatings can maintain aerodynamic smoothness in icing climates. In desert regions, CNF-enhanced coatings greatly improve particle erosion resistance.

Over a 25-year life, blade tips are likely to suffer multiple lightning strikes, progressing from superficial scorching to deep cracks or adhesive failures. Non-conductive blade materials like GFRPs and CFRPs typically use embedded copper or metallic lightning arresters to route strikes safely to ground, bypassing the blade and gearbox. Conductive carbon fibres can also help by integrating the strike path directly into the blade's structural elements, avoiding extra deadweight.

Lightning Protection System

A wind turbine's lightning receptor cable, also known as a down conductor, is a key part of the turbine's Lightning Protection System (LPS). It is typically made of copper or a copper-based alloy and serves as a low-resistance electrical pathway that connects the metallic lightning receptor at the blade tip to the ground termination system within the tower. Its function is to safely direct the immense electrical energy from a lightning strike away from the turbine's structural and electrical components and into the earth, preventing flashover, arcing, or catastrophic damage.

The LPS operates through a sequence of controlled electrical processes. First, attraction occurs when the metallic lightning receptors on the blade tips — and sometimes at mid-blade locations — act as intentional strike points. These receptors are designed to attract lightning discharges, safely guiding them into the system. Next, during the conduction phase, the electrical current travels along a low-resistance copper down conductor embedded within the blade. This conductor extends from the receptor tip to the blade root, ensuring the charge remains within the designated path.

At the transfer stage, the electrical current moves from the blade's internal conductor to the hub or nacelle grounding system. This is achieved through specially designed brush contacts or coupling systems that maintain electrical continuity even as the blades rotate. Finally, in the dissipation stage, the current travels down the turbine's internal grounding cable through the tower and is released safely into the earth (or seabed, in offshore installations) via a ground termination system that disperses the energy over a wide area.

The LPS is composed of several integral elements. Lightning receptors are metallic points installed on the blade surface, typically near the tip, to provide a safe and predictable attachment point for lightning strikes. The down conductor — a large copper cable — runs through the interior of the blade to carry the electrical charge to the root. The root connection or transfer system then ensures the smooth handover of the current from the rotating blade to the stationary nacelle and tower grounding system. Finally, the ground termination system dissipates the electrical energy harmlessly into the surrounding soil or ocean, completing the protection circuit.

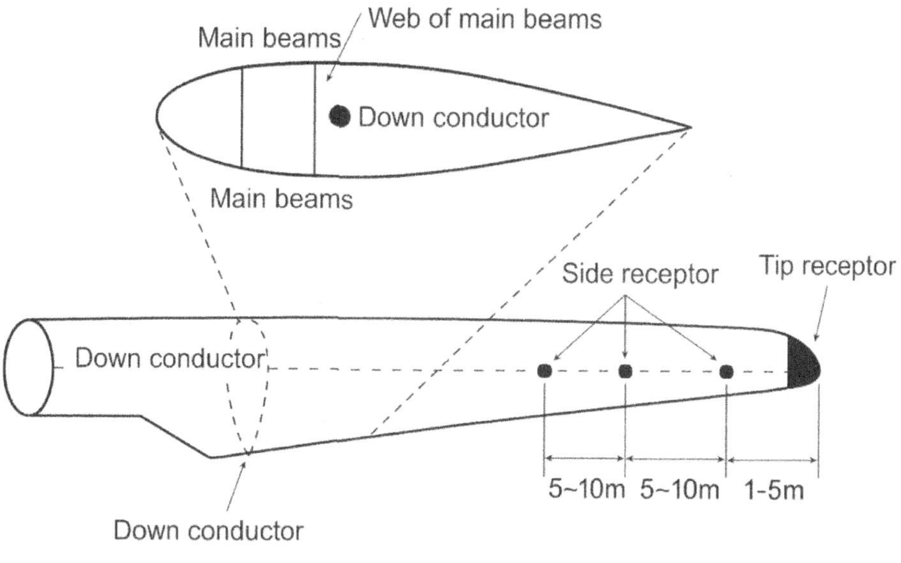

Figure 82: Lightning protection system of wind turbine blade.

A well-functioning lightning protection system is essential for turbine safety, reliability, and performance. By safely channelling high-voltage discharges through designated conductors, the LPS prevents severe damage to the blades, hub, and electrical systems. It also reduces operational downtime, since turbines without adequate protection can be disabled for extended periods following a strike. Additionally, the system minimizes fire and thermal risks, as uncontrolled lightning currents can cause intense localized heating that may ignite internal components or damage structural materials.

In essence, the lightning receptor cable and its associated components form a critical defence mechanism, ensuring that even when lightning strikes a wind turbine — a frequent occurrence given their height and exposed locations — the structure remains protected, operational, and safe.

Blade Repair

Wind turbine blades are complex structures that require careful consideration in maintenance and repair due to the various conditions they endure during operation. Typically, these blades necessitate repair after 2 to 5 years of service because of damage from manufacturing defects, transportation issues, assembly, and environmental factors, including thermal cycling and lightning strikes [127-129]. Erosion and structural fatigue further compound the damage, highlighting the essential role of maintenance strategies in extending blade lifespan and efficiency [130].

Effective repair approaches depend on the nature of the damage. For example, addressing leading edge erosion may involve specialized coatings or protective tapes, while structural repairs may utilize bonding or fastening techniques with new materials [130, 131]. Non-structural issues, such as matrix cracks, often require resin injections or fills to prevent minor damage from escalating into more serious structural failures [128]. Distinguishing between structural and non-structural repairs is critical, as neglecting minor cracks can significantly reduce blade performance and longevity, thereby increasing operation and maintenance (O&M) costs as turbines age [127].

To inform repair strategies, four distinct zones have been identified on the blade, each requiring tailored repairs. Zone 1, the leading edge, often necessitates interventions for erosion or cracks. Zone 2, located close to the tip, may require semi-structural repairs to manage aeroelastic responses. The middle area (Zone 3) commonly suffers from erosion, necessitating similar treatments as Zone 1. Finally, Zone 4, encompassing the root and near-root of the blade, may require more complex structural repairs due to higher shear and bending stresses experienced during operation [131, 132].

The urgency for effective blade repair methodologies is underscored by economic factors within the wind energy sector, where O&M costs can account for 20-25% of the total levelized cost of energy from wind [127, 128]. As wind turbines age, these costs are projected to rise, emphasizing the importance of timely and effective maintenance. Ignoring minor damages not only affects power production but can also accelerate the decline in blade functionality and increase repair costs over time [130].

Considering the composite materials used in blade construction, repair techniques prevalent in the aerospace industry are often adapted for wind turbine blade maintenance [133]. This cross-industry exchange highlights technological advancements and methodologies available for addressing the specific challenges associated with wind turbine blades. Innovation in repair technologies, including methods based on dynamic analysis and modular designs for ease of maintenance, presents pathways for improved efficiency and reduced costs in the repair process [130, 131].

Pitch Control

Pitch control systems are designed to optimize the efficiency and safety of wind turbines by actively adjusting the angle (or "pitch") of the turbine blades relative to the wind. By changing blade pitch, the system maximizes energy capture at low and medium wind speeds, limits output at high wind speeds, and helps stop or slow the rotor during emergencies. This makes pitch control one of the most important systems in modern turbines for balancing performance and protection.

Although the basic principle—turning the blades to the best angle for the current wind—is simple, the execution is complex. A typical pitch control system includes:

- **Sensors and Controllers:** Sensors on the nacelle and hub continuously monitor wind speed, wind direction, and turbine output. This data is sent to a controller, which uses programmed algorithms to calculate the optimal blade angle for the conditions.
- **Actuators and Bearings:** The controller sends commands to electric or hydraulic actuators inside each blade hub. These actuators physically rotate each blade around its longitudinal axis. Bearings in the hub allow the blade to rotate smoothly and handle the large aerodynamic and gravitational loads.
- **Feedback Loop:** Because wind conditions can change quickly, the system uses continuous feedback from sensors to verify that the blades are at the commanded angle. If conditions change or a discrepancy is detected, the controller issues new commands to adjust the pitch.

This closed-loop feedback ensures the turbine can respond dynamically to gusts, turbulence, or grid-control demands.

Because pitch control systems consist of sensors, actuators, bearings, and software working together, any malfunction can quickly lead to lost efficiency, uneven loads, or unplanned downtime. For example, faulty sensors may give incorrect wind-speed data, causing the controller to miscalculate blade angles, while worn actuators may fail to reach the commanded position.

Regular inspections and preventive maintenance are therefore essential to ensure these systems remain reliable. This includes checking the health of actuators, lubricating bearings, testing sensor calibration, and ensuring control software is updated.

Some common challenges and their solutions include:

1. Mechanical Stress: Pitch-controlled turbines undergo frequent blade-angle adjustments, which place significant cyclic loads on the blade roots, hub, and pitch bearings. Misalignment or errors in the pitch system can cause uneven stresses, leading to accelerated wear or even blade damage.

Solution: Start with high-quality turbine components and sensors. Once in operation, implement regular inspections and early corrective actions to prevent misalignment. Early detection of mechanical or electrical faults can prevent costly breakdowns.

2. Maintenance Needs: The moving parts and sensitive electronics of pitch control systems require regular maintenance. Faulty sensors can compromise the whole system's accuracy, and actuator wear can cause slow or incomplete blade movement.

Solution: Use advanced inspection technologies such as drones or autonomous systems to perform blade and hub inspections more frequently and with less disruption than traditional rope-access methods. Plan maintenance schedules based on real data from sensors to reduce downtime.

3. System Complexity: Despite its simple goal, a pitch control system is a highly complex integration of mechanical, electrical, and software components. Control algorithms may become outdated, reducing efficiency or responsiveness. Additionally, failures in one component can disable the entire system.

Solution: Consider upgrading control algorithms to more advanced or adaptive versions to improve responsiveness. Where possible, add redundancies or fail-safes so the turbine can continue to operate safely even if one part of the pitch system fails. Some manufacturers already integrate backup power supplies and emergency pitch mechanisms to maintain safety and energy production during faults.

Pitch control systems dramatically improve turbine performance by ensuring optimal blade angles for power production and protection. However, their benefits only materialize with proper maintenance and inspection. Poorly maintained systems can reduce output, increase loads, and shorten turbine life.

By combining high-quality design, regular inspections, advanced monitoring technologies, and updated algorithms, operators can ensure pitch control systems deliver maximum efficiency and reliability throughout the turbine's lifespan.

Independent Pitch Systems (one actuator per blade)

In an independent pitch system, each blade has its own pitch bearing and its own actuator + controller, so the three blades can be moved together (synchronously) or individually (asynchronously) when the control system needs finer load control. This is the dominant architecture on modern utility-scale turbines because it gives the controller maximum authority over power, loads, and braking.

Two drive types are common:

Hydraulic independent pitch: A pitch bearing (four-point contact ball or similar) is bolted between the hub and each blade root. Inside the hub, a hydraulic cylinder connects via a pitch lever on the blade flange. As the piston extends/retracts, it rotates the blade about its longitudinal axis. The cylinder is trunnion-mounted so it can swing as the lever follows a circular path. A linear position transducer in each cylinder provides precise blade-angle feedback. Pressure and command signals pass through the hollow main shaft via a rotary union/slip ring to the nacelle controller. Accumulators allow fail-safe feathering if grid power is lost.

Electric independent pitch: Each blade root has a pitch motor + gearbox that turns a ring gear integrated with the pitch bearing. Designs come in external-gear (teeth on the outer race) and internal-gear (teeth on the inner race). A blade-angle encoder or geared transducer closes the loop, and a holding brake in each motor locks the blade when stationary. Power and signals for the three motor/inverter "pitch cabinets" in the hub are fed via the slip ring; batteries or ultracapacitors in each cabinet provide emergency feather power.

Control behaviour across operating states:

- **Start:** from ~90° (feather, "full-wind") the blade turns to ~45° to build starting torque, then toward ~0° as rpm rises.
- **Normal operation:** hold ~0° below rated; above rated wind, modulate 0–30° to cap generator power.
- **Shutdown:** drive to ~90° (feather) to aerodynamically stop the rotor; in any power-loss/fault, the system auto-feathers from stored energy (accumulator or ultracap).

Strengths/limitations: Superb control (individual-blade load mitigation, precise power regulation, reliable emergency feather). Electric systems are cleaner and simpler to service; hydraulic systems deliver very high torque density but need leak-free plumbing and fluid care. Both require robust slip-ring management and periodic actuator/gearbox inspections.

<u>Drive-Rod Unified Variable Pitch (mechanically linked, synchronous)</u>

A drive-rod system uses one central linear drive (often a hydraulic cylinder behind the gearbox) to move all three blades together via a forked pitch drive rod and linkages.

Mechanics: Each blade sits in a pitch bearing whose inner ring is integrated with the blade flange and carries a pitch lever. A three-armed fork (one arm per blade) at the hub connects to each lever via link rods. The fork attaches to a long, axially moving rod that passes through the hollow main shaft and gearbox low-speed shaft. When the central actuator pushes/pulls the rod, the fork advances/retracts, the three links drive all blade flanges to rotate together, and the pitch angles change synchronously. The main rotor bearing is sleeved so the rod can slide through it.

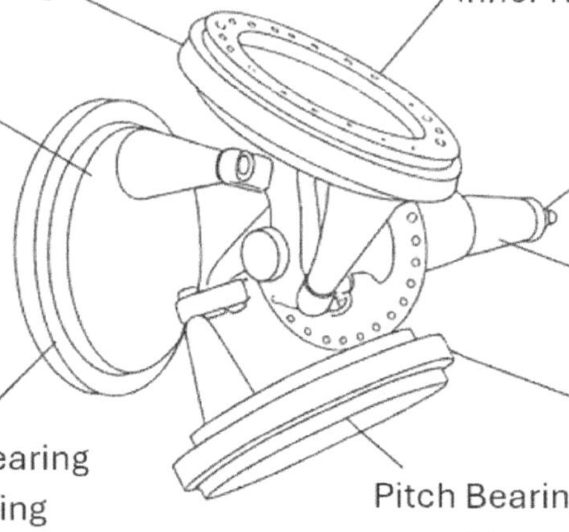

Figure 83: Pitch drive mechanism of three blades.

Control behaviour: Functionally the same operating states as above (start, regulate, feather), but all blades move together—there's no individual-blade control.

Strengths/limitations: Very simple, robust, and cost-effective with minimal hub equipment—good for small and some medium turbines. Fewer in-hub electronics, fewer drives to maintain. However, the lack of independent pitch means less capability for asymmetric load reduction and advanced control strategies, and a single mechanical chain is a larger single point of failure. Stroke limits must cover full 0–90° travel, and linkage geometry must keep angles favourable across the range to avoid force spikes.

<u>Toothed-Belt Pitch Systems (independent, belt-driven)</u>

A toothed-belt system is an independent electric pitch variant that swaps a gear-mesh at the blade for a timing belt drive wrapped around the pitch bearing's rotating member.

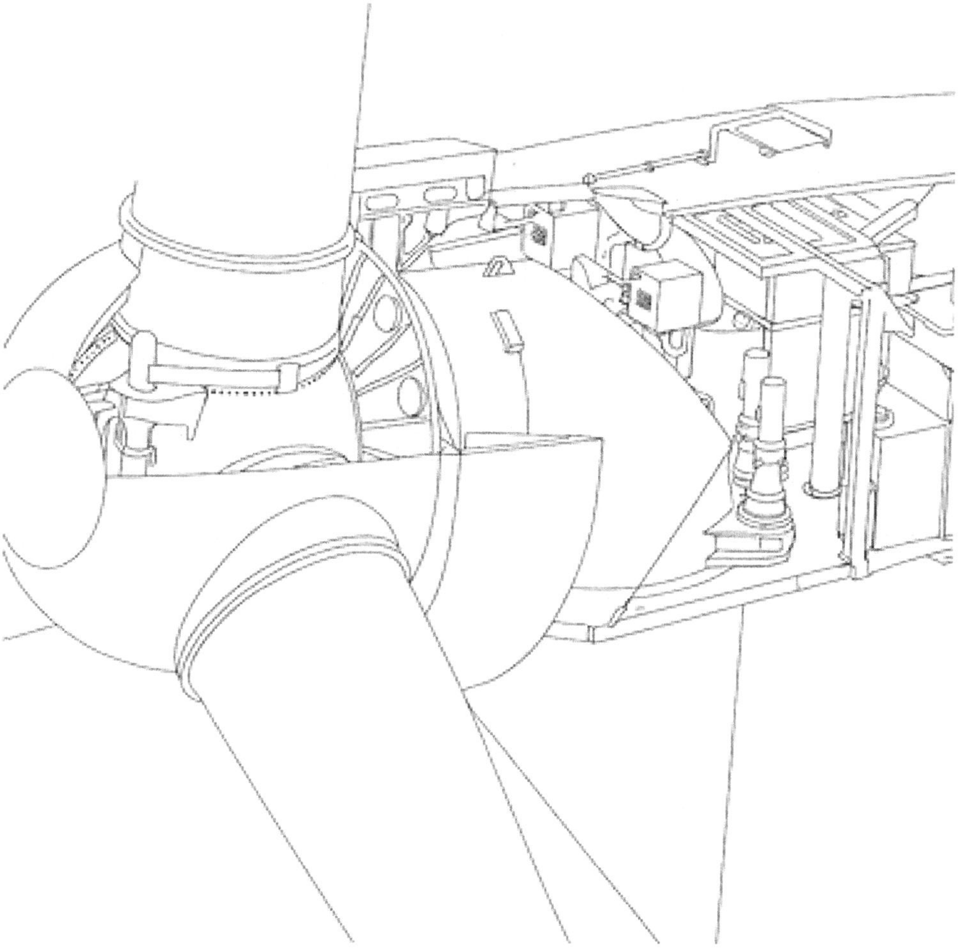

Figure 84: Toothed belt pitching mechanical structure.

Mechanics: The blade mounts to the rotating ring of the pitch bearing (some designs mount blade on the outer ring, others on the inner). A timing belt (steel/fibre tension members with polyurethane/neoprene teeth) is anchored at both ends to the rotating ring via clamping plates and routed over a drive pulley (on the pitch motor/gearbox) and one or two idler/tension pulleys on a bracket in the hub. When the motor turns the drive pulley, the belt walks around the ring, rotating the blade. Tensioners ensure positive, no-slip engagement and adequate belt wrap. Each blade has its own pitch cabinet (dc supply, inverter, controller, encoders, end-limits, motor brake) in the hub; power/control come through the slip ring. Ultracapacitors commonly provide emergency feather energy.

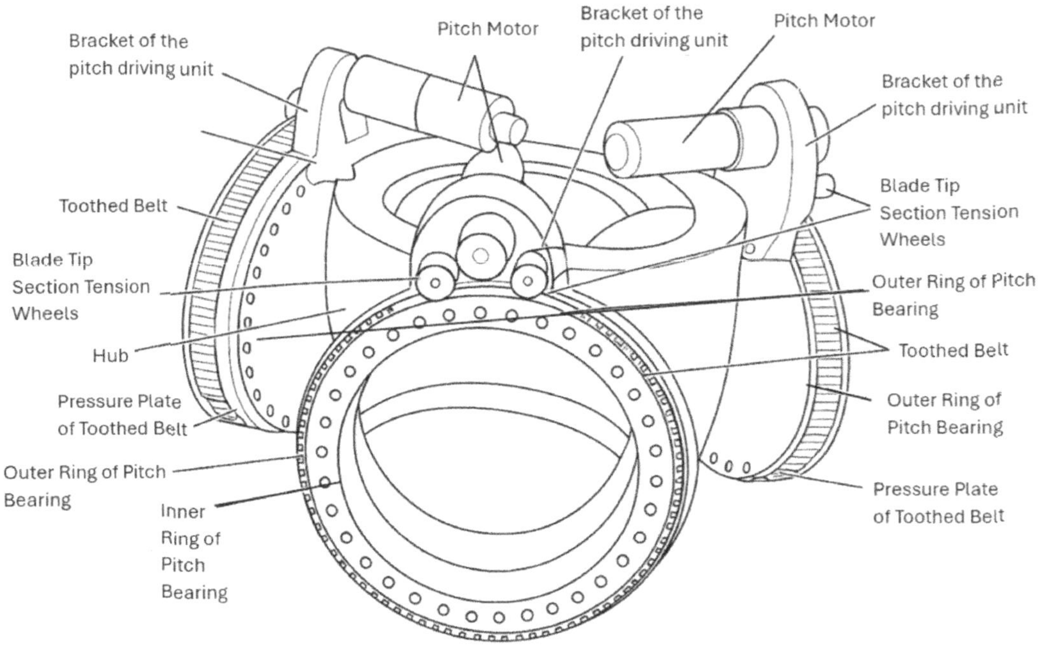

Figure 85: Driving structure of toothed belt pitch system.

Control behaviour: Identical to other independent systems: individual or synchronous blade control, full-feather on trip, continuous regulation above rated wind.

Strengths/limitations: Belts bring smooth, low-backlash, low-noise torque transmission, no gear grease at the blade interface, and easy belt replacement. They also tolerate slight misalignment and distribute tooth contact, reducing local wear. Limitations include belt life management (tension, tooth wear, temperature, contamination) and ensuring the clamp plates, belt wrap, and pulley sizing can transmit full emergency-feather torque without tooth shear. Good guarding keeps dust/oil off the belt; periodic re-tensioning and inspection are part of the routine.

Chapter 7

Electrical and Control Systems

Electricity is the lifeblood of a wind turbine's operation, and the systems that generate, condition, and control that power are among the most sophisticated in modern renewable energy technology. *Chapter 7: Electrical and Control Systems* provides an in-depth exploration of how mechanical energy captured from the wind is transformed into usable electrical energy and how advanced control networks keep turbines operating safely, efficiently, and in harmony with the grid.

The chapter begins by examining the generator and power conversion process, the point where kinetic energy from the rotating blades becomes electrical energy through electromagnetic induction. Readers are introduced to key turbine configurations—geared and direct-drive—and the types of generators used, such as induction and permanent magnet synchronous machines. The section explains how variable-speed operation improves efficiency and how converters and inverters condition the raw power to match grid frequency and voltage requirements.

From there, the chapter delves into the fundamental physics of energy transformation, clarifying the principle that energy is not created but converted. Concepts like translational and rotational kinetic energy, Faraday's Law, and the Lorentz Force are explained to show how magnetic fields and motion interact to produce electricity. Different generator types—DC, synchronous AC, and induction—are compared to illustrate their respective advantages and applications across turbine sizes and operating environments.

The next section addresses protection and reliability within electrical systems. Wind turbines face continuous exposure to lightning, switching surges, and grid disturbances. To mitigate these risks, the chapter explains the use of Surge Protective Devices (SPDs), grounding systems, and international standards such as IEC 61400-24 and IEC 62305 that govern lightning and overvoltage protection. Readers learn how SPDs are strategically installed throughout the turbine—from the base to the nacelle and blade roots—to safeguard generators, converters, and control electronics.

The chapter then transitions to Supervisory Control and Data Acquisition (SCADA) systems—the digital command centres of modern wind farms. These systems continuously monitor parameters such as rotor speed, power output, vibration, and temperature, enabling predictive maintenance, fault diagnosis, and performance optimization. SCADA networks connect individual turbines to central control rooms, allowing operators to oversee entire wind farms remotely and respond swiftly to changes in wind conditions or equipment status.

the chapter concludes with industry standards and cybersecurity frameworks that underpin the design and protection of electrical and control systems. Standards like IEC 62443, ISA-95, and NERC CIP ensure interoperability, reliability, and resilience against cyber threats in an increasingly digitalized energy sector. This section highlights how the integration of automation, data communication, and cybersecurity has become essential to safe and sustainable turbine operation.

Together, these topics give readers a complete understanding of the electrical and control foundations of wind power. *Electrical and Control Systems* bridges the gap between mechanical engineering and digital intelligence—showing how modern wind turbines not only generate electricity, but also think, communicate, and protect themselves in real time to keep renewable energy flowing safely to the grid.

Generators and Power Conversion

Generator Purpose and Function

Wind turbines convert the kinetic energy of moving air into mechanical energy. As wind flows across the rotor blades, the airfoil-shaped blades generate lift (like an airplane wing) and begin to rotate. The amount of energy captured depends on blade size, wind speed, and air density.

The generator is one of the most critical components of a wind turbine because it is responsible for converting mechanical energy from the rotor into electrical energy. Without a properly functioning generator, wind power production cannot take place.

In large horizontal-axis turbines, the blades are connected to a central hub that spins the low-speed shaft. This shaft usually turns at 10–25 revolutions per minute (rpm), which is far too slow for electrical generation.

The low-speed shaft enters the nacelle, which contains a gearbox and generator.

- **Gearbox (in geared turbines):** The gearbox increases the rotational speed from the slow rotor shaft to the much faster high-speed shaft needed by the generator (typically 1,000–1,800 rpm).
- **Direct-drive turbines**: Skip the gearbox. Instead, they use a large-diameter, low-speed generator (often a permanent magnet synchronous generator) to produce electricity at variable frequencies.

The generator converts mechanical energy into electrical energy by electromagnetic induction.

- **Basic principle**: As the rotor (or armature) spins inside a magnetic field (from permanent magnets or an electromagnet), it cuts through magnetic flux lines. This induces an electric current in the copper windings of the stator.
- Induction generators (commonly used in geared turbines): They produce alternating current (AC) whose frequency varies with rotor speed. Controllers and converters are used to adjust this power to match grid requirements.
- Permanent magnet synchronous generators (PMSGs) (common in direct-drive turbines): They use strong magnets to create the field, eliminating the need for external excitation.

Wind turbines produce variable-frequency AC, because rotor speed changes with wind speed. The electricity must be conditioned before it enters the grid:

1. AC from the generator is rectified to DC.
2. Power electronics (converters) smooth and control the DC.
3. An inverter converts the DC back into stable, synchronized 50 or 60 Hz AC for the grid.

This full conversion allows variable-speed operation, which maximizes energy capture and reduces mechanical stress on the drivetrain.

A closed-loop control system keeps the turbine safe and efficient:

- Yaw system rotates the nacelle so the rotor faces the wind.
- Pitch system adjusts the blade angle to regulate power output and protect the turbine in high winds.
- Braking system can stop the rotor in emergencies or for maintenance.
- Sensors monitor wind speed, direction, rotor speed, temperatures, vibration, and more, feeding data to the main controller.

Once the electricity is synchronized, it flows through transformers inside or near the turbine to step up the voltage for efficient transmission to the grid.

As such, a wind turbine generator's role is to take the rotational mechanical power produced by the wind-driven rotor (also called the prime mover) and convert it into useful electrical power. This electricity can be used to supply homes and businesses directly, feed into the grid, or charge batteries for later use. In small-scale systems, a low-speed, low-rpm generator is typically used, because the rotor blades in small turbines do not rotate quickly. For large-scale turbines, the generator is a high-capacity electrical machine that can handle significant power outputs.

Energy Conversion, Not Creation

A fundamental law of physics — the conservation of energy — states that energy cannot be created or destroyed, only transformed from one form to another. Wind turbines do not "make" energy; rather, they convert the energy that already exists in moving air into electricity. Understanding this requires looking at the two main forms of energy involved: translational kinetic energy (in the moving wind) and rotational kinetic energy (in the spinning turbine components), which is then converted into electrical energy.

Air in motion carries translational kinetic energy, meaning the energy of movement from one place to another. The faster the air moves (i.e., the stronger the wind), the more kinetic energy it carries. A wind turbine is designed to intercept part of this moving air and transfer its energy to a rotating system. This is accomplished by the rotor blades, which are shaped like airfoils to extract as much energy as possible from the wind.

As the wind flows over and past the blades, differences in air pressure and flow direction generate lift forces that cause the blades to spin. This action converts the wind's translational kinetic energy into rotational kinetic energy — the energy of spinning motion — in the rotor. This is similar to holding a pencil and twisting it in your hand: the pencil does not move through space, but it rotates about its axis, gaining rotational energy.

The rotating blades are bolted to a hub, which is connected to the main shaft inside the nacelle. This shaft is what physically transmits the rotation into the rest of the drivetrain. Depending on the turbine design, the shaft may directly drive the generator (direct-drive turbine) or first pass through a gearbox to increase the rotational speed to the level required by the generator. Either way, at this point the wind's energy has been successfully transformed from moving air into a spinning shaft.

The second major transformation is from rotational kinetic energy to electrical energy. This happens in the generator, which is essentially a machine that uses electromagnetic induction to make electrons move. In a typical wind turbine generator, the rotating shaft turns a rotor containing magnets or electromagnets inside coils of wire (the stator). As the magnetic field moves past the coils, it induces an electric current. In other words, the mechanical rotation is turned into the motion of electrons — an electrical current — which can then be sent through transformers and power lines for use on the grid.

In summary, a wind turbine involves two key energy transformation processes:

1. Rotor blades convert the wind's translational kinetic energy into rotational kinetic energy of the shaft.
2. Generator converts the shaft's rotational kinetic energy into electrical energy.

Both processes are essential. The blades "capture" and transform the wind's motion, and the generator "translates" that rotational motion into usable electricity. Without either step, the system would not function. This is why the efficiency and maintenance of both the rotor and generator are crucial for reliable wind power production.

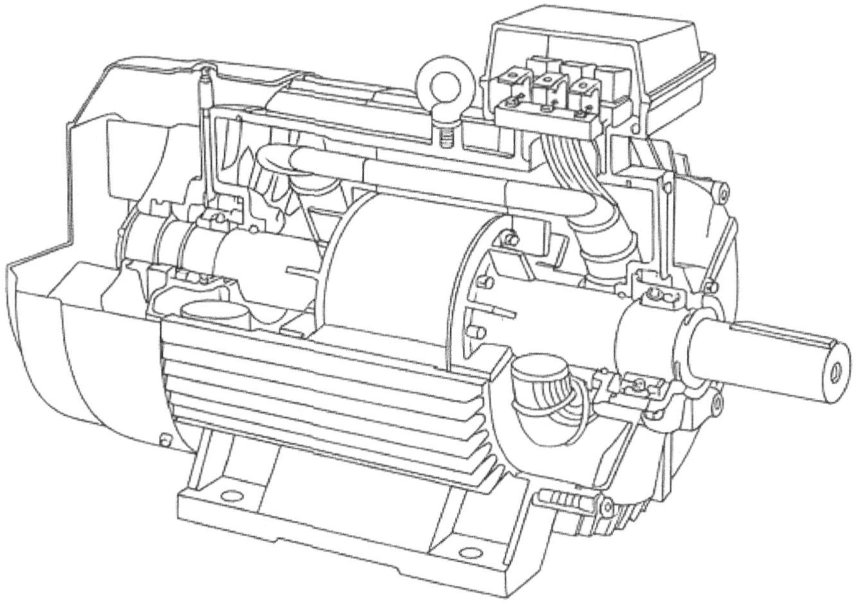

Figure 86: Generator structure.

Types of Electrical Machines Used in Wind Turbines

Several types of rotating electrical machines can be used to achieve this conversion, each with its own characteristics:

1. **Direct Current (DC) Machines or Dynamos** – These produce a direct current output and are common in smaller or homemade wind turbines. They are inexpensive, easy to wire up, and well suited to charging batteries directly.

2. **Alternating Current (AC) Synchronous Generators** – These are also known as AC generators. They produce an AC output at a fixed frequency when connected to the grid. They are often used in larger, grid-connected wind turbine systems.

3. **Alternating Current (AC) Induction Generators (Alternators)** – Induction generators are also common in larger wind turbines. They are robust, require less maintenance than synchronous generators, and can operate effectively over a range of speeds.

Despite their differences, all these machines work on the same principle: Faraday's Law of Electromagnetic Induction. This law states that an electric current is induced in a conductor when it experiences a changing magnetic field. In practical terms, a wind turbine generator has coils of wire and magnetic fields arranged so that when the rotor turns, it induces electricity in the coils.

Electricity from motion is possible because of a physical phenomenon called electromagnetic induction. This is the principle at the heart of all electrical generators, including wind turbines.

In everyday life, we already see the same concept in use – for example, induction cooking appliances also rely on electromagnetic fields to transfer energy. In a generator, however, instead of heating pots, the aim is to produce an electric current.

The principle is straightforward: when a conductor (such as a wire) moves through a magnetic field, an electric current is induced in that conductor. Conversely, when an electric current flows through a wire, it creates its own magnetic field. This mutual relationship between magnetism and electricity was discovered by Michael Faraday in 1831, and his work forms the foundation of all modern power generation.

To understand how electromagnetic induction works inside a generator, it helps to start with the Lorentz Force Law, which is often demonstrated using the "Right Hand Rule." Stretch your right-hand thumb, index finger, and middle finger so they are perpendicular to each other. In this model:

- The thumb represents the direction of motion of an electric charge carrier (an electron).
- The index finger represents the direction of the magnetic field.
- The middle finger shows the direction of the resulting force acting on the moving charge.

This rule tells us that when electrons move through a magnetic field, a force pushes them sideways, causing them to drift collectively in one direction. When you move a wire (which contains many electrons) through a magnetic field, the same force acts on all of them, pushing them along the wire. This organized movement of electrons is what we call an electric current.

A generator exploits this effect by moving conductors (wires) inside magnetic fields. Instead of a straight wire, the conductor is bent into loops called coils to increase the area of interaction with the magnetic field. A light bulb or other electrical load can be connected to this coil to make the current visible. The coil is rotated mechanically — for example, by a hand crank or by the rotor of a wind turbine — while a magnet (often a horseshoe magnet) creates a steady magnetic field around it. In practice, magnets inside modern generators are far stronger than simple bar magnets, but the principle is the same.

As the coil rotates, different parts of it cut across the magnetic field in opposite directions. Using the Right Hand Rule, we can predict the direction of current flow in each segment of wire at any instant. In one part of the coil, the electrons are pushed one way ("backwards"), while in the opposite part, they are pushed the other way ("forwards").

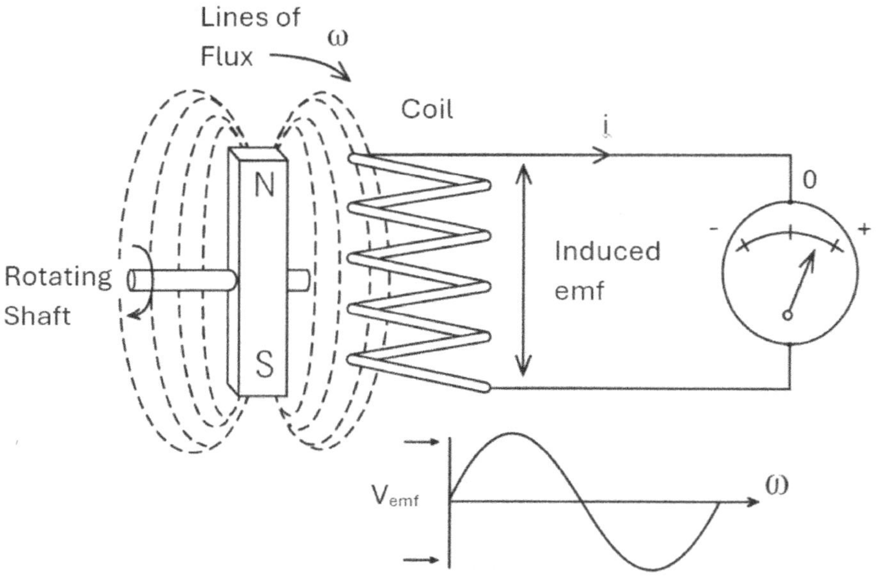

Figure 87: Simple Generator using Magnetic Induction.

Because the coil keeps rotating, each side of the wire alternately moves left and right through the magnetic field. This reverses the direction of the force on the electrons every half turn, which means the current also reverses direction. The result is an alternating current (AC) — electrons oscillate back and forth rather than flowing steadily in one direction. This reversal happens once per half-rotation of the coil, so a complete rotation produces two full reversals of current flow.

This is why the simplest generator design produces AC power naturally. As long as the coil continues to rotate, the current alternates at a frequency determined by the rotation speed.

It's also possible to generate direct current (DC), where electrons flow in only one direction. This requires additional components such as commutators or rectifiers that switch the coil's connections at exactly the right moment or convert AC into DC electronically. These additions make DC generation slightly more complex, but still entirely feasible. In small wind turbines or portable devices, DC generators are often used because they can directly charge batteries without further conversion.

In short, electromagnetic induction underlies all wind turbine generators:

1. The rotor blades capture wind energy and rotate a shaft.
2. The shaft turns a coil of wire within a magnetic field inside the generator.
3. The motion of the coil through the field causes electrons to move, creating an electric current.

4. The current naturally alternates direction (AC) with each half-turn of the coil, but with additional technology it can also be made into DC if needed.

This two-step process — converting the wind's translational motion into rotation, and then converting that rotational motion into electrical current — is how moving air ultimately produces electric energy.

Stator and Rotor

The term stator comes from "stationary," meaning all the generator parts that remain still during operation. In a very simple generator, like the one described earlier with a rotating coil inside a horseshoe magnet, the magnet itself would be considered the stator because it does not move.

In most modern wind turbine generators, however, the design is reversed: the stator holds the coils of wire that form the electric circuit. These coils are mounted securely in a frame so they stay completely still while the rotor moves inside. As the rotor's magnetic field sweeps past the stator's coils, electricity is induced in the coils according to Faraday's Law.

The stator is therefore the electrical "output" side of the generator. It's connected to the power electronics (like converters or inverters) that condition the electricity for use in the grid. Its design focuses on providing a strong, stable environment for the coils while dissipating heat efficiently, since the coils can get hot during operation.

The rotor is the moving part of the generator. It rotates inside (or sometimes outside) the stator depending on the design. In most modern wind turbine generators, the rotor contains the magnets — either permanent magnets or electromagnets — which produce a strong magnetic field.

As the wind turbine shaft turns the rotor, its magnetic field sweeps across the stator's stationary coils, inducing an electric current. The magnets on the rotor can be:

- Mounted on the surface of the rotor,
- Embedded just below the surface,
- Or installed inside the rotor (called an interior magnet design).

Each option affects cost, performance, and how much rare-earth material is needed. The rotor must also be precisely balanced to handle the high rotational speeds and avoid vibration.

The mechanical input to the rotor is provided by the wind turbine's main shaft, which turns as the blades capture wind energy. However, wind turbine blades typically rotate slowly — often only 10–20 RPM for large turbines. This is far too slow for most generators, which are designed to produce electricity efficiently at much higher rotational speeds (hundreds or thousands of RPM).

To solve this, most wind turbines use a gearbox between the rotor shaft and the generator. This gearbox increases the rotational speed to the level the generator needs. Many wind turbines use planetary gear sets because they can achieve high speed-increase ratios (like 1:20 or higher) in a compact space. In direct-drive turbines, which have no gearbox, the generator itself is much larger and designed to produce power at low speeds, but this requires heavier, more expensive magnets.

One of the critical details in generator design is the air gap — the small distance between the rotor and stator. Keeping this gap as small as possible improves the generator's performance for two main reasons:

1. Maximising Efficiency: A small gap means the magnetic field crosses the coils more effectively, reducing energy losses and unwanted eddy currents in the stator.

2. Reducing Magnet Requirements: With a small gap, the magnets can be smaller or weaker while still producing the same magnetic flux. This reduces the amount of rare-earth material (like neodymium) needed, which lowers cost and environmental impact.

Maintaining a tight but uniform gap requires extremely precise manufacturing and alignment, which is one reason wind turbine generators are high-precision machines.

Reversible Nature of Electromechanical Machines

Because of how they are built, these machines can work in both directions:

- As motors, they take electrical power and produce mechanical rotation.
- As generators, they take mechanical rotation and produce electrical power.

In wind turbines, we always use them as generators. However, during certain maintenance procedures, the same components can be energized as motors to perform controlled rotations.

Generator Components

As outlined, the simplest form of a generator consists of a two-pole rotor — one north pole and one south pole — rotating inside or alongside a stator that is connected to the electrical grid. As the rotor turns, its magnetic field passes through the stator's coils and induces an alternating current (AC). For grid connection, this AC must meet strict requirements for voltage, phase, and frequency. For example, in Europe, the standard is a single-phase supply at 50 Hz. This means the generator's design has to ensure the output current oscillates exactly 50 times per second, regardless of how the wind behaves.

Poles and Frequency: Two key characteristics describe an AC waveform: amplitude and frequency. Amplitude refers to the voltage level (how "tall" the sine wave peaks and valleys are), while frequency indicates how many cycles the current completes per second.

A major challenge in wind turbine design is that the generator's frequency is directly tied to the rotational speed of the rotor shaft. Wind speeds fluctuate constantly, which means the rotor blades (and therefore the generator shaft) rotate at varying speeds. Without intervention, the generator would produce electricity at a variable frequency, which cannot be fed directly into the grid.

One way to address this is to increase the number of magnetic poles in the rotor. Instead of a simple two-pole setup, designers can use six or more poles. With more poles, the stator coils experience more magnetic field reversals per revolution of the rotor, effectively increasing the electrical frequency without increasing rotational speed. This allows the generator to produce 50 Hz AC even at slower shaft speeds, which is particularly useful for wind turbines.

Coils and Phases: Another key factor is the number of phases of AC electricity. While a single-phase sine wave is possible, most grid systems and large turbines use three-phase AC, which consists of three sine waves offset in time from one another by 120°. This arrangement delivers smoother power with less vibration and more efficient transmission over long distances.

Creating multiple phases inside a generator is straightforward: instead of a single coil, multiple coils are placed around the stator at specific angular offsets. As the rotor's magnetic field sweeps past, each coil generates its own voltage wave, and because of the spacing, these waves are naturally phase-shifted from one another. This setup yields the three-phase current required by modern electrical systems.

Slip Rings: Because a generator has both rotating and stationary parts, it must have a way to transmit electricity from the moving components without tangling wires. This is achieved using slip rings.

Slip rings are conductive cylindrical rings attached to the rotating part (usually the rotor). Stationary brushes press against these rings, allowing electrical current to pass between the rotating and stationary circuits without twisting wires. This makes it possible to maintain a closed circuit while the rotor spins freely. Without slip rings, cables would quickly twist, wear, and break.

Braking System: Wind turbine generators also include a brake to protect the machine from overload damage. The most common are "power-off" brakes — disc brakes that automatically clamp shut when power is lost. They can only be released when a specific voltage is supplied. This ensures that, during maintenance, transport, or assembly (when no electricity is being produced), the shaft remains locked in place for safety.

Additionally, the brake can be automatically actuated by the internal control electronics if conditions exceed permissible limits, such as excessive wind speeds, temperatures, or rotational speeds. This is part of the turbine's overall safety and control system.

Power Electronics: Modern wind turbine generators integrate sensors that continuously monitor temperature, current, voltage, and rotational speed. When preset limits are reached, the control system can reduce torque to zero and activate the brake. These sensors feed data to power electronics — converters and inverters — that condition the generator's output for grid connection. They can also trigger emergency stops and manage how the generator interacts with the grid during disturbances.

Control Systems: Finally, the control system coordinates all these elements to protect the drive train and extend the turbine's service life. The drive train is subjected to fluctuating aerodynamic and structural loads from the wind. Up to a certain wind speed (the "cut-off" speed), the structure can withstand these loads, but above that, the control system must constrain rotation.

By adjusting pitch, torque, and braking, the control system limits mechanical stress and keeps power generation stable. This closed-loop control ensures the turbine operates efficiently, safely, and in compliance with grid requirements even under rapidly changing wind conditions.

Generator Choice Depends on Application

Whether you use a DC dynamo, an AC synchronous generator, or an AC induction generator depends on the application, system size, and desired output:

- A low-voltage DC output (from a dynamo) is ideal for charging batteries in off-grid or small-scale applications.
- A higher-voltage AC output (from an alternator) can be directly connected to the grid, eliminating the need for intermediate conversion.

The choice also depends on the wind resource at the location. If the wind speed fluctuates frequently, the generator's output will also vary. Some generator types handle variable speed better than others, and some systems add power electronics (like inverters or converters) to smooth out the power output.

For smaller or home-built wind turbines, a low-speed permanent magnet DC generator is common because it's simple, cost-effective, and easier to integrate into battery-based systems. For larger commercial turbines, synchronous and induction generators dominate due to their ability to produce higher power levels and their compatibility with grid standards.

The type of wind turbine generator (WTG) chosen for a particular site depends on two main factors:

1. **The energy content of the wind at that location** — which is a function of the local wind speeds and their variability over time.
2. **The electrical and mechanical characteristics of the generator itself** — including its cut-in speed, rated output, and ability to withstand loads.

Because wind speed varies from site to site, the generator must be matched to the conditions so that it can produce electricity efficiently and reliably without being over-stressed or frequently shut down.

Cut-In Wind Speed and Power Production

Every wind turbine generator has a cut-in wind speed. This is the minimum wind speed at which the rotor blades can generate enough torque to overcome internal friction and losses in the system (such as bearings, gearbox, and generator drag). Below the cut-in speed, the rotor may spin slowly but will not produce usable electricity because the generator's rotational speed is too low.

Once the wind reaches or exceeds the cut-in speed, the blades accelerate, the rotor reaches an adequate speed, and the generator begins to produce usable electrical power. This is the point at which the turbine transitions from idle to active energy production.

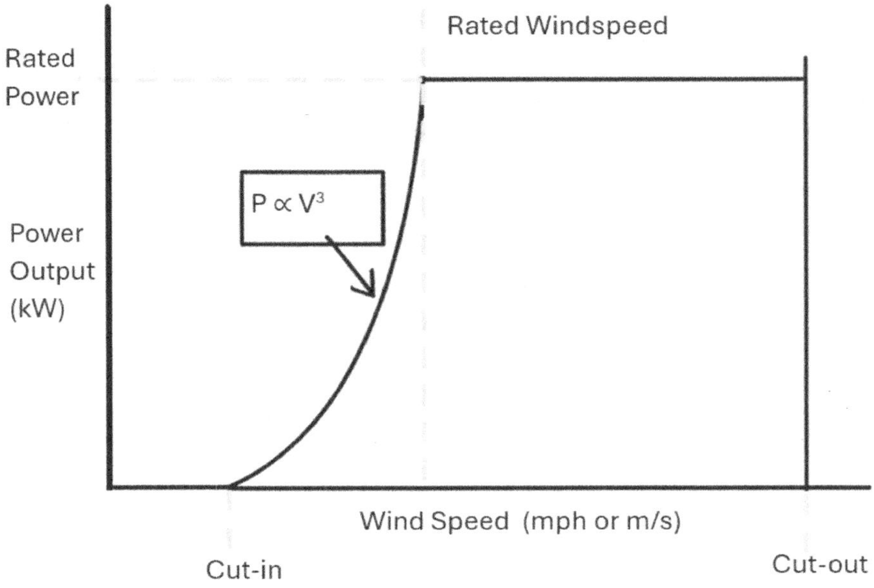

Figure 88: Wind Turbine Generator Output Curve.

Between the cut-in speed and the rated wind speed, a well-designed wind turbine generator produces power roughly proportional to the cube of the wind speed (often expressed as $k{\cdot}V^3$, where V is the wind speed). This cubic relationship reflects the physics of wind energy: doubling the wind speed increases the available power by a factor of eight.

During this operating window, the generator ramps up its power output naturally as the wind increases, and the turbine captures as much of that energy as its aerodynamic and electrical design allows.

Once the wind reaches the rated speed, the generator achieves its maximum or rated power output. This point marks the limit of what the electrical and mechanical components can safely handle. Beyond this point, if the generator continued to produce power proportional to wind speed, it would exceed its structural and thermal limits.

Therefore, above rated speed, the turbine uses control systems — typically blade pitch control or other mechanisms — to limit the power output and maintain it at a steady rated value despite stronger winds. This protects both the rotor blades and the generator.

If the wind continues to rise beyond a certain threshold, the cut-out speed, the wind turbine must stop completely to avoid mechanical or electrical damage. At this point, the loads on the blades, drive train, and generator would exceed safe design limits.

To protect the system, a braking mechanism is engaged. This can be a mechanical governor (such as a brake or pitch feathering system) or an electrical speed sensor tied to control electronics. These systems work together to arrest rotation and shut down the generator safely until the wind drops back into an acceptable range.

Wind Power

The notation $P \propto V^3$ means:

"Power (P) is proportional to the cube of the wind speed (V^3)."

This relationship comes from the basic physics of wind energy.

The kinetic energy contained in moving air is given by:

$$E_{\text{kinetic}} = \frac{1}{2} \rho A V^2$$

where:

- **ρ** = air density (kg/m^3)
- **A** = swept area of the rotor (m^2)
- **V** = wind speed (m/s)

The power available from wind is the rate at which this energy passes through the rotor's swept area. That means multiplying by the wind speed again (because energy per unit time depends on flow rate):

$$P_{\text{wind}} = \frac{1}{2} \rho A V^3$$

So you can see that the power in the wind increases with the cube of the wind speed. Double the wind speed → $2^3 = 8$ times the power available.

For a wind turbine:

- Below the rated speed, the turbine tries to capture as much of this power as possible, so its output power also rises roughly with V^3.
- This is why even small increases in wind speed make a big difference in energy production.

For example:

- At 5 m/s wind speed, the available power might be 100 kW.
- At 10 m/s (twice the speed), the available power is about 800 kW (eight times more).

This is what $P \propto V^3$ expresses: power grows dramatically as wind speed increases.

Generator Maintenance

It is essential to keep the generator in optimal working condition at all times. The most effective way to achieve this is through regular maintenance and prompt repairs.

A well-planned preventive maintenance program not only extends the lifespan of the wind turbine's equipment but also reduces overall costs. By following proper preventive maintenance practices, operators can closely monitor the generator's performance and address potential problems before they escalate into full component failures.

Common wind turbine generator problems include:

Excessive Vibration: Excessive vibration is one of the leading causes of generator failure in wind turbines. It usually occurs in the head or body of the turbine due to factors such as imbalance in the rotor, misalignment of components, or wear of critical parts. Over time, these vibrations stress the mechanical and electrical systems, loosening fasteners, damaging insulation, and accelerating fatigue in bearings, couplings, and windings. If not detected early, excessive vibration can lead to complete generator failure and unplanned shutdowns. Regular vibration monitoring and balancing of rotating parts are crucial to prevent this issue.

Ineffective Yaw Systems: Wind turbines rely on yaw systems to keep the nacelle and rotor facing directly into the wind. When wind speed and direction change frequently, the yaw system is under constant demand, which can degrade its performance over time. An ineffective yaw system struggles to rotate the turbine correctly, resulting in suboptimal wind capture and reduced power output. In some cases, the yaw mechanism may overshoot or rotate too quickly (overspeeding), impairing the turbine's stability and placing additional loads on the generator and drivetrain. Regular inspection and maintenance of yaw motors, gearboxes, and sensors are essential to ensure reliable performance.

Other Generator Problems: Wind turbine generators may also suffer from a range of other issues that affect their performance and lifespan:

- **Irregular Voltage:** Fluctuations in voltage output can result from electrical faults, control system errors, or inconsistent wind conditions. Irregular voltage stresses both the generator and connected grid systems, increasing wear on electrical components.
- **Cooling System Failures:** Generators produce significant heat during operation. If cooling systems (air or liquid-based) fail due to clogged filters, pump malfunctions, or coolant leaks, overheating can occur, damaging insulation and windings.
- **Wind Loading:** High and variable wind loads put cyclic stress on generator bearings and support structures, increasing the risk of mechanical fatigue.
- **Bearing Failures:** Bearings support the rotating parts of the generator. They are highly susceptible to wear, contamination, and electrical erosion, leading to friction, heat build-up, and eventual breakdown.
- **Harsh Weather Conditions:** Turbines exposed to salt spray, dust, ice, or extreme temperatures face additional corrosion and degradation of components, especially insulation and seals.
- **Thermal Cycling:** Repeated heating and cooling during operation causes expansion and contraction of generator parts, which can lead to cracks, loosening of connections, and premature failure of electrical insulation.

Maintenance Frequency

The location and type of wind turbine greatly influence the frequency and depth of maintenance required. Turbines in offshore or high-wind environments experience more severe loads, harsher weather, and higher corrosion risk than those onshore. In such cases, operators often need to exceed the manufacturer's maintenance schedule to address accelerated wear and reduce the likelihood of catastrophic failures. Proactive inspections, condition monitoring, and timely repairs can significantly extend the generator's service life and improve reliability.

Wind Turbine Converter

A wind turbine converter is an essential part of a modern wind turbine's electrical system. It's a type of power electronics device—like an inverter but designed for much higher power and for dynamic wind conditions—that manages the flow of electrical energy between the generator and the grid.

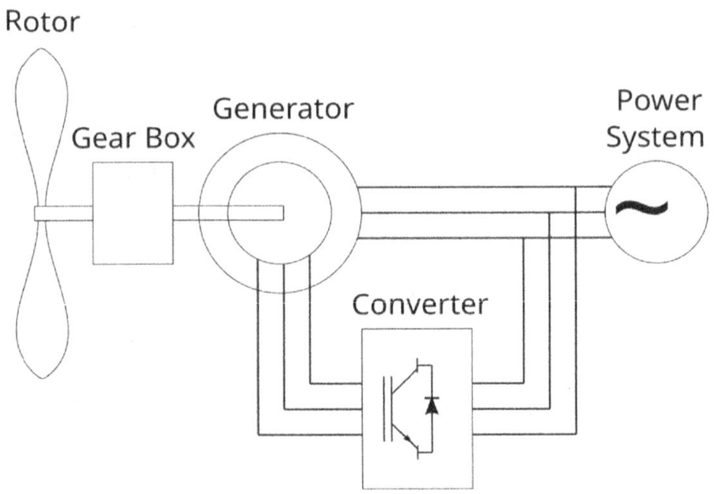

Figure 89: DFIG in Wind Turbine. Funkjoker23, CC0, via Wikimedia Commons.

Because the generator in a wind turbine produces variable-frequency, variable-voltage AC (depending on wind speed), this power cannot be directly fed into the grid, which requires a stable 50 Hz (in Europe) or 60 Hz (in North America) AC at a fixed voltage. The converter acts as an intermediary:

1. It rectifies (converts) the generator's variable AC into DC.
2. It processes the DC with power electronics and controls to regulate voltage, current, and power factor.
3. It inverts the DC back into stable, grid-synchronized AC at the required frequency and voltage.

In short, the converter allows variable-speed operation of the turbine while still meeting the strict requirements of the electrical grid.

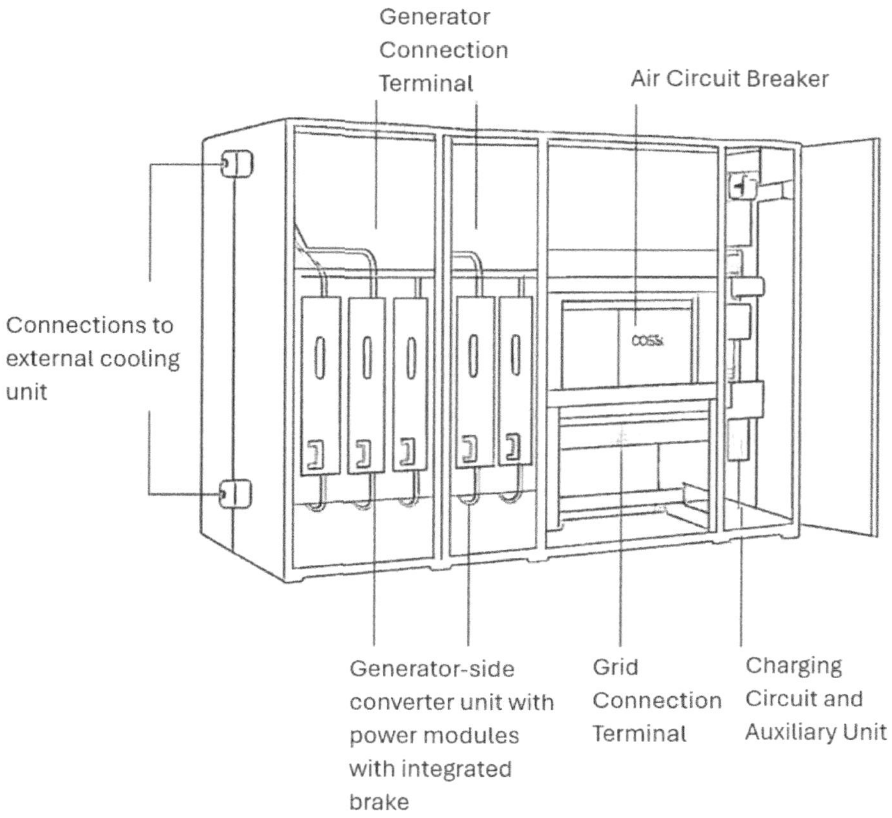

Figure 90: Wind Turbine Converter components.

<u>Two Main Types of Wind Turbine Converters</u>

1. Doubly-Fed Converter (Partial-Scale Converter)

This is typically used with a Doubly-Fed Induction Generator (DFIG). In this setup, the generator stator is connected directly to the grid, and only the rotor circuit passes through a power converter.

Key Characteristics:

- **Bidirectional power flow:** The converter allows power to flow both into and out of the rotor circuit, enabling fine control of active and reactive power and making the turbine more flexible.
- **Partial power conversion:** Only a fraction of the total turbine power (typically 20–30%) passes through the converter because the stator connects directly to the grid. This makes the converter physically smaller and cheaper.

- **Smaller capacity and lower cost:** Because the converter handles only part of the power, it requires fewer high-power electronic devices and less rare material, reducing cost.
- **Improved voltage quality and grid support:** With control over rotor voltage and frequency, the converter can provide reactive power and help stabilize voltage fluctuations in the grid.

This configuration has been very popular in onshore turbines up to a few MW because of its cost efficiency and proven reliability.

2. Full-Power Converter (Full-Scale Converter)

This approach is used with Permanent Magnet Synchronous Generators (PMSG) or other synchronous machines. Here, the generator is decoupled from the grid; all the turbine power flows through the converter.

Key Characteristics:

- **Full power conversion:** The entire output of the turbine passes through the converter, which completely controls the voltage and frequency before feeding it into the grid.
- **High efficiency:** Modern full-power converters use advanced semiconductor switches (IGBTs, SiC devices) and sophisticated control algorithms to maximize conversion efficiency and minimize losses.
- **High stability and grid adaptability:** Because the converter sits between the generator and grid, it isolates the generator from grid disturbances and can maintain stable output under fluctuating grid conditions or faults.
- **Network support functions:** Full-power converters can actively regulate reactive power, voltage, and frequency, acting like a "smart" grid device to support stability and reliability.

This design is increasingly dominant for large offshore wind turbines where full control, fault ride-through, and grid-code compliance are critical.

<u>Practical Comparison</u>

- **Doubly-Fed Converter:** Cheaper, smaller, handles partial power, but the stator is directly tied to the grid so it's more exposed to grid disturbances.
- **Full-Power Converter:** More expensive and heavier but offers total control over output power quality, better fault ride-through, and easier integration with permanent magnet generators.

In both cases, the converter is the "brain" of the electrical interface, allowing variable-speed wind turbines to deliver steady, grid-compliant electricity. Without it, a turbine's generator could not effectively supply power to the grid.

Doubly-Fed Wind Turbine

Modern wind turbines operate under highly variable wind conditions, so to harvest the maximum possible energy it is ideal for the rotor to run at variable speed. However, the electrical grid requires a fixed frequency, such as 50 Hz in Europe. The doubly-fed asynchronous generator (DFIG) overcomes this challenge by allowing the mechanical speed of the turbine to vary while still producing a constant-frequency electrical output. This capability makes DFIGs the mainstream configuration for large, variable-speed, constant-frequency wind turbines.

In a conventional alternator, the voltage frequency is directly tied to rotor speed: if the rotor spins slower or faster, the output frequency changes accordingly. The DFIG introduces an innovative solution by feeding its rotor winding with controlled AC power via a converter. This ensures that the generator's internal rotating magnetic field remains constant relative to the stator, even as the mechanical rotor speed fluctuates. In other words, the rotor's magnetic field "slips" relative to the shaft, keeping the stator output fixed at grid frequency.

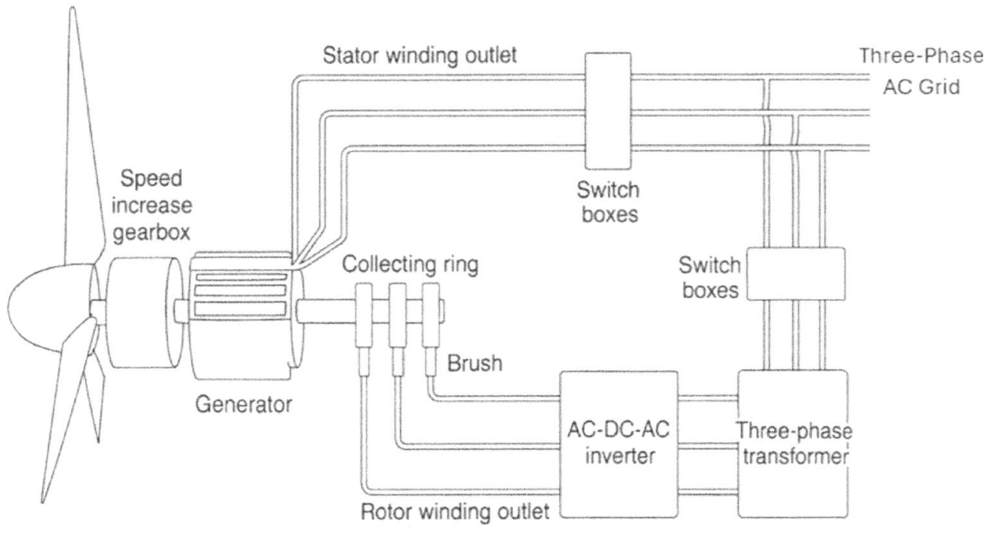

Figure 91: Composition of the doubly-fed wind turbine.

To understand how the rotating magnetic field works, consider a 2-pole generator designed to produce 50 Hz AC. Inside, the rotating magnetic field must always move at 50 revolutions per second relative to the stator. In synchronous operation, when the rotor turns at 50 revolutions per second, the magnetic field is stationary relative to the rotor and rotates at 50 revolutions per second relative to the stator. This is standard synchronous operation. In sub-synchronous operation, if the rotor turns at 30 revolutions per second, the converter injects an extra 20 Hz AC into the rotor windings. This creates a rotating magnetic field at 20 revolutions per second relative to the rotor. Combined with the rotor's 30 revolutions per second, the stator still "sees" 50 revolutions per second and outputs 50 Hz AC. Conversely, in super-synchronous operation, if the rotor turns at 70 revolutions per second, the converter injects a reverse-phase 20 Hz AC into the rotor windings. This reverse rotating field combined with the rotor's higher speed again sums to a 50 revolutions per second field relative to the stator, still producing 50 Hz AC. This is why the system is called "doubly fed": both the stator and the rotor receive electrical power, with the stator connected directly to the grid and the rotor supplied via the converter.

A typical DFIG wind turbine includes several key components. The wind rotor (blades) and gearbox step up the low blade speed to generator speed. The doubly-fed asynchronous generator's stator connects directly to the grid, while its rotor is fed through slip rings by the converter. The AC-DC-AC converter consists of a generator-side converter and a grid-side converter that manage the rotor's voltage, frequency, and phase. A transformer and switchgear match voltages and safely connect or disconnect from the grid. Because the converter only needs to handle a fraction of the turbine's total power (about 20–30% to excite the rotor windings), the system is significantly cheaper than a full-power converter arrangement.

In terms of operation, the DFIG can work in three modes. In sub-synchronous operation, when the rotor speed is below synchronous, power flows from the grid through the converter into the rotor to create the necessary rotating magnetic field, while the stator still outputs fixed-frequency AC. In synchronous operation, when the rotor speed matches synchronous speed, the converter supplies DC to the rotor, producing a fixed magnetic field relative to it so that the stator emits grid-frequency AC. In super-synchronous operation, when the rotor speed is above synchronous, the rotor windings induce AC of opposite phase; the converter converts this to grid-frequency AC and sends it to the grid. In this mode, power flows from both the rotor and stator to the grid.

Finally, the AC-DC-AC converter plays a crucial role. These back-to-back PWM converters are bi-directional: the generator-side converter injects or extracts current from the rotor to control its field, while the grid-side converter conditions power to or from the grid to maintain the correct frequency, voltage, and phase. Because only the rotor power is processed, the converter size is typically about 25% of the machine's rated power, offering a major cost advantage over full-power converter systems. This combination of efficiency, flexibility, and cost-effectiveness explains why DFIG technology dominates in large-scale wind power generation.

Electronic Devices in the Nacelle and Heat

Although wind turbines sit in naturally windy environments, the electronic equipment inside their nacelles still generates considerable heat and requires active thermal management. The most important devices are the generator and the power conversion systems. The generator is the heart of a wind turbine, converting the rotor's mechanical energy into electrical energy that can be fed into the grid. This energy conversion process always produces losses in the form of heat, especially in the copper windings of the generator. As turbines get larger, the challenge grows because the thermal losses scale with the cube of the generator's dimensions. Without proper cooling, efficiency declines and components risk damage.

To keep temperatures within safe limits, most turbines use ducted enclosures with large fans to blow air over the generator. Some manufacturers go further and provide water-cooled generators, where liquid coolant circulates around the stator and rotor windings. These water-cooled models include a radiator installed inside the nacelle to reject the absorbed heat. Both air and liquid cooling aim to protect generator efficiency and extend its service life, but water cooling handles much greater heat loads than air, making it necessary for multi-megawatt machines.

Wind turbines can be designed with either synchronous or asynchronous generators and may connect directly or indirectly to the power grid. In a direct grid connection, the generator's output frequency matches the grid (usually three-phase AC). In an indirect connection, the generator output first passes through a power converter. These AC–AC converters (often AC/DC/AC) rectify the generator's AC to DC and then invert it back to grid-quality AC.

During conversion, the current passes through a series of Insulated Gate Bipolar Transistors (IGBTs). These power switches are rapidly turned on and off to create an artificial sine wave. The more frequently they switch, the closer the waveform is to an ideal sine wave—and the cleaner the electricity injected into the grid. However, high switching frequencies also mean higher heat production. Because wind inputs are variable, IGBTs in turbine converters cycle very frequently, which generates substantial heat that can dramatically reduce efficiency unless well cooled.

Even with modern efficiency improvements, wind turbine power electronics must manage hundreds of kilowatts of waste heat inside a confined nacelle. Air cooling works well for small turbines but is impractical for large multi-megawatt units due to air's low thermal capacity. It becomes difficult to blow enough air across a generator or converter to maintain safe operating temperatures. This is why water cooling is widely adopted for large turbines.

Medium-voltage wind turbine converters (up to 12 MW) typically use closed-loop systems with a mix of deionized water and glycol to cool electronics. However, water-cooled systems are large and their thermal limitations force the size and weight of the power-generation subsystems to scale with power output. For example, a 10 MW wind turbine's power equipment is nearly twice as large and heavy as that of a 5 MW model because water cooling alone cannot remove the additional heat without increasing the system footprint.

To overcome these constraints, some suppliers have developed advanced liquid cooling systems. Parker Hannifin's Vaporizable Dielectric Fluid (VDF) system provides much higher heat-transfer capability than water. It requires less fluid and lower pump rates, allowing a much smaller cooling system. For instance, the same heat dissipation achieved with 6 liters per minute of water can be reached with only 1 liter per minute of VDF. The VDF is non-conductive and housed in a hermetically sealed assembly; if a leak occurs, it will not damage electronic components.

This high-efficiency cooling also minimizes thermal cycling, keeping the electronics at a stable temperature range. Avoiding frequent temperature swings removes a major source of stress on sensitive components like IGBTs, control boards, and sensors. In turn, this extends the service life of the wind turbine's electronic systems, improves reliability, and reduces maintenance costs.

Converter Maintenance

The wind turbine converter is a critical part of the turbine's power electronics. It manages the conversion of variable-frequency, variable-voltage power from the generator into grid-compliant AC. Because it operates under high electrical and thermal stresses, it requires regular maintenance to avoid failures that could cause costly downtime or damage to other turbine components.

Routine inspections and timely servicing also help maintain grid-code compliance and prolong the converter's life. The exact schedule depends on environmental conditions, such as dust, humidity, and temperature at the installation site.

Core Maintenance Items

Radiator Temperature and Cleaning: Converters generate a lot of heat, which is dissipated through radiators and cooling fans. Checking radiator temperature ensures that the cooling system is working effectively. Radiators should be cleaned every 6–12 months depending on site contamination (dust, insects, or salt spray). For dusty or poorly sealed towers, cleaning before the hot summer months is essential to prevent overheating.

Filter Screens: Dust filters prevent airborne particles from entering the converter cabinet. Over time these filters clog and restrict airflow, reducing cooling performance. Filters should be cleaned or replaced annually, but in particularly dusty conditions (e.g. northwest wind farms) every six months is recommended.

Circuit Breaker Inspection: Circuit breakers protect the converter and connected equipment from overcurrent and faults. They must be checked annually or after a specified number of operations to ensure proper tripping characteristics and contact integrity.

Power Section Connections: The converter's power section contains busbars, connectors, and terminals carrying high voltages and currents. Loose or contaminated connections can lead to arcing, overheating, or insulation failure. Routine inspection and cleaning prevent such problems.

Temperature and Humidity Controller: The internal environment of the converter cabinet must be controlled. Adjusting the temperature/humidity controller and verifying heater operation prevents condensation and corrosion. Each radiator and fan should also be manually tested through the converter's software interface to confirm correct function.

Optical Fibre Connections: Modern converters rely on fibre optics for high-speed, noise-free communication between control boards. Loose or damaged fibres can cause miscommunication or faults. Inspect fibre connections for damage, proper seating, and minimum bend radius.

UPS Power Supply Maintenance: Many converters include a UPS (uninterruptible power supply) to power control systems during grid disturbances. The UPS's internal batteries should be replaced every three years, and if the UPS remains in standby mode for long periods, it should be discharged and recharged every 3–6 months to extend battery life.

Safety Precautions

Isolate Power Before Work: Before any installation or service work, disconnect both the generator stator and the converter power supply. Engage the mechanical brake to lock the rotor to prevent rotation.

Discharge Capacitors: Converters contain large intermediate DC capacitors. Even after disconnection, they can hold dangerous voltages for several minutes. Always wait at least 5 minutes for discharge, and confirm with a multimeter between UDC+ and UDC– terminals that the voltage is zero before touching.

Avoid Insulation Tests Without Disconnecting Cables: Applying external test voltages without disconnecting converter cables can damage sensitive power modules.

High-Voltage Areas: Many parts inside the cabinet—such as brake control terminals (UDC+, UDC–, R+, and R–)—carry dangerous DC voltages. Depending on the wiring, relay outputs may also contain 24 V or 230 V. Always treat these as live until proven otherwise.

Handling the Converter Module: The module's weight distribution may be top-heavy. When moving it, support it from a high point to prevent tipping or injury.

Handling Optical Fibers: Fibers are fragile and should be disconnected by holding the connector, not the fibre itself. Maintain at least a 35 mm bending radius to avoid damage or signal loss.

Transformers, Switchgear, and Grid Connection

Wind turbine transformers are essential components in the power conversion chain of a wind energy system. Their primary role is to step up the voltage produced by the wind turbine generator to a higher voltage level, suitable for transmission to the grid. Because wind turbines are often located far from demand centres, efficient transmission at higher voltages is necessary to minimize energy losses. Without transformers, the electricity produced at the relatively low generator voltage could not be economically transmitted over long distances.

Transformers play a central role in the success and reliability of wind power systems. While wind turbines are often seen as the most visible part of renewable energy infrastructure, the electricity they produce cannot be used directly by homes, industries, or the grid without conditioning and transformation. This is where wind power transformers—specialized devices designed for the unique demands of wind generation—come in. They not only step up the voltage to transmission levels but also stabilize and protect the system, ensuring that the energy harvested from the wind can be delivered efficiently and reliably [134].

Figure 92: Wind turbine transformer back. Patrick Finnegan, CC BY 2.0, via Flickr.

Inside a wind turbine nacelle, the generator produces electricity at a voltage typically in the range of a few hundred volts. This electricity is first stabilized and conditioned by power converters, which ensure that its frequency and voltage are compatible with the grid. After this, a step-up transformer increases the voltage to a medium level, usually several kilovolts (kV). From there, a second, larger substation transformer at the wind farm site may step up the voltage again—often to tens or hundreds of kilovolts—so it can be transmitted efficiently through high-voltage transmission lines.

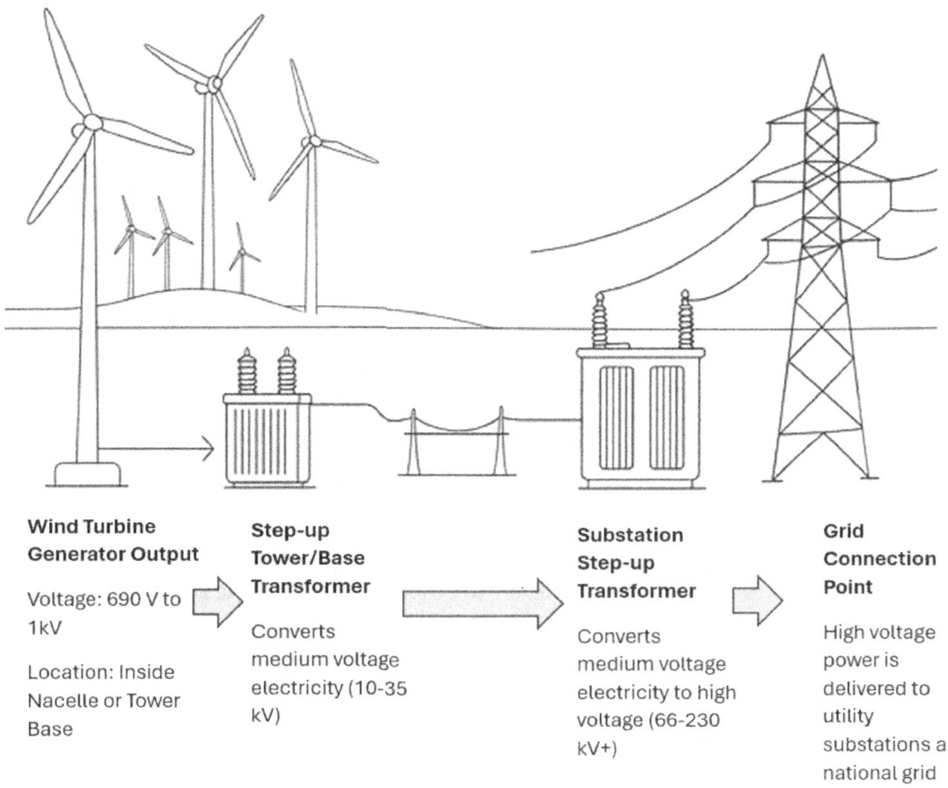

Figure 93: Role of transformers in the system [134].

Unlike standard utility transformers used in conventional power distribution, wind power transformers are engineered to handle highly variable conditions. Wind is not constant; it fluctuates with gusts, storms, diurnal cycles, and seasonal changes. These fluctuations cause frequent changes in load and output voltage, meaning the transformer must adapt dynamically. Offshore turbines add further challenges, as their transformers must withstand corrosive salty air, while onshore turbines may face high altitudes, dust, and extreme thermal stresses. Reliability under these conditions is crucial, because a transformer failure can bring an entire turbine—or in some cases, multiple turbines in a farm—offline [134].

A wind power transformer, sometimes called a wind turbine transformer or renewable energy step-up transformer, is a specialized electrical device that converts the generator's low-voltage output into a higher voltage suitable for transmission or distribution. Wind turbine generators typically produce power in the range of 690 volts to 1.2 kilovolts (kV), which is far too low to be transmitted efficiently over distance. Transformers step this voltage up—commonly to 10 kV to 220 kV or higher—so that the electricity can enter medium- and high-voltage collection networks or transmission lines. This process is essential for minimizing line losses and ensuring efficient delivery of renewable energy to the grid [134].

Key functions of wind power transformers include:

1. **Voltage Transformation:** The primary role is stepping up the turbine's low-voltage output into medium or high voltage levels. Without this process, long-distance transmission would be impractical due to excessive energy losses.

2. **Power Flow Regulation:** Because wind energy output is irregular, the transformer helps stabilize and regulate the flow of power. It ensures that short-term fluctuations in production do not destabilize the grid, acting as part of the broader control and protection system.

3. **Insulation and Protection:** Wind power transformers provide electrical isolation between voltage levels, which helps protect equipment and personnel. They are equipped with protection mechanisms against overloads, faults, and short circuits, thus preventing damage and maintaining grid integrity.

4. **Grid Compatibility:** Grid operators demand electricity that is synchronized in terms of frequency, voltage, and phase. Wind power transformers are designed with customized impedance and tuning features to support synchronization. This is especially important in wind farms with multiple turbines, where dozens of generators must work together as a single, stable power source for the grid.

There are different transformer designs used in wind turbines depending on size, placement, and cooling requirements. Some are located inside the nacelle close to the generator, reducing cable losses between the generator and the transformer. Others are located at the base of the tower to simplify maintenance and reduce weight in the nacelle. Large wind farms also use pad-mounted or substation transformers to collect power from multiple turbines before feeding it into the transmission grid.

Because wind turbines operate in variable and sometimes harsh conditions, transformers must be designed to handle fluctuating loads, frequent switching, and environmental challenges. They often use advanced insulation systems, forced air or oil cooling, and compact designs to fit within limited nacelle or tower spaces. The reliability of these transformers is critical: failures can lead to long turbine downtimes and costly repairs.

A wind farm typically uses several types of transformers, each with a distinct role in handling the flow of power from the turbine generator all the way to the high-voltage transmission system.

Turbine-Level Transformer: The first transformer in the chain is located either inside the nacelle, at the base of the tower, or in close proximity to the turbine. Its role is to step up the low voltage produced by the wind turbine generator—typically between 690 V and 1.2 kV—into a medium-voltage range that can be efficiently transferred to collector systems. In addition to stepping up voltage, it stabilizes output and helps reduce losses over short cable runs. These units are often compact and designed to withstand vibrations, variable loads, and harsh conditions directly at the turbine site.

Collector Transformer: After the turbine-level step-up, multiple turbines feed their output into collector transformers located at small local substations. These transformers aggregate power from several turbines, smoothing out fluctuations caused by wind variability and helping ensure a more consistent flow into the main collection system. They act as a buffer between dozens of turbines and the central substation, playing a vital role in minimizing electrical imbalances within the wind farm.

Main Substation Transformer: At the heart of the wind farm is the main substation transformer, which performs a major voltage step-up to high transmission levels—often from medium voltage (e.g., 30–66 kV) to very high voltage (110–400 kV or higher). This step-up is critical for long-distance transmission with minimal losses. The main transformer ensures that the electricity produced by the entire farm is compatible with the regional or national grid and can be delivered efficiently to demand centres. Its reliability and capacity are crucial, since any failure here affects the entire wind farm's output.

Auxiliary Transformer: Supporting the operation of turbines and substations is the auxiliary transformer, which provides low-voltage power for essential systems. These include station lighting, cooling systems, control circuits, emergency backup systems, and other operational equipment. Auxiliary transformers can be found at both substations and individual turbine sites, ensuring that the infrastructure has a continuous supply of AC or DC power to keep monitoring and safety systems functioning.

Modern wind power transformers are no longer just passive devices. They are increasingly equipped with smart grid features that enable predictive maintenance and improved operational performance.

- **SCADA-Compatible Sensors:** Allow remote monitoring of temperature, load levels, and fault conditions, ensuring operators can respond quickly to anomalies.
- **Tap Changers:** Enable voltage adjustments under fluctuating load conditions, keeping output stable and grid-compliant.
- **Partial Discharge Monitoring:** Detects early insulation degradation, preventing catastrophic transformer failures.
- **IoT Integration:** Provides real-time analytics and predictive maintenance data, helping operators extend transformer lifespan and reduce unplanned downtime.

These smart features and specialized transformer roles combine to give wind farms greater stability, efficiency, and reliability. By stabilizing voltage at each stage, providing grid-compatible output, and enabling real-time monitoring, transformers ensure that fluctuating wind energy can be delivered as a dependable power source. They also make it possible to operate large, geographically dispersed wind farms with high levels of automation and minimal on-site supervision.

Wind power transformers are designed with unique characteristics that set them apart from conventional transformers. This is because wind energy presents highly variable operating conditions, unlike the stable and predictable environments found in most urban or industrial grid applications. A wind farm transformer must not only step up voltage but also handle unpredictable load changes, remote installation challenges, and harsh environmental conditions.

Conventional transformers operate under stable, predictable loads from industrial plants, commercial facilities, or city grids. These loads change gradually and within defined patterns. By contrast, wind power transformers deal with highly variable loads because wind speeds fluctuate minute by minute. The power generated can swing from very low to full capacity within short intervals, meaning that wind transformers must be engineered to withstand constant changes in loading without overheating or degrading prematurely.

Traditional transformers are typically placed in indoor substations or controlled environments, often close to demand centres with easy access for maintenance. Wind power transformers, however, are usually installed in remote and exposed sites such as coastal areas, deserts, mountain ridges, or offshore platforms. These settings require designs that are compact, rugged, and modular to simplify installation and servicing in areas where space is limited and access is difficult. Many are pad-mounted, pole-mounted, or integrated into turbine towers to minimize footprint and streamline deployment.

Because wind farm locations are often exposed to extreme weather, wind transformers require enhanced cooling and sealing solutions. Unlike conventional ONAN (Oil Natural Air Natural) or ONAF (Oil Natural Air Forced) cooling, wind transformers may use corrugated fins, sealed radiators, forced ventilation, or even heat exchangers to dissipate heat efficiently in compact housings.

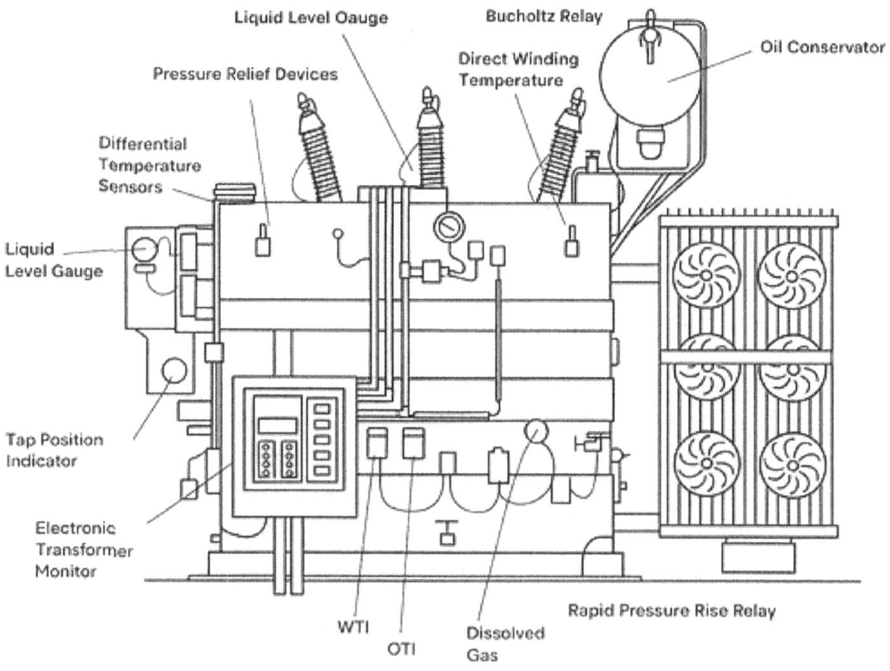

Figure 94: Oil filled transformer.

They must also endure environmental hazards such as:

- High humidity and salt spray in coastal and offshore wind farms, which cause corrosion.
- Sandstorms and UV exposure in desert regions.
- Freezing temperatures and ice buildup in high-altitude or northern sites.

To counter these stresses, manufacturers use sealed oil tanks, hermetically designed conservators, stainless steel or galvanized enclosures, and corrosion-resistant coatings to extend service life.

Because turbine output fluctuates with wind speed, wind power transformers must be capable of managing frequent and dynamic voltage swings. They are often equipped with on-load tap changers (OLTC), which allow real-time voltage regulation, ensuring that power delivered to the grid remains within acceptable limits. They also employ fault-tolerant insulation systems and surge arrestors to handle voltage spikes and protect against grid instability.

Unlike older conventional designs, modern wind power transformers are increasingly hybrid-ready. They can integrate with solar arrays, battery storage systems, and microgrids, supporting bidirectional power flow in smart grids. This allows wind farms to not only deliver power but also absorb or store it, improving overall grid flexibility and reliability. Many of these transformers are equipped with IoT-based monitoring and SCADA integration, enabling real-time diagnostics, predictive maintenance, and better performance management.

In essence, wind power transformers differ from conventional transformers in their design philosophy and operational resilience. They are built to withstand harsh climates, manage rapid voltage fluctuations, and support integration with hybrid renewable systems. While conventional transformers emphasize efficiency and stability in predictable environments, wind power transformers focus on adaptability, durability, and smart functionality, ensuring that renewable energy can be fed reliably into modern power grids.

Oil Filled/Immersed Transformers

Oil-filled transformers are essential for the efficient operation of electrical power systems. These transformers utilize insulating oils not only for electrical insulation but also as a cooling medium to manage the heat generated during operation. The various cooling methods employed in these transformers include Oil Natural Air Natural (ONAN), Oil Forced Air Forced (OFAF), and Oil Forced Water Forced (OFWF) configurations. ONAN relies on the natural convection of both oil and air for cooling, while OFAF employs pumps and fans to enhance circulation. OFWF, commonly used in larger transformers, integrates water cooling to effectively manage thermal loads [135-137].

It is important to note that all oil-filled transformers fall under the category of oil-immersed transformers, as they utilize oil as a cooling and insulating medium. However, the term "oil-immersed transformers" specifically refers to those where the core and winding are completely submerged in oil. This immersion supports optimal thermal management and significantly improves the electrical insulation strength compared to air, facilitating compact designs which are particularly beneficial in applications like wind turbines [136, 138, 139].

The thermal performance characteristics of oil-filled transformers are further augmented by employing alternative fluids, such as natural and synthetic esters, which provide enhanced environmental profiles and lower fire hazards compared to conventional mineral oils [140, 141].

In wind turbine applications, the terminology can often be confusing, as "oil-filled" and "oil-immersed" are frequently used interchangeably to describe transformers designed to handle the fluctuating loads generated by wind energy. The cooling properties of these oils are advantageous, effectively managing the high heat loads generated under variable wind conditions. Additionally, the compactness of oil-filled designs is critical for integration into the turbine nacelle or tower base, where space constraints are prevalent [142, 143].

Special considerations arise in offshore wind farms, where the adoption of synthetic esters or silicone-based fluids is becoming standard practice. These alternatives present advantages such as biodegradability and reduced flammability, making them safer choices in the event of oil spills. Furthermore, offshore transformers are typically hermetically sealed to guard against moisture ingress, which is crucial in marine environments susceptible to salt and oxidation [143-145].

Oil-filled transformers provide several advantages in wind turbine applications, particularly in terms of performance, reliability, and adaptability to demanding environments. One of the key benefits is high efficiency, as transformer oil has far better thermal conductivity than air. This allows for effective heat transfer, enabling compact designs with superior cooling performance. As a result, oil-filled transformers can handle higher loads without overheating, making them ideal for the fluctuating demands of wind power.

Another important advantage is their compact size. Because oil cooling supports higher power density, these transformers can be built smaller than equivalent dry-type units. This is critical in wind turbines where installation space is limited, such as inside the nacelle or at the tower base. In addition, oil not only cools the transformer but also helps to absorb and distribute heat evenly, reducing thermal stresses on windings and insulation. This contributes to longer service life and improved durability under variable operating conditions. Finally, oil-filled transformers offer a wide rating range, making them suitable for both turbine-level step-up transformers and larger collector or substation transformers within a wind farm.

Despite these strengths, oil-filled transformers also come with specific challenges in the wind energy sector. Leakage risks are a major concern, as any oil spill can cause environmental damage—particularly offshore or in ecologically sensitive areas. To mitigate this, many modern wind farms use biodegradable synthetic esters or natural ester oils instead of mineral oil. Fire safety is another issue, since mineral oil is flammable. For higher-risk applications, fire-retardant fluids such as esters are often specified.

Maintenance requirements also need careful attention. Oil must be regularly monitored for moisture, dissolved gases, and particulate contamination, as these can degrade insulation and compromise transformer performance. In addition, transformers deployed in harsh environments—such as high-altitude sites, desert regions, or offshore installations—face temperature variability and environmental stresses like extreme cold, intense heat, or salt spray. In these cases, specially formulated oils and sealed tank designs are used to ensure reliable performance and long-term protection.

Switchgear

Switchgear is a critical part of a wind turbine's electrical system because it ensures safe operation, protection, and control of power flow from the turbine to the grid. In simple terms, switchgear refers to a combination of electrical disconnect switches, fuses, and circuit breakers used to isolate equipment, protect against faults, and allow maintenance without disrupting the wider power system. In wind turbines, switchgear plays an even more important role because of the highly variable and sometimes harsh operating conditions.

The main function of switchgear is protection and isolation. If a fault such as a short circuit, overload, or insulation failure occurs, switchgear will detect it and disconnect the affected turbine or section of the wind farm from the grid. This prevents damage to costly equipment like the generator, transformer, and converter. It also ensures personnel safety by creating a safe, de-energized environment for maintenance.

Switchgear also provides control. It allows operators to connect or disconnect a turbine from the grid as needed, such as during commissioning, maintenance, or in response to grid conditions. In many wind farms, switchgear is integrated with Supervisory Control and Data Acquisition (SCADA) systems, allowing remote monitoring and control.

Wind turbine switchgear can be medium-voltage (MV) or high-voltage (HV), depending on the turbine rating and grid requirements:

- **Turbine-Level Switchgear (MV 10–36 kV):** Usually located at the base of each turbine, this switchgear connects the turbine's step-up transformer to the collector network. It isolates the turbine during faults or service.
- **Collector Substation Switchgear (MV/HV 36–132 kV):** Used at substations where multiple turbines feed into a collector system. It aggregates outputs and provides sectionalizing protection.
- **Main Substation Switchgear (HV 132–400 kV):** Located at the grid interconnection point, providing grid synchronization and large-scale protection.

Wind turbine switchgear must be compact, reliable, and environmentally resilient because it is often installed in remote or offshore locations. Common types include:

- **Air-Insulated Switchgear (AIS):** Traditional design using air as the insulating medium. More common onshore but bulkier.
- **Gas-Insulated Switchgear (GIS):** Uses SF_6 or alternative gases for insulation, allowing compact designs and higher reliability—ideal for offshore or space-constrained towers.
- **Vacuum Circuit Breakers:** Frequently used at the turbine level, offering high reliability, minimal maintenance, and strong fault interruption capability.

Wind turbine switchgear must withstand harsh conditions such as salt spray (offshore), temperature extremes, humidity, and dust. Therefore, switchgear enclosures are often weatherproof and corrosion-resistant. Fire safety and arc-flash protection are also essential, since confined tower spaces present risks during faults.

Modern wind turbine switchgear has undergone significant evolution to keep pace with the increasing operational demands for efficiency, safety, and sustainability in renewable energy systems. A cornerstone innovation is the integration of remote monitoring and diagnostics through SCADA (Supervisory Control and Data Acquisition) systems. This technology allows operators to monitor switchgear performance in real time, swiftly identifying faults or irregularities, ultimately facilitating better decision-making without physical site presence. Such capabilities are particularly essential for wind farms located in remote or offshore areas, as they minimize downtime and maintenance costs while enhancing system reliability [146, 147].

The incorporation of advanced condition monitoring techniques into SCADA systems enhances fault detection and management. These methods can pre-emptively address potential issues, which is critical because undetected faults can lead to severe downtime and substantial repair costs [146]. A literature review indicates that effective condition monitoring and early diagnostics are vital for maintaining high availability and operational efficiency in wind energy systems [147].

Another noteworthy advancement in modern switchgear design is the integration of arc flash detection systems. These systems significantly enhance safety for both equipment and personnel by detecting dangerous electrical discharges and isolating faults almost instantaneously. Research indicates that implementing arc-flash sensors and fast protective relays minimizes the risk of catastrophic equipment damage and protects technicians from hazards associated with electrical failures [147]. The ability to quickly intervene in the event of an arc flash incident is essential for upholding safety protocols in high-voltage environments typical in wind energy systems.

Sustainability also stands out as a critical focus area in the development of wind turbine switchgear. Traditionally, SF_6 (sulphur hexafluoride) was used as an insulating medium in high-voltage switchgear; however, its classification as a potent greenhouse gas has prompted a shift toward alternatives with lower environmental impact. New designs are increasingly utilizing vacuum interrupters or eco-friendly gases, which maintain superior insulation and arc-quenching performance while supporting global carbon reduction initiatives [148]. The push for environmental sustainability aligns with the broader trend in the energy industry toward reducing carbon footprints and minimizing ecological disruption [149].

Additionally, the design of modern wind turbine switchgear is increasingly leaning towards compact and modular structures. Given the spatial constraints within turbine towers or nacelles, compact designs facilitate efficient installation and operation [150]. Modular systems further allow for easy assembly, transport, and maintenance, significantly enhancing the ability to replace or extend components without extensive downtime. This modularity is exceptionally beneficial in offshore wind farms, where logistical challenges often complicate servicing and technology upgrades [151]. The trend towards modularity answers the critical need for flexibility, enabling operators to adapt systems for evolving energy demands efficiently.

Air-Insulated Switchgear

Air-Insulated Switchgear (AIS) is an essential component in electrical power systems, particularly within the context of renewable energy, such as wind power. AIS employs ambient air as its insulation medium to separate and isolate conductive components, in contrast to gas-insulated switchgear (GIS), which utilizes insulating gases like sulphur hexafluoride (SF_6) or liquid dielectrics [152, 153]. This fundamental distinction contributes to the cost-effectiveness and reliability of AIS, making it an attractive option for medium- and high-voltage applications [154].

One major advantage of AIS is its simplicity and ease of maintenance. Since air serves as the dielectric barrier, this design allows for straightforward servicing, enabling operators to visually inspect components, and facilitating the detection of faults more easily [155, 156]. This reduces operational complexity, which is particularly beneficial in maintaining wind farm infrastructures. Additionally, the materials used in AIS systems tend to be more affordable than those required for GIS, leading to lower overall installation costs [157]. The AIS configuration is commonly employed at collector stations or main substations, rather than inside wind turbine nacelles, due to space requirements; onshore wind farms, where land availability is less restricted, particularly favour this type of switchgear [153].

However, AIS is not without drawbacks. The need for larger physical clearances between live parts to prevent electrical arcing results in bulkier installations, making AIS less suitable for compact applications, such as offshore wind farms, where space is limited [157, 158]. Moreover, exposure to environmental conditions can degrade AIS performance; factors like dust, humidity, and pollution may require proper maintenance and sometimes the application of special coatings to mitigate these issues [152, 159].

Despite these challenges, the safety and operational reliability of AIS are commendable. Circuit breakers in AIS typically utilize vacuum or air-blast technology to effectively interrupt fault currents, thereby protecting both the wind turbine systems and the grid connection (Komiya et al., 2018). With consistent maintenance, AIS systems can exhibit long service lifespans, proving to be a cost-efficient solution for various wind power projects [158]. Furthermore, modern advancements in digital monitoring systems, including Supervisory Control and Data Acquisition (SCADA) technologies, enhance real-time performance tracking and fault detection, improving the reliability of AIS and solidifying its role in contemporary electrical infrastructure [159, 160].

Gas-Insulated Switchgear

Gas-Insulated Switchgear (GIS) represents a significant advancement in the design and functionality of high-voltage switchgear systems, primarily utilizing sulphur hexafluoride (SF_6) for insulation. Unlike traditional air-insulated switchgear (AIS), which depends on expansive air clearances to maintain operational integrity, GIS encapsulates its components—conductors, busbars, and circuit breakers—within sealed metallic enclosures filled with insulating gas. This innovative design facilitates the same voltage ratings as its AIS counterparts but in a notably smaller spatial footprint, making GIS particularly advantageous in constrained environments such as offshore wind turbine platforms, nacelles, and urban substations [161, 162].

In the context of wind power systems, GIS is frequently implemented in turbine-level switchgear and offshore substations. Offshore platforms are characterized by limited space and face demanding environmental conditions, including saline exposure, high humidity, and strong winds. GIS offers a corrosion-resistant solution that effectively shelters its interior from moisture, dust, and various pollutants, ensuring reliable and uninterrupted operation in these challenging settings [162]. The compact and robust nature of GIS reduces land usage and construction costs while maintaining high performance [161].

Reliability and safety are paramount attributes of GIS. The completely enclosed design minimizes risks related to arc faults and short circuits that may result from environmental contaminants or animal interference. Additionally, GIS systems are typically lower maintenance than AIS due to their sealed compartments, which protect internal components from degradation over extended periods. Consequently, the frequency of visual inspections and part replacements is significantly reduced, offering operators enhanced service continuity and safety [162, 163].

However, the reliance on SF_6 for insulation in GIS poses substantial environmental concerns due to its classification as a potent greenhouse gas, with a global warming potential reported to be thousands of times greater than that of carbon dioxide [162]. This growing environmental awareness has prompted the industry to explore environmentally benign alternatives; manufacturers are investigating vacuum switchgear options and other gas combinations such as nitrogen and fluoronitrile-based gases, which promise similar performance advantages while mitigating ecological impacts [162, 163].

From a technical standpoint, GIS systems provide several performance benefits, such as the capacity to function safely at higher voltage ratings (up to several hundreds of kV) and the integration of advanced digital monitoring systems that encompass sensors for gas pressure, temperature, and partial discharge detection. These features enable predictive maintenance capabilities that can be integrated into SCADA systems, thereby enhancing operational efficiency and responsiveness [162, 164].

Vacuum Circuit Breakers

The Vacuum Circuit Breaker (VCB) represents an advanced solution in circuit protection technology, utilizing a vacuum as an arc-extinguishing medium to efficiently interrupt electrical faults in power systems, particularly in medium-voltage applications. Unlike traditional methods that rely on air, oil, or gas as insulation mediums, VCBs operate within a sealed vacuum chamber, which inherently possesses a high dielectric strength. This unique feature is crucial as it prevents ionization; hence, any electrical arc formed between the breaker contacts is rapidly extinguished upon separation of the contacts, often within a single half-cycle of current flow [165, 166]. The resulting rapid interruption capability provided by VCBs enhances the safety of electrical equipment and ensures reliable operation under various fault conditions commonly encountered in wind power systems.

In the context of wind energy, VCBs are integral components of medium-voltage switchgear, managing connections between turbines, collector systems, and substations. Their design is particularly beneficial in wind farms, where they must reliably handle fluctuating loads and environmental conditions. The compact nature of VCBs facilitates installation in space-restricted environments, such as turbine nacelles, thereby optimizing the available space for critical electrical infrastructure [165, 167]. Moreover, VCBs are celebrated for their resilience against various environmental contaminants, contributing to their longevity and reducing maintenance requirements, making them suitable for harsh offshore or desert conditions [166, 168].

The advantages of VCBs extend beyond functionality; they demonstrate noteworthy reliability and performance consistency over extensive operational periods. Reports indicate that VCBs can undergo thousands of switching operations with minimal wear and tear [165, 169]. This distinguishes VCBs in terms of longevity compared to traditional circuit breakers, which may require frequent maintenance and replacement. Additionally, their composition negates the use of environmentally harmful gases like SF_6, positioning VCBs as a more sustainable option in modern energy projects, aligning with global efforts to mitigate environmental impacts associated with power generation and distribution [170].

Despite their notable strengths, certain limitations of VCBs must be acknowledged. Primarily, they are suited for medium-voltage applications, typically up to 72.5 kV, and are not as common in high-voltage environments where gas-insulated systems are preferred due to their superior insulation capabilities [165, 169]. A critical consideration lies in the integrity of the vacuum within the interrupter. If the seal is compromised, the breaker fails, necessitating replacement. The precision engineering involved in maintaining such integrity is paramount for ensuring long-term reliability [171].

Modern advancements continue to enhance the application and performance of VCBs, particularly in wind energy contexts. Integration with digital monitoring technologies, arc-flash detection systems, and SCADA (Supervisory Control and Data Acquisition) allows for improved operational control and maintenance of wind farm electrical systems, underscoring a trend toward increasingly intelligent power management solutions [166, 168]. Consequently, VCBs stand out not only for their innovation in arc-extinguishing capabilities but also for their role in fostering the growth of efficient, environmentally responsible energy infrastructures.

Grid Connection

A wind turbine grid connection is the process by which electricity generated by a wind turbine is integrated into the local or national electrical power grid. Since the grid operates at a constant frequency and voltage (for example, 50 Hz in Europe or 60 Hz in North America), while wind energy is variable and intermittent, specialized systems and components are required to ensure compatibility, reliability, and safety.

Wind turbines generate electricity at relatively low voltages (typically 690 V – 1.2 kV). This output is stepped up using transformers to medium voltage (10–35 kV) for collection in wind farms, and then to high voltage (110–400 kV) at substations for transmission across long distances. The generator output also needs to be synchronized in frequency, voltage, and phase angle with the grid to ensure seamless integration.

Direct vs. Indirect Grid Connection

When it comes to integrating wind turbines into the power grid, two main approaches are used: direct connection and indirect connection. The method chosen depends largely on the turbine design, the size of the installation, and the grid requirements.

In a direct connection, which is more typical of older or smaller wind turbines, the generator—often a synchronous type—is connected straight to the grid. For this system to work, the generator's rotor must spin at a constant speed to keep the output frequency in line with the grid's fixed frequency (such as 50 Hz in Europe or 60 Hz in North America). Because the frequency of the generated electricity is directly tied to the rotor speed, any fluctuations in wind speed can disrupt the match between turbine output and grid requirements. This rigidity limits the turbine's efficiency since wind is naturally variable and does not always provide a steady rotational speed. As a result, direct connection systems are becoming less common in modern wind power technology.

In contrast, indirect connection has become the standard for modern wind turbines, especially larger installations. This approach uses power converters—typically AC-DC-AC systems—to manage the link between the generator and the grid. First, the generator's variable-frequency AC output is converted into DC, and then it is inverted back into AC at the precise frequency, voltage, and phase required by the grid. By using this method, the turbine's generator is free to operate at variable speeds, which allows it to capture more energy from changing wind conditions. At the same time, the converter ensures that the electricity delivered to the grid remains stable and compliant with grid codes.

Overall, the shift from direct to indirect connection reflects the growing need for efficiency, flexibility, and reliability in wind power systems. While direct connection is simple, it cannot take advantage of variable wind speeds. Indirect connection, though more complex, allows turbines to maximize energy capture while still delivering clean, stable power to the grid.

Role of Power Electronics

The power converter system is the central technology that makes modern wind turbine grid connection possible. Without it, the variable and unpredictable nature of wind energy could not be effectively integrated into a stable power grid. The converter's role is to act as an intelligent interface between the wind turbine's generator and the grid, ensuring that the electricity produced is both usable and reliable.

At its core, the converter works in two main stages. First, it performs rectification, where the variable-frequency alternating current (AC) produced by the generator is converted into direct current (DC). Since the generator's output frequency depends on wind speed and rotor speed, it can fluctuate significantly. By turning this into DC, the converter creates a stable, controllable intermediate form of energy that is easier to manage.

The second stage is inversion, where the DC is converted back into AC—but this time at the exact frequency, voltage, and phase required by the power grid. This process allows the wind turbine to operate at variable speeds, capturing the maximum possible energy from changing wind conditions, while still delivering perfectly synchronized power to the grid.

Beyond simply converting energy, modern converters provide essential grid-supporting functions. They can regulate voltage levels, control frequency and phase angles, and even supply or absorb reactive power to help stabilize the grid. Another critical feature is fault ride-through capability, which enables turbines to remain connected during short disturbances or faults in the grid instead of shutting down. This resilience is increasingly important as more renewable energy sources are integrated into power networks.

In practice, these advanced functions are enabled by the use of Insulated Gate Bipolar Transistors (IGBTs), which switch power rapidly and efficiently within the converter. By controlling how often and how precisely these transistors turn on and off, the system can produce a high-quality AC output that closely mimics an ideal sine wave. This combination of precision, adaptability, and reliability makes the power converter system the true backbone of wind turbine grid connection technology.

Grid Codes and Compliance

When wind turbines are connected to the electrical grid, they cannot simply deliver power in any form they generate. Instead, they must comply with a set of rules known as grid codes, which are established by each country or region. These codes are designed to ensure that renewable energy integrates smoothly with the existing grid, maintaining both reliability and stability. Without such standards, large-scale wind power could disrupt grid operations instead of supporting them.

One key requirement in most grid codes is maintaining voltage and frequency limits. Power grids operate at fixed frequencies—such as 50 Hz in Europe or 60 Hz in North America—and within strict voltage ranges. Wind turbines must ensure their output stays within these ranges to avoid destabilizing the grid. If frequency or voltage fluctuates too much, it could lead to blackouts or equipment failures across the network.

Another important requirement is Fault Ride-Through (FRT) capability. In the past, older turbines disconnected from the grid whenever a short disturbance occurred. However, as wind power became a larger share of total generation, this practice created risks of cascading failures. Modern turbines must now stay connected during short faults, such as voltage dips, and even help support the grid by injecting reactive power until the disturbance clears.

Grid codes also require turbines to provide reactive power control. Unlike active power, which delivers real energy to consumers, reactive power supports the grid by maintaining voltage stability. Wind turbines must be able to adjust their reactive power output dynamically, just like traditional power plants, to help stabilize the system when demand or conditions fluctuate.

Finally, wind turbines must be capable of active power control, which means adjusting how much power they deliver based on instructions from grid operators. For example, during periods of excess generation or low demand, turbines may be curtailed to prevent overloading the grid. Conversely, during shortages, they may be required to ramp up their output quickly if wind conditions allow.

Together, these compliance requirements ensure that wind turbines are not just passive energy producers but active participants in grid stability. Meeting grid code standards makes renewable energy reliable, scalable, and capable of functioning alongside conventional power plants in modern energy systems.

Challenges of Grid Connection

One of the main challenges of connecting wind power to the grid is intermittency. Unlike conventional power plants that can operate continuously at a controlled output, wind turbines depend on wind speed, which changes unpredictably. This means that the electrical output of a wind farm can fluctuate from moment to moment. Without proper balancing systems, such as flexible backup generation, demand-side management, or energy storage, these variations can cause instability in the grid, especially as the proportion of wind energy increases.

Another issue arises from the use of power electronics, which are essential in modern wind turbines for variable-speed operation and grid compliance. While these converters and inverters enable turbines to deliver high-quality, grid-compatible electricity, they can also generate electrical disturbances known as harmonics. Harmonics are unwanted frequency components that distort the smooth sinusoidal waveform of AC electricity. If left unchecked, they can reduce efficiency, overheat equipment, and affect sensitive electronic devices on the grid. To mitigate this, wind turbines often include harmonic filters or advanced control strategies to maintain power quality.

A further complication comes from the remote locations where many wind farms are built. Offshore sites and rural regions are often chosen because of their stronger and more consistent wind resources, but they are far from major population centres and industrial loads. As a result, the electricity must travel long distances through high-voltage transmission lines before reaching consumers. This increases infrastructure costs, requires grid reinforcement, and introduces transmission losses. Offshore wind farms, in particular, need specialized submarine cables and substations, adding to the technical and economic challenges of grid connection.

Finally, as the energy transition accelerates, there is a growing emphasis on integration with storage and hybrid systems. Wind power is increasingly combined with battery energy storage systems (BESS) or other renewables like solar PV to create hybrid plants. Storage helps smooth the output of wind farms by absorbing excess generation during high winds and releasing it during calm periods. Similarly, coupling wind with solar can balance daily and seasonal variations, creating a more stable supply before injecting power into the grid. While this improves reliability, it also requires advanced control systems, additional investment, and careful coordination between technologies to optimize performance.

Together, these challenges highlight that successful wind power deployment is not just about building turbines but also about creating a resilient and flexible grid infrastructure that can accommodate renewable variability while maintaining stability and reliability.

Surge Protection for Wind Turbines

Wind turbines are designed to capture wind energy in open, elevated, and often remote environments — conditions that also make them highly susceptible to lightning strikes and transient overvoltages. The combination of tall structures, conductive materials, and exposed locations means that wind turbines are struck by lightning far more frequently than most other electrical installations. Lightning and switching surges can cause severe damage to turbine blades, generators, transformers, sensors, and control systems, leading to costly repairs and significant downtime. To mitigate these risks, a coordinated Lightning Protection System (LPS) combined with Surge Protective Devices (SPDs) is essential for modern wind turbine design and operation.

Wind turbines are frequently installed in areas with high wind availability — such as mountain ridges for onshore systems and coastal regions for offshore farms. However, these exposed environments also have higher lightning flash densities. The height of modern turbines, often exceeding 100 metres, further increases their likelihood of being struck by lightning. When lightning strikes a turbine, it can inject millions of volts and hundreds of kiloamperes of current into the structure within microseconds. This energy, if not safely dissipated, can damage electronic and electrical systems within the nacelle, control units, and sensors.

Lightning strikes also induce lightning electromagnetic impulses (LEMP), powerful transient waves that travel along conductive cables and equipment housings. LEMP is often more destructive than the direct strike itself because it propagates through control and communication lines, overwhelming insulation and damaging sensitive electronics. To counter these effects, coordinated surge protection systems are installed throughout the turbine. These systems channel excess electrical energy through low-resistance paths, preventing voltage spikes from reaching critical components.

The need for surge protection in wind turbines arises from three main factors:

1. **High Risk of Lightning Strikes:** Due to their exposed locations and tall structures, wind turbines experience a much higher frequency of lightning strikes than other facilities. Blades, made from composite materials with embedded conductors, are particularly prone to triggering upward lightning leaders during thunderstorms.

2. **High Maintenance and Downtime Costs:** Lightning strikes can cause severe blade ablation, electrical component failures, and control system malfunctions. For offshore turbines, where access is limited and maintenance is costly, repairs can take weeks or months, resulting in significant lost revenue from downtime.

3. **Threat from Lightning Electromagnetic Pulses (LEMP):** Even if a turbine avoids a direct hit, nearby lightning activity can induce powerful electromagnetic fields that disrupt operations. Control systems such as pitch and yaw modules, which contain delicate electronic circuits, have low immunity to transient overvoltages. LEMP can cause insulation breakdown, component burnout, or even data loss in communication systems.

The installation of Surge Protective Devices (SPDs) is one of the most effective methods for protecting wind turbines from lightning electromagnetic impulses (LEMP) and transient overvoltages. SPDs function as voltage limiters, safely redirecting excess electrical energy to ground before it can reach or damage sensitive electrical and electronic components. They are installed at key points within the turbine's electrical system to provide comprehensive protection against overvoltage surges.

At the turbine base and control cabinets, SPDs protect both power and communication lines as they enter the tower, preventing surges from propagating through the control system. Within the nacelle and generator units, SPDs safeguard vital equipment such as transformers, converters, and control electronics that regulate turbine operation. At the blade root and hub junctions, SPDs play a critical role in preventing voltage surges caused by lightning currents traveling through blade conductors.

To ensure reliable performance, SPDs must be carefully selected and rated according to recognized international standards. In wind turbine applications, this includes IEC 61400-24, which outlines lightning protection requirements specific to turbines, and IEC 62305, which defines general lightning protection principles. SPDs should be installed at the boundaries between lightning protection zones, ensuring that surge energy is discharged in a controlled manner consistent with the endurance capability of the protected equipment.

All surge protection systems in wind turbines must comply with established international and national standards to guarantee both safety and performance consistency. IEC 62305 (Parts 1–4) defines the general framework for lightning protection, including risk assessment, system design, and implementation principles. IEC 61400-24 provides detailed guidance specific to wind turbines, addressing the protection of blades, structural components, and internal systems against both direct and indirect lightning effects. IEC 61643 sets the testing methods, classification criteria, and performance standards for SPDs themselves.

These standards also account for the challenging environmental conditions faced by wind turbines, such as high humidity, temperature fluctuations, salt spray, and mechanical vibration. SPDs must be robust enough to maintain their performance under these demanding conditions, ensuring long-term reliability and minimal maintenance requirements.

The IEC 62305 standard establishes four Lightning Protection Levels (LPL I–IV) to categorize systems based on their exposure and the maximum lightning current they must withstand. LPL I represents the highest level of protection and is designed for severe lightning conditions, capable of withstanding a peak current of up to 200 kA and an impulse charge of 100 C. Lower levels—LPL II, III, and IV—are suitable for environments with reduced lightning risk, offering proportionally lower protection thresholds.

For wind turbines, especially large onshore and offshore units exposed to frequent storms, LPL I is generally recommended. Adopting the highest level of protection ensures maximum resilience against direct strikes and induced surges, reducing the risk of catastrophic failure and prolonged downtime.

Control Systems Power

A wind turbine's control systems and SCADA devices—including the sensors, computers, and communication equipment that monitor and operate the turbine—require a constant and reliable source of electrical power, even when the blades are not turning or the turbine is shut down. The power that supports these systems can come from several sources depending on the turbine's operational state, location, and design.

During normal operation, when the turbine is generating electricity, it supplies its own control and auxiliary systems directly from the power produced by its generator. A portion of the generated energy is routed through low-voltage auxiliary circuits within the nacelle or tower to feed essential internal systems. These include the control panels and PLCs (Programmable Logic Controllers), yaw motors and pitch systems, lubrication pumps, cooling fans, heaters, and SCADA computers that handle monitoring and communication. The remainder of the electricity is stepped up through the transformer and exported to the grid. In essence, when a turbine is producing power, it is self-sufficient—running its electrical, control, and monitoring equipment from its own output.

When the turbine is idle, experiencing low wind conditions, or undergoing maintenance, it cannot generate power. However, its control systems, safety circuits, sensors, and communication links must remain active to allow for continuous monitoring and quick reactivation when conditions improve. In such cases, the turbine draws a small amount of external electrical power from the grid or site substation through its transformer in reverse. This external feed is typically a three-phase low-voltage supply (400–690 V), stepped down through internal transformers to 230 V AC or 24 V DC for control circuits. The power keeps vital systems such as SCADA communication, yaw brakes, hydraulic systems, dehumidifiers, and safety alarms operational. This continuous external feed is critical, allowing the turbine to "wake up" automatically when wind speeds increase or after a controlled shutdown.

In the event of a grid outage—caused by storms, faults, or maintenance work—the turbine must still maintain a minimal level of power for safety and control. To manage this, backup systems are installed to provide short-term or emergency power. The most common form of backup is an uninterruptible power supply (UPS), which uses batteries to provide power for several minutes or hours, giving the system time to safely shut down, maintain communication, and hold yaw or pitch mechanisms in a secure position. Some remote or offshore turbines also have small auxiliary generators—diesel or battery-powered—located in the tower base or at a nearby substation. In modern pitch systems, supercapacitors are used to store sufficient energy to feather the blades safely in the event of total power loss. These layered backup measures ensure that even when all external power is lost, the turbine can safely stop and remain in a stable, protected state until power is restored.

Offshore turbines and microgrid systems present additional complexity in power management. In offshore wind farms, SCADA and control systems usually receive their power through the array cable network that links multiple turbines to an offshore substation. If the farm is offline, auxiliary supplies from the substation or from service vessels—connected via temporary power cables—are used to maintain critical systems. In isolated microgrids or hybrid renewable systems combining wind, solar, and battery storage, standby power is often provided by a battery bank or secondary generator. These arrangements ensure continuous operation of monitoring, safety, and communication systems regardless of the turbine's operational status or environmental conditions.

Together, these power sources—self-generation, grid supply, and backup systems—allow wind turbines to maintain constant control, safety, and communication functionality. This layered approach ensures operational resilience and technician safety across all wind energy environments, whether onshore, offshore, or remote.

SCADA and Control Systems Overview

SCADA Overview

Supervisory Control and Data Acquisition (SCADA) systems are the digital backbone of modern wind power plants. They provide operators with the ability to monitor, control, and optimize wind turbines and entire wind farms in real time. Because wind turbines are often located in remote or offshore locations, SCADA plays a crucial role in ensuring reliable operation, early fault detection, and efficient maintenance planning.

At the most basic level, a wind turbine SCADA system continuously collects operational data from each turbine. This includes rotor speed, generator output, wind speed and direction, blade pitch angle, nacelle orientation, temperature (in components like the gearbox, generator, or bearings), and electrical parameters such as voltage, current, and power factor. This data is transmitted to a central control room where operators can visualize the performance of individual turbines and the farm as a whole.

Beyond simple monitoring, SCADA enables remote control functions. Operators can start or stop turbines, change pitch angles, adjust yaw orientation, or apply braking systems—all from a distance. This minimizes the need for on-site personnel and ensures quick responses to unexpected events such as storms, faults, or grid operator requests.

One of the most valuable roles of SCADA is in predictive maintenance and fault detection. The system's sensors and alarms detect anomalies, such as abnormal vibrations, temperature rises, or power fluctuations, which could indicate a developing failure. By identifying these issues early, operators can schedule maintenance before a minor issue escalates into a costly breakdown, reducing downtime and extending component life.

SCADA is also essential for grid compliance and integration. Grid operators often require wind farms to provide specific support functions, such as reactive power control, active power curtailment, or frequency regulation. SCADA ensures that turbines respond quickly to these commands, helping maintain grid stability.

In modern wind farms, SCADA systems are increasingly being enhanced with advanced analytics, IoT integration, and cloud platforms. These allow for big data processing, machine learning-based performance optimization, and even fleet-wide benchmarking across multiple sites. The result is improved efficiency, reduced operational costs, and better long-term planning for wind energy production.

SCADA in wind turbines is more than just a monitoring tool—it is a comprehensive control and optimization system. By enabling real-time data collection, remote operation, predictive maintenance, and compliance with grid requirements, SCADA ensures that wind farms operate safely, efficiently, and profitably under highly variable wind conditions.

A SCADA system is essentially the central nervous system of an industrial or energy operation, combining both hardware and software to provide seamless monitoring, control, and automation. At its core, it has three main components: Human Machine Interfaces (HMIs), Programmable Logic Controllers (PLCs), and Remote Terminal Units (RTUs). Each of these plays a different but interdependent role in ensuring the system functions smoothly.

The PLCs and RTUs are microcomputers that act as the "field brains" of the system. They collect raw data from sensors, motors, and other devices located throughout the facility—in the case of wind turbines, this might include data from rotor blades, generators, gearboxes, or yaw systems. They don't just collect data but also interact with these devices by executing automated commands, such as adjusting a motor's operation or activating safety mechanisms. This allows SCADA to perform real-time control and not just passive observation.

The Human Machine Interfaces (HMIs) act as the visual and interactive layer for human operators. Through screens and dashboards, the HMI translates the complex streams of data collected by PLCs and RTUs into clear, understandable graphics, charts, and alerts. This gives operators a real-time overview of what's happening in the system. In a wind farm, for example, an HMI might show live wind speed, turbine power output, and whether a particular turbine is experiencing overheating or abnormal vibration. Operators can then use the HMI to issue commands, such as shutting down a turbine or adjusting blade pitch.

In the renewable energy sector, SCADA plays an especially vital role because it helps manage resources that are inherently variable, like wind, water, or sunlight. SCADA systems in wind energy monitor wind speeds, turbine performance, and power output, allowing operators to optimize energy production while protecting equipment from overload or faults. Similarly, in solar and hydro systems, SCADA monitors irradiance or water pressure and ensures that power production is efficient and stable.

Finally, SCADA is not limited to the turbines or generation units alone—it is also deeply integrated into the power distribution and control centres. Here, it helps manage the balance of supply and demand by monitoring grid output, coordinating turbine responses, and ensuring that energy flows consistently and reliably into the grid. Without SCADA, large-scale renewable energy projects would be far less efficient and far more difficult to manage, as operators would lack the ability to oversee and control complex systems in real time.

Principles and Architecture

A Supervisory Control and Data Acquisition (SCADA) system is a vital part of any offshore wind farm, acting as the *nerve centre* of the project. It provides operators with the "handles" they need to oversee, control, and optimize the operation of the entire plant [172]. Without a SCADA system, running a professional wind power project efficiently and reliably would be impossible. It serves as the main interface between the operator and the wind farm, ensuring that data is continuously collected, analysed, and used for decision-making.

At its core, the SCADA system connects the individual turbines, the substation, and meteorological stations to a central computer. This central hub, supported by a robust communication network, enables real-time supervision of both individual turbines and the wind farm as a whole. Through this setup, the operator can monitor performance, detect faults, and track overall system health. The SCADA system automatically records activity such as energy output, turbine availability, and error signals. These records are not only crucial for operational management but also form the basis of warranty calculations and claims, as they provide evidence of downtime, failures, or underperformance [172].

A critical role of SCADA is ensuring grid compliance. Modern connection agreements require wind farms to contribute to grid stability by controlling both active and reactive power output, supporting voltage and frequency regulation, and sometimes limiting power output when instructed by grid operators. The SCADA system enforces these requirements automatically, adjusting turbine behaviour and farm output to keep the wind farm synchronized with grid demands. This makes it a key enabler of large-scale wind integration into national power systems [172].

Communication within a SCADA network almost always relies on optical fibre, which provides fast, reliable, and high-capacity data transfer between turbines and the central system. The SCADA software and hardware are typically provided by the turbine manufacturer for simplicity and contractual efficiency. However, some independent SCADA suppliers offer systems that can integrate data from multiple turbine types into standardized reporting formats. This is especially useful for owners and operators who manage diverse projects with turbines from different manufacturers. Independent systems also improve transparency in calculations related to availability and warranty issues, reducing reliance on proprietary manufacturer data [172].

To further enhance accuracy and performance evaluation, large offshore wind farms typically install permanent meteorological masts equipped with instruments to measure wind speed, direction, and other key weather variables. These measurements allow operators to compare actual turbine output against the expected wind resource. If energy production is lower than projected, high-quality meteorological data makes it possible to determine whether the shortfall is due to equipment underperformance or simply weaker-than-expected wind conditions. This ensures that underperformance is diagnosed correctly and that maintenance, upgrades, or financial claims can be managed with confidence.

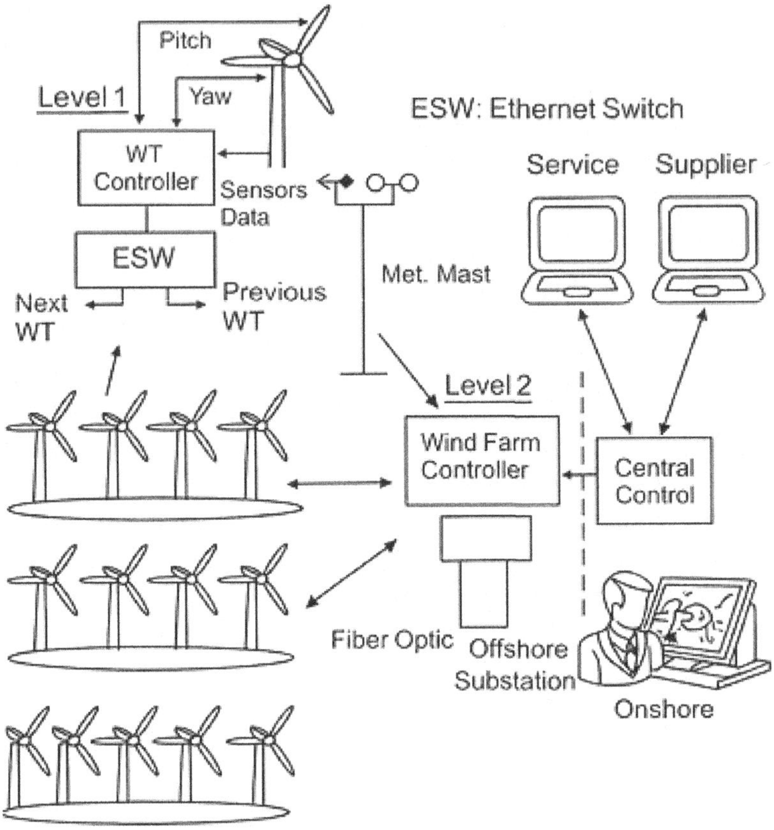

Figure 95: Sample SCADA architecture.

SCADA systems in offshore wind farms provide centralized monitoring, grid compliance, operational optimization, and data transparency. When coupled with permanent meteorological instrumentation, they allow operators to fully understand the performance of their assets and take corrective actions when needed, ensuring the long-term efficiency and reliability of the wind farm [172].

Sensors

In a wind turbine, the SCADA system depends on a network of sensors that constantly monitor both the turbine's critical components and the surrounding environment. These sensors send continuous data to the turbine controller, local PLCs, and the central SCADA server, giving operators the ability to track performance, identify potential faults, and optimize power generation. Each type of sensor is strategically placed to monitor specific parts of the turbine and its systems.

Meteorological sensors are mounted on the nacelle and meteorological masts. These include anemometers to measure wind speed, wind vanes to determine wind direction, temperature sensors for ambient conditions, and barometers to measure air pressure. Together, they provide the vital input needed for yaw control, pitch regulation, and estimating the available wind resource.

Rotor and blade sensors are fitted to monitor the moving parts of the turbine. Pitch angle sensors, typically mounted at the blade root, measure blade positioning. Strain gauges inside the blades detect bending loads, while vibration sensors identify abnormal oscillations that might indicate imbalance or icing. Dedicated ice detection sensors are also often installed on blades or nacelles to identify hazardous icing conditions. These devices protect the blades from overloading, improve aerodynamic efficiency, and prevent damage.

Within the nacelle itself, yaw position encoders track rotation relative to the tower base, yaw brake pressure sensors monitor braking forces, and nacelle vibration sensors detect movement from wind gusts or drivetrain issues. This allows the turbine to maintain optimal alignment and prevents excessive stress on structural components.

The gearbox and drivetrain, being some of the most failure-prone components, are heavily instrumented. Oil temperature and pressure sensors ensure lubrication is effective, while oil particle and moisture sensors monitor for contamination or wear debris. Torque sensors on the low-speed shaft measure transmitted forces, and vibration or acoustic sensors on the gearbox casing detect misalignments or early damage. These sensors are critical for predictive maintenance.

Generator sensors focus on preventing overheating and mechanical failure. Winding temperature sensors and bearing temperature sensors measure heat levels, while vibration sensors detect imbalance. Voltage and current sensors track electrical output, ensuring the generator is operating efficiently and safeguarding against catastrophic breakdowns.

Tower and structural sensors are also vital for long-term monitoring. Strain gauges at critical points measure bending stresses, accelerometers detect vibrations from gusts or resonance, and displacement sensors track sway or tilt. Together, they provide valuable insights into tower integrity and fatigue over time.

Electrical components such as converters and transformers are monitored with specialized sensors. IGBT temperature sensors track the heat of switching devices, while cooling system sensors measure coolant flow, pressure, and temperature. Partial discharge sensors in transformers detect insulation breakdown before failures occur, helping to maintain reliable grid connection.

Finally, auxiliary sensors play a support role. Hydraulic system sensors monitor pressure and flow in pitch and brake systems, UPS and battery condition sensors ensure emergency power readiness, and humidity or moisture sensors prevent condensation inside nacelles and control cabinets.

Table 2: SCADA sensors in wind turbines by category, function, and location.

Category	Sensor Type	Function	Typical Location
1. Meteorological Sensors	Anemometers	Measure wind speed	Nacelle top, meteorological mast
	Wind Vanes	Measure wind direction	Nacelle top, meteorological mast
	Temperature Sensors	Monitor ambient air temperature	Outside nacelle, meteorological mast
	Barometers	Measure air pressure for power curve analysis	Meteorological mast, nacelle
2. Rotor & Blade Sensors	Blade Pitch Angle Sensors (Encoders/Resolvers)	Measure blade angle for pitch control	Blade root
	Strain Gauges	Detect bending loads and stress	Inside blades
	Vibration Sensors	Detect abnormal oscillations, imbalance, or ice	Blades, hub
	Ice Detection Sensors	Identify icing conditions	Blades, nacelle
3. Nacelle Sensors	Yaw Position Encoders	Measure nacelle rotation	Nacelle base
	Yaw Brake Pressure Sensors	Monitor hydraulic/brake pressure	Yaw system
	Nacelle Vibration Sensors	Detect tower-top or drivetrain vibrations	Inside nacelle
4. Gearbox & Drivetrain Sensors	Oil Temperature & Pressure Sensors	Monitor lubrication condition	Inside gearbox
	Oil Particle & Moisture Sensors	Detect wear debris/contamination	Gearbox oil system
	Torque Sensors	Measure torque transfer	Low-speed shaft
	Vibration & Acoustic Sensors	Detect wear, misalignment, or bearing faults	Gearbox casing, bearings
5. Generator Sensors	Winding Temperature Sensors (RTDs/Thermocouples)	Prevent overheating	Generator windings
	Bearing Temperature Sensors	Monitor lubrication issues/wear	Generator bearings
	Vibration Sensors	Detect imbalance or faults	Generator housing
	Voltage & Current Sensors	Monitor electrical output	Generator stator & rotor (DFIGs)

Category	Sensor Type	Function	Typical Location
6. Tower & Structural Sensors	Strain Gauges	Measure bending loads	Tower walls/critical points
	Accelerometers	Detect vibrations from gusts or resonance	Tower structure
	Displacement Sensors	Track sway or tilt	Tower base or mid-sections
7. Converter & Transformer Sensors	IGBT Temperature Sensors	Monitor heating of power electronics	Converter modules
	Cooling System Sensors	Track coolant flow, pressure, temperature	Converter/transformer cooling circuits
	Partial Discharge Sensors	Detect insulation breakdown	Transformer units
8. Auxiliary Sensors	Hydraulic System Pressure & Flow Sensors	Monitor pitch and brake hydraulics	Pitch system, braking system
	UPS & Battery Condition Sensors	Ensure emergency backup readiness	UPS cabinets
	Humidity & Moisture Sensors	Prevent condensation and shorts	Nacelle, control cabinets

SCADA sensors are carefully distributed across the turbine: meteorological sensors on the nacelle and masts, blade and hub sensors for aerodynamic performance, drivetrain and gearbox sensors for mechanical health, generator sensors for electrical safety, tower sensors for structural monitoring, and electrical system sensors for converter and transformer protection. Together, these devices give operators a comprehensive real-time picture of turbine performance, condition, and safety, enabling efficient energy production and proactive maintenance.

Data Flow Within a Wind Turbine

Modern wind turbines are increasingly reliant on sophisticated data management systems to ensure optimal performance, safety, and reliability. The continuous flow of information begins with an array of sensors embedded in the turbine framework, measuring critical parameters such as wind speed, rotor speed, generator temperature, gearbox oil pressure, yaw and pitch angles, vibration levels, and grid connection status. These sensors transmit data as raw signals, typically in both analogue and digital formats, to the turbine's internal control systems [173, 174].

At the control level, local controllers, which can include Programmable Logic Controllers (PLCs) or dedicated embedded control systems, process the incoming sensor data in real time. This processing is crucial for adjusting turbine functionalities such as blade pitch control, yaw positioning, braking systems, and generator loading, thus ensuring that turbines operate efficiently and within prescribed safety standards [175, 176]. Research indicates that advancements in pitch control significantly enhance energy capture and overall turbine performance [177, 178].

Once the data is processed by the local controllers, it is disseminated into the Supervisory Control and Data Acquisition (SCADA) system. This pivotal component gathers operational data—ranging from status updates and alarms to performance metrics—and forwards it to SCADA nodes located strategically within the turbine's nacelle and tower base. These nodes serve as crucial intermediaries, relaying information to central wind farm control centres where operators can conduct remote monitoring, analyse performance trends, and implement operational commands such as turbine curtailments or restarts [179-181].

In addition to operational monitoring, SCADA systems integrate data from power electronics and grid interface systems, capturing metrics related to temperatures, voltages, and harmonics. This integration is vital for ensuring that the turbine's electrical output aligns with grid requirements, thereby adhering to established grid codes and maintaining operational standards [182, 183]. Such data management not only supports immediate operational control but also aids in predictive maintenance strategies by enabling anomaly detection and fault diagnostics based on historical and real-time SCADA data [184-186].

Finally, the processed data flows out of individual turbines through various communication systems, including fibre optics, copper cabling, and wireless communication methods. This data travels to a central substation within the wind farm, from where it can be transmitted through wide-area networks (WANs) to remote monitoring centres and utility grid operators, ensuring comprehensive oversight of turbine functionality and health [187-189].

Network Requirements for Wind Turbines

The operation of wind turbines relies heavily on well-structured communication networks that are robust and reliable. This is particularly crucial given that many turbines operate in harsh, remote, or offshore environments where maintaining uninterrupted connectivity is essential for operational safety and the efficient transmission of critical data. In these settings, high bandwidth and low latency are paramount due to the substantial amount of data generated by modern turbines, including Supervisory Control and Data Acquisition (SCADA) telemetry, condition monitoring data, video surveillance outputs, and meteorological information. Fiber optic networks are favoured for their capability to provide gigabit-level bandwidth with extremely low latency, thereby ensuring that data transmission occurs in real-time and is reliable under challenging conditions [190].

Reliability and redundancy in communication networks are also vital, as these systems must function seamlessly to maintain control and safety of the turbines at all times. In wind farms, a common strategy involves deploying redundant network paths using robust topological structures, such as ring topologies supported by self-healing protocols like Rapid Spanning Tree Protocol (RSTP) or Ethernet Ring Protection Switching (ERPS). This setup allows the network to automatically reroute communication in the event of a failure, thereby minimizing downtime [190].

Redundant network paths are a critical part of wind turbine and wind farm communications because the network must operate continuously without single points of failure. In practice, this means that if one cable or switch fails, the system automatically re-routes traffic along an alternative path so that SCADA data, monitoring signals, and control commands can still flow without interruption. Robust network topologies, particularly ring-based structures, are commonly used to achieve this.

In a ring topology, each network device (such as switches in turbine towers, substations, or control centres) is connected to two others, forming a closed loop. This provides two potential paths for data: clockwise and counterclockwise around the ring. Under normal operation, only one of these paths actively carries traffic while the other remains on standby as a backup. If a fault occurs—for example, a cable is cut between two turbines—the system immediately activates the backup path, and data flows in the opposite direction around the ring. This prevents downtime and isolates the fault without impacting the entire network.

The effectiveness of a ring topology depends on the self-healing protocol used. Two widely implemented protocols are:

1. **Rapid Spanning Tree Protocol (RSTP)** – RSTP is an evolution of the original Spanning Tree Protocol (STP). It detects looped paths and keeps one link in a blocking state to prevent broadcast storms. If a failure occurs, RSTP quickly reconfigures the network by unblocking the backup path, restoring connectivity typically within a few hundred milliseconds. Although not designed exclusively for rings, RSTP is widely supported by industrial switches and works well in smaller to medium-sized wind farms.

2. **Ethernet Ring Protection Switching (ERPS, ITU-T G.8032)** – ERPS was specifically developed for Ethernet ring networks and offers faster recovery than RSTP. ERPS can restore communications within less than 50 milliseconds after a failure, making it ideal for critical infrastructures such as wind power systems, where SCADA data and control signals must remain reliable. Unlike RSTP, ERPS scales better in large industrial networks and supports multiple interconnected rings without introducing complex topology changes.

By combining ring structures with these self-healing protocols, wind turbine networks achieve high availability, fault tolerance, and resilience against harsh environmental conditions. This ensures that even in remote or offshore wind farms, operators maintain uninterrupted monitoring and control of turbines, substations, and grid connections.

Additionally, considering the complexity and scale of modern wind farms, which may consist of hundreds of interconnected turbines, scalability is a major concern. Communication networks must accommodate simultaneous data transmission from multiple sources without creating bottlenecks. To facilitate this, industrial Ethernet-based protocols—such as Modbus TCP/IP, IEC 61850, and OPC UA—are commonly employed to ensure interoperability among devices while maintaining effective scalability [191, 192].

Given the critical nature of wind turbines within national energy infrastructure, cybersecurity is a pressing issue that must not be overlooked. Strong protective measures, including firewalls, data encryption, and intrusion detection systems, are essential to guard against possible cyber intrusions that could disrupt operations or jeopardize safety [186, 193]. Furthermore, precise time synchronization through Network Time Protocol (NTP) or GPS synchronization is necessary for accurate event logging, fault analysis, and regulatory compliance, which ensures operational consistency across all turbines [194].

Lastly, environmental hardening of networking equipment is crucial, as devices installed in turbine nacelles and towers must endure extreme conditions such as temperature fluctuations, vibrations, humidity, and electromagnetic interference. Therefore, ruggedized industrial switches and routers designed specifically to maintain long-term reliability in these challenging operational environments are preferred [192, 195].

Industry Standards and Regulations

Standards and guidelines are a critical foundation for securing and designing Industrial Automation and Control Systems (IACS), but they are not prescriptive. Each system is unique, meaning that while standards provide valuable frameworks and common practices, they must be tailored to the specific needs, risks, and environments of the business. Rather than serving as rigid blueprints, they act as adaptable best practice references that help ensure consistency, security, and interoperability across industrial systems. Below is a detailed explanation of some of the most widely referenced and applied standards in the industrial and energy sectors [196].

ISA-95/PERA (Purdue Enterprise Reference Architecture): ISA-95 and the Purdue Enterprise Reference Architecture (PERA) define a general architecture for all types of industrial systems. These standards are especially useful because they establish common terminology, concepts, and building blocks for IACS. They help create structured layers of automation and control, making it easier for organizations to manage integration across business planning, manufacturing execution, and control systems. ISA-95 and PERA are often used as a foundation for modelling complex industrial networks and aligning IT with operational technology (OT) [196].

IEC 62443/ISA-99: The IEC 62443 series (formerly ISA-99) extends IT security principles into the industrial domain. While the ISO/IEC 27000 standards focus on general-purpose IT systems, IEC 62443 addresses the unique requirements of industrial control systems (ICS). Cybersecurity in ICS is not only about protecting data—it has direct implications for health, safety, and the environment (HSE). For this reason, IEC 62443 emphasizes integrating cybersecurity into broader risk management practices. It provides guidelines for secure design, development, and operation of automation systems, covering everything from policies to technical controls [196].

NIST Cybersecurity Framework: The NIST Cybersecurity Framework, introduced in 2014, is not a mandatory standard but a best practice guideline. It was developed to provide a voluntary, industry-led set of practices, processes, and methodologies to reduce cyber risks in critical infrastructure. Its strength lies in flexibility: organizations can adapt the framework to their specific environments. It organizes cybersecurity activities into five core functions—Identify, Protect, Detect, Respond, and Recover—helping organizations build resilience against evolving threats while aligning with business objectives.

NIST 800 Series: The NIST 800 Series provides a comprehensive set of documents detailing U.S. government security policies and practices. While originally developed for federal agencies, these guidelines have been adopted globally by private companies and governments alike. Of particular importance to the industrial sector is NIST SP 800-82: Guide to Industrial Control Systems Security, which directly addresses the protection of SCADA systems, Distributed Control Systems (DCS), and related technologies. It covers ICS architectures, known vulnerabilities, threats, and recommended countermeasures, making it a cornerstone reference for organizations securing critical infrastructure [196].

NERC CIP (Critical Infrastructure Protection): The North American Electric Reliability Corporation's Critical Infrastructure Protection standards (NERC CIP) were initially developed for utilities but have since been applied more broadly across industries dealing with critical infrastructure. NERC CIP focuses on protecting energy delivery systems, and its terminology and requirements align closely with the realities of IACS environments. It is particularly relevant for the energy sector as it provides concrete requirements for asset identification, security management, and incident response, ensuring both reliability and security in power systems [196].

IEEE 1588 – Precise Time Protocol (PTP): IEEE 1588, known as the Precise Time Protocol (PTP), was developed to synchronize clocks across distributed devices in packet-based networks.

Precise time synchronization is essential in industrial systems where coordination across multiple devices is required, such as substation automation, outage monitoring, or renewable power generation. Enhanced in 2008 with PTPv2, this protocol has been adapted into specific "profiles" for different applications: the Default Profile for general industrial use, the Power Profile for IEC 61850 substation communications, and specialized profiles for telecom and time-sensitive networking (TSN). PTP allows industrial systems to operate with sub-microsecond synchronization accuracy while requiring minimal bandwidth and processing overhead, making it indispensable for advanced control and monitoring systems [196].

Together, these standards—ISA-95/PERA, IEC 62443, NIST frameworks, NERC CIP, and IEEE 1588—form the backbone of best practices in industrial system security, design, and operation. They provide organizations with structured approaches to architecture, cybersecurity, risk management, and synchronization. However, they are not "one-size-fits-all" solutions. Their real value lies in being adaptable frameworks that can be customized to the operational context of each organization, ensuring resilient, secure, and efficient industrial automation systems.

Wind Farm Security Architecture

A wind farm security architecture is the integrated framework of physical, network, and cybersecurity measures that safeguard turbines, substations, and control centres from potential threats. Because wind farms are classified as critical infrastructure and are often located in remote or offshore environments, their security systems must address a wide range of risks, including harsh weather conditions, cyber-attacks, vandalism, and operational disruptions. To ensure resilience, a robust security architecture combines layered defence strategies, network segmentation, and continuous monitoring.

The first line of defence is physical security. Wind farms protect their assets through perimeter fencing, gated access, and surveillance systems. Turbine towers are fitted with secure access doors and electronic locks, while offshore turbines may require vessel access control. Video surveillance, motion detectors, and intrusion alarms provide real-time monitoring, with all signals integrated into the SCADA or security operations centre. Physical protections also include environmental hardening, such as anti-corrosion materials, reinforced housings, and fire suppression systems to withstand demanding site conditions.

Beyond physical measures, network segmentation forms the backbone of wind farm security. This involves separating operational networks from business IT systems, typically following frameworks such as ISA-95 (Purdue Model) or IEC 62443. These define hierarchical layers, from field devices and sensors at the base, through control systems and SCADA operations, up to enterprise IT and cloud platforms. To reduce risks, firewalls and demilitarized zones (DMZs) isolate operational technology from corporate IT, while secure VPNs, multi-factor authentication, and session logging regulate vendor or remote access.

Cybersecurity measures are equally essential. Core protections include encrypting communications between turbines, substations, and control centres using TLS or IPSec. Intrusion detection and prevention systems (IDS/IPS) monitor traffic for anomalies, while role-based access control (RBAC) ensures only authorized personnel can access sensitive systems. Patch and vulnerability management keep SCADA systems and controllers secure, while Security Information and Event Management (SIEM) platforms centralize monitoring and incident response. Adherence to standards such as IEC 62443, NIST SP 800-82, and NERC CIP helps ensure compliance with global best practices.

Resilience and redundancy are critical to maintaining continuous operations. Wind farm networks often employ redundant communication paths and self-healing topologies, such as ring networks using RSTP or ERPS, to keep SCADA data flowing even during equipment failures. Backup SCADA servers, redundant power supplies, and disaster recovery plans further protect against outages, whether caused by cyber incidents or physical disruptions.

Monitoring and incident response complete the architecture. Centralized SCADA systems, supported by cybersecurity dashboards, allow operators to monitor turbine performance, alarms, and potential threats in real time. Automated alerts can trigger incident response playbooks, which may include shutting down turbines, isolating compromised networks, or dispatching on-site teams.

Integration with the wider grid and smart systems is essential. Wind farms must comply with grid code requirements and utility cybersecurity standards while synchronizing with national transmission systems. Security architectures incorporate protocols such as IEC 61850 and IEEE 1588 PTP while safeguarding against risks such as spoofing, data manipulation, or man-in-the-middle attacks.

Use Case Example

The following sample use case shows how SCADA and telemetry systems work in a modern wind farm, from the turbine level through to central operations and the grid operator:

Use Case: Detecting and Managing a Gearbox Fault in a Wind Farm

1. Sensor and Local Monitoring

Inside one of the turbines, a set of vibration and oil-particle sensors in the gearbox detect abnormal patterns—slightly elevated vibration and an increase in metallic debris. These raw signals are picked up by the turbine's PLC (Programmable Logic Controller). The PLC processes the data in real time and compares it against predefined thresholds.

2. Turbine Controller Response

The turbine controller determines that the gearbox is showing early signs of wear but not yet at critical failure levels. It automatically adjusts turbine operation by slightly reducing rotor torque to reduce mechanical stress, keeping the turbine online but at a lower output.

3. SCADA Data Transmission

The processed event (gearbox vibration alarm, oil contamination warning, torque reduction action) is sent via the turbine's SCADA node to the wind farm's local SCADA server located at the substation. The data flows over a fibre optic link using an industrial protocol (e.g., IEC 61850 or Modbus TCP/IP).

4. Centralized SCADA Monitoring

At the wind farm control centre, operators view the alarm on their HMI (Human Machine Interface) dashboard. They see live turbine performance indicators (rotor speed, generator temperature, vibration trends) alongside the gearbox fault. The SCADA system also logs the alarm in the event database with timestamps synchronized via GPS/NTP.

5. Telemetry to Grid Operator

Simultaneously, telemetry data (active power, reactive power, voltage, frequency) is sent from the substation to the national grid operator's control centre. The grid operator sees that the turbine is producing slightly less power than expected but is still stable. The wind farm continues to meet its grid code compliance requirements, including reactive power support and frequency stability.

6. Predictive Maintenance Scheduling

The SCADA system correlates the gearbox vibration data with historical trends and flags the event to the Condition Monitoring System (CMS). An automated work order is generated in the maintenance management platform, scheduling a service visit in two weeks—before the gearbox risk becomes critical. Spare parts are ordered in advance, reducing downtime.

7. Operator Actions

Operators use the SCADA interface to:

- Confirm the torque reduction applied automatically by the turbine controller.
- Place the turbine in a "curtailed" mode to further limit loads until service.
- Notify the maintenance team and record the event for warranty tracking.

8. Outcome

Thanks to SCADA/telemetry:

- The turbine remained operational, avoiding an immediate shutdown.
- Maintenance was scheduled proactively, reducing the risk of catastrophic gearbox failure.

- The grid operator received continuous power data, ensuring no disruption to grid stability.
- Long-term reliability of the turbine was preserved through early fault detection.

This use case illustrates how SCADA/telemetry integrates sensors, controllers, communications, operators, and grid interaction into a seamless loop. It highlights how data moves from a single component (gearbox sensor) to local control (PLC), then through SCADA to central monitoring, and finally to the grid operator—all while enabling predictive maintenance and compliance with grid codes.

Control Methods

Pitch Control – Adjusting the Blade Angle

Pitch control refers to changing the angle of the turbine blades relative to the incoming wind. The goal is to maintain the optimum angle of attack so that the blades generate maximum aerodynamic lift while minimizing drag. At moderate wind speeds, blades are pitched to maximize energy capture. However, at very high wind speeds, too much aerodynamic force could damage the turbine or overload the generator.

In such cases, pitch control becomes a safety and regulation tool:

- Stalling occurs when the blade angle is increased, causing the flat side of the blade to face further into the wind. This disrupts smooth airflow, reducing lift and limiting rotation speed.
- Furling does the opposite—by decreasing the angle of attack, the blade edges face more into the wind, reducing surface exposure and power capture.

Pitch adjustment is the most effective way to limit power output, as it directly regulates aerodynamic forces acting on the rotor.

Yaw Control – Orienting the Turbine to the Wind

Yaw control refers to rotating the entire nacelle (the housing for the gearbox, generator, and control systems) on top of the tower so that the rotor always faces into the wind. This rotation occurs around the turbine's vertical axis.

Wind direction changes constantly, and if the turbine is not aligned properly, its rotor intercepts less wind energy. This misalignment causes power losses, which can be estimated using the formula:

$$\Delta P = \alpha \cos(\varepsilon)$$

Here, ΔP is the lost power, and ε is the yaw error angle. Even small misalignments reduce efficiency significantly. Automatic yaw motors, guided by wind vane data, correct the nacelle orientation to minimize these losses and maintain high power output.

Generator and Electrical Control

Beyond mechanical adjustments, wind turbines also use electrical subsystem control to optimize performance. This is achieved through power electronics and electronic converters linked to the generator. By controlling current flow in either the stator (stationary part) or rotor (rotating part), operators can decouple generator speed from the fixed voltage and frequency of the grid.

This means the turbine can run at variable rotor speeds while still delivering electricity that matches grid requirements. Controlling generator speed is especially effective at low wind speeds, as it allows the turbine to maximize power capture without being constrained by synchronous grid speed.

System-Level Integration

At a system-wide level, the most powerful combination for wind energy optimization is:

1. Pitch control to adjust aerodynamic efficiency and protect the turbine from overload.
2. Generator speed control to fine-tune power conversion and maximize energy capture across varying wind speeds.

Together with yaw alignment, these control methods form the backbone of wind turbine operation, balancing efficiency, reliability, and safety in variable wind conditions.

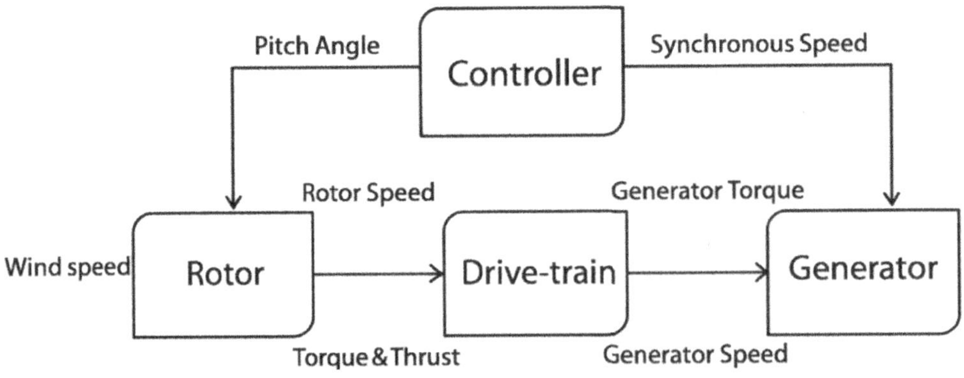

Figure 96: System-Level Layout of a Wind Energy System.

Control System

Wind turbine control is one of the most critical aspects of ensuring both efficiency and safety in modern wind energy systems. The overall objective of the control system is twofold: to extract the maximum possible energy from the wind by maximizing the coefficient of power (Cp), and to maintain safe operation under all wind conditions, from very low to extreme winds. To achieve these goals, wind turbine control systems are structured into different levels, each addressing different operational needs.

At the wind farm level, a central controller manages groups of turbines, coordinating their output to respond to environmental conditions and grid demands. This function is often called power management, as the controller can initiate or shut down turbines, balance output across the farm, and ensure compliance with grid operator instructions.

At the individual turbine level, the supervisory controller governs each turbine's basic operation. It decides when to start or shut down the turbine depending on wind speeds (cut-in, rated, and cut-out thresholds) and load conditions. The supervisory controller also provides inputs to the dynamic controllers that directly regulate the rotor speed, blade pitch, and braking systems. These dynamic controllers ensure the turbine maintains an optimum tip-speed ratio (λ), usually between 6 and 8 depending on turbine design, to maximize efficiency.

One way to control turbine performance is by adjusting the aerodynamic torque on the rotor. This is achieved by managing the rotor's lift-to-drag ratio (L/D), which can be controlled in two main ways:

1. Stall-Regulated Designs – These turbines are built with blade shapes and angles that cause the rotor to stall at high wind speeds. As the blades stall, lift decreases and drag increases, naturally limiting power without requiring active adjustment.
2. Pitch-Regulated Designs – Here, blade pitch angle is actively changed. Reducing pitch lowers the angle of attack, decreases lift, and limits aerodynamic torque. Pitch control starts once wind velocity reaches rated power and continues to adjust blade angles to maintain constant power output until the cut-out speed is reached.

Pitch regulation is generally more effective and flexible, especially for large turbines, as it actively adapts to varying wind conditions.

Another approach to controlling the tip-speed ratio is through electrical torque control. Most large wind turbines use synchronous generators. These machines allow adjustment of reactive power by controlling the rotor's field current, making them suitable for grid support as well as power generation.

In a synchronous machine with P poles, the relationship between mechanical speed and electrical frequency is:

$$\omega = \frac{P}{2}\omega_m$$

Or in terms of frequency (*f*) and rotor speed (*n* in rpm):

$$n = \frac{120f}{P}$$

With a 4-pole generator, achieving 60 Hz output (in the U.S.) requires the rotor to spin at 1800 rpm. Fixed-speed turbines therefore need a gearbox to ensure rotor blades run at the correct speed for synchronous generation. However, this approach is restrictive. A more flexible solution is to convert AC output to DC and then back to grid-standard AC using power electronic converters, enabling variable-speed operation while still maintaining grid frequency.

Here are five short, fully-worked examples that show how electrical torque control and synchronous-machine basics play out in real wind-turbine design and operation.

1) Pole count ↔ grid frequency ↔ generator speed

For a synchronous machine, electrical frequency f(Hz), pole count P, and mechanical speed n(rpm) are linked by

$$n = \frac{120\,f}{P}.$$

Example A (U.S. grid, 60 Hz):

- $P = 4$ poles $\rightarrow n = \frac{120 \times 60}{4} = 1{,}800\ \mathrm{rpm}.$
- $P = 6$ poles $\rightarrow n = \frac{120 \times 60}{6} = 1{,}200\ \mathrm{rpm}.$
- $P = 8$ poles $\rightarrow n = \frac{120 \times 60}{8} = 900\ \mathrm{rpm}.$

Example B (Europe, 50 Hz):

- $P = 4$ poles $\rightarrow n = \frac{120 \times 50}{4} = 1{,}500\ \mathrm{rpm}.$
- $P = 6$ poles $\rightarrow n = \frac{120 \times 50}{6} = 1{,}000\ \mathrm{rpm}.$

Takeaway: To get 60 Hz with a 4-pole generator you need 1,800 rpm; with 6 poles you only need 1,200 rpm. This is why pole count and gearbox ratio are design levers.

2) From wind and rotor size to gearbox ratio

Tip-speed ratio λ(kept near its optimum by control) relates wind speed V, rotor radius R, and rotor speed ω(rad/s):

$$\lambda = \frac{\omega R}{V} \Rightarrow \omega = \frac{\lambda V}{R}, \qquad n_{\text{rotor}} = \frac{\omega}{2\pi} \times 60.$$

Assume a large turbine: $R = 50$ m, $V = 9$ m/s, $\lambda^{\star} = 7$.

- $\omega = \dfrac{7 \times 9}{50} = 1.26 \text{ rad/s}.$
- $n_{\text{rotor}} = \dfrac{1.26}{2\pi} \times 60 \approx 12.0 \text{ rpm}.$

Match to a 50 Hz, 4-pole synchronous generator at 1,500 rpm:

Required gearbox ratio

$$G = \frac{1{,}500}{12.0} \approx 125{:}1.$$

If instead you pick 6 poles (1,000 rpm at 50 Hz),

$$G = \frac{1{,}000}{12.0} \approx 83{:}1.$$

Takeaway: Higher pole count relaxes the gearbox ratio; variable-speed + converter can avoid "hard-locking" to these exact speeds.

3) Electrical torque control for MPPT (the $T = K\omega^2$ law)

At optimum tip-speed ratio $\lambda^{\star}$and power coefficient $C_p^{\star}$, aerodynamic power is

$$P = \frac{1}{2}\rho A C_p^{\star} V^3, \quad \text{with } V = \frac{\omega R}{\lambda^{\star}} \Rightarrow P = \left(\frac{1}{2}\rho A C_p^{\star} \frac{R^3}{(\lambda^{\star})^3}\right)\omega^3.$$

Therefore the maximum-power torque command is

$$T_{\text{ref}} = \frac{P}{\omega} = K\,\omega^2, \quad K = \frac{1}{2}\rho A C_p^{\star} \frac{R^3}{(\lambda^{\star})^3}.$$

Numerical example (large onshore turbine):

$$\rho = 1.225 \text{ kg/m}^3,\ R = 50 \text{ m},\ A = \pi R^2 \approx 7{,}854 \text{ m}^2,\ C_p^{\star} = 0.45,\ \lambda^{\star} = 7.$$

Compute K:

- $\frac{1}{2}\rho A \approx 0.5 \times 1.225 \times 7{,}854 \approx 4{,}815$.
- $4{,}815 \times C_p^\star \approx 2{,}167$.
- $R^3 = 125{,}000 \text{ m}^3$; $2{,}167 \times 125{,}000 \approx 270.9 \times 10^6$.
- $(\lambda^\star)^3 = 343$; $K \approx \dfrac{270.9 \times 10^6}{343} \approx 7.90 \times 10^5 \text{ N . pm}/(\text{rad/s})^2$.

Now pick two rotor speeds:

- **Case A:** $\omega = 1.20 \text{ rad/s}$ (~11.5 rpm)
  $T_{\text{ref}} = K\omega^2 \approx 7.90 \times 10^5 \times 1.44 \approx 1.14 \text{ MN . pm}$.
  $P = T\omega \approx 1.14 \times 10^6 \times 1.20 \approx 1.37 \text{ MW}$.
- **Case B:** $\omega = 1.50 \text{ rad/s}$ (~14.3 rpm)
  $T_{\text{ref}} \approx 7.90 \times 10^5 \times 2.25 \approx 1.78 \text{ MN . pm}$.
  $P \approx 1.78 \times 10^6 \times 1.50 \approx 2.67 \text{ MW}$.

How it's used: The converter commands generator torque $T_e$to track $T_{\text{ref}} = K\omega^2$, keeping λnear $\lambda^\star$for maximum energy capture in Region II.

4) Reactive power via field current (synchronous generator support)

Synchronous generators can adjust power factor by changing rotor field current (excitation), supplying or absorbing vars to support voltage.

Example: A 2 MW turbine must operate at 0.95 lagging PF.

- Apparent power $S = \dfrac{P}{\text{PF}} = \dfrac{2.0}{0.95} = 2.105 \text{ MVA}$.
- Reactive power $Q = \sqrt{S^2 - P^2} = \sqrt{(2.105)^2 - (2.0)^2} \text{ MVAr}$.
- $S^2 \approx 4.433$, $P^2 = 4.000 \Rightarrow Q \approx \sqrt{0.433} \approx 0.658 \text{ MVAr}$.

By increasing excitation, the machine can supply ~+0.66 MVAr (lagging PF).

By decreasing excitation, it can absorb vars (leading PF), within thermal limits.

Takeaway: Field current is a direct lever for grid-supporting vars (voltage control), independent of real-power MPPT torque control.

5) Variable-speed with a full converter (frequency decoupling)

With a full-scale AC–DC–AC converter, the generator can run at whatever speed optimizes aerodynamics; the converter synthesizes grid-quality 50/60 Hz.

Example with a multi-pole PM generator:

Suppose the rotor runs at $n_{\text{rotor}} = 15$ rpmand the generator has $P = 64$poles (common for direct-drive or medium-speed designs). For a synchronous PM machine,

$$f_{\text{electrical}} = \frac{P}{120} n_{\text{mech}} = \frac{64}{120} \times 15 = 8.0 \text{ Hz}.$$

The generator produces **8 Hz variable-amplitude AC**. The **converter**:

1. **Rectifies** to DC,
2. **Inverts** to **exactly 50/60 Hz**, phase-locked to the grid, with commanded voltage and reactive power.

If wind rises and the controller speeds the rotor from 12 rpm to 18 rpm (to hold λ^*), the raw generator frequency slides from 6.4 Hz to 9.6 Hz—but the output of the converter stays rock-solid at grid frequency and setpoint PF.

Takeaway: Full converters (and DFIG rotor converters) decouple mechanical speed from grid frequency, enabling variable-speed MPPT while meeting grid codes.

What these examples show:

- **Pole count & gear ratio** set the synchronous-speed target if you're directly tied to the grid.
- **Electrical torque control** (via the converter) enforces the $T = K\omega^2$law for maximum power capture.
- **Synchronous excitation** lets you **push/pull vars** for voltage support.
- **Converters** (full-scale or DFIG rotor-side) **decouple** rotor speed from grid frequency, so you can run at the **aerodynamic optimum** and still deliver compliant 50/60 Hz power.

Wind Turbine Operation Strategies

Wind turbines are designed with four strategic objectives:

1. Maximize energy production within safe speed and load limits.

2. Minimize extreme loads and fatigue damage from cyclic stresses on the blades and drivetrain.

3. Deliver acceptable power quality that meets grid connection standards.

4. Ensure overall safe operation under all conditions.

Control approaches differ depending on wind speed:

- **Between cut-in and rated wind speeds ($U_{cut\text{-}in} < U_\infty < U_{rated}$):** the goal is to maximize power output.

- **Between rated and cut-out wind speeds ($U_{rated} < U_\infty < U_{cut\text{-}out}$):** the goal is to limit power to the turbine's rated capacity.

Fixed Speed vs. Variable Speed Designs

Fixed-speed turbines are either stall regulated or active pitch regulated.

- In stall-regulated designs, blades are fixed at a certain angle and naturally stall at higher wind speeds, limiting power passively.

- In active pitch regulated designs, blades remain fixed below rated speed but actively adjust pitch above rated speed to smooth power output and prevent overload.

Variable-speed turbines, on the other hand, use electrical torque control to continuously maintain the optimum tip-speed ratio (λopt).

- In stall-regulated variable speed turbines, blade stall is used to limit power above rated speeds, but rotor speed is actively adjusted at lower speeds to maximize efficiency.

- In pitch-regulated variable speed turbines, torque control maintains constant power while pitch control fine-tunes rotor speed and load.

Variable speed designs are more advanced, enabling adaptive torque control that allows the rotor to accelerate or decelerate as needed to stay near the optimum λ. This results in higher efficiency and smoother power regulation.

Advanced Control Concepts: Axial Induction and Lift Control

Axial Induction Control builds on aerodynamic theory to optimize how much energy is extracted from the wind stream. By controlling induction factors (how much the wind slows down through the rotor), operators can regulate torque and improve efficiency beyond simple pitch or speed controls.

Lift Control applies aerodynamic modifications similar to those used in aircraft. Techniques like trailing edge flaps, Gurney flaps, trailing edge blowing, or even plasma actuators can dynamically alter lift characteristics of the blade airfoil. For instance, a Gurney flap at the trailing edge can increase or decrease lift depending on placement, allowing fine control of blade performance.

Chapter 8

Structural Components

Wind turbines look simple from a distance: a tall tower, a box on top, three blades turning in the wind. Up close, they're finely tuned structures that have to do a very hard job—carry hundreds of tonnes of moving machinery a hundred meters in the air, ride out storms and waves, and keep working for decades. This chapter opens the tower and foundation "black box," showing how height, stiffness, mass, and soil interact to keep a turbine upright and quiet under constantly changing loads.

We begin with towers: why they're the height they are, what shapes and materials are used (tubular steel, lattice, concrete and hybrids), and how they handle gravity, thrust, and vibration without drifting into resonance with the rotor. From there we step into the ground (and seabed), comparing spread footings, piles, monopiles, jackets, suction caissons, and floating substructures—and how site geology and water depth drive those choices. Worked examples translate wind and seismic forces into base moments and stresses to make the numbers tangible.

Because technicians have to reach everything we build, we cover access systems—ladders, lifts, platforms, and fall-arrest gear—and how they're integrated for safe, repeatable work at height. The chapter then tackles durability: corrosion protection and coatings onshore and offshore, cathodic protection, and the inspection/repair routines that keep steel and composites alive in salt, sun, and spray. Finally, we map lightning and surge protection from blade tip to earth grid, explaining LPZ zoning, SPD selection, and ride-through demands on modern electronics.

By the end, you'll see how structural choices ripple into installation, maintenance, and reliability—and how good engineering turns a tall, slender tower into a stable, serviceable, long-lived asset.

Towers and Foundations

A wind turbine tower and foundation are critical components that provide stability, height, and structural support to the turbine. They must withstand extreme mechanical loads, harsh environments, and the dynamic forces created by wind and rotor operation.

Wind Turbine Tower

The tower is the vertical structure that supports the nacelle (containing the gearbox, generator, and control systems) and the rotor (blades + hub). Its main functions are to elevate the blades into stronger, more consistent wind streams and to transmit loads safely into the foundation.

Design Considerations:

- **Height:** Modern onshore towers range from 80–120 m, while offshore towers can exceed 150 m to access stronger winds.
- **Loads:** Towers must resist vertical loads (from turbine weight), horizontal thrust (from wind on blades), and dynamic vibrations (from rotor rotation and gusts).
- **Aerodynamics:** The tower must be designed to minimize drag and vortex shedding, which can cause structural fatigue.

Wind turbine towers come in several different types, each with unique design characteristics and applications.

The most common design is the tubular steel tower, which is constructed in cylindrical or conical sections that are transported to the site and bolted together during assembly. These towers house internal ladders or elevators that provide access to the nacelle for maintenance personnel. To withstand weathering and environmental exposure, tubular steel towers are typically painted or coated for corrosion resistance.

Figure 97: Top section of a turbine tubular steel tower being lifted into position. Paul Anderson, CC BY-SA 2.0, via Wikimedia Commons.

Another design is the lattice tower, built from interconnected steel trusses. This structure is lighter and generally less expensive than tubular towers, but it has become less common in modern, large-scale wind farms due to its higher maintenance needs and less favourable visual appearance.

Figure 98: Lattice towers. Vyacheslav Argenberg , CC BY 4.0, via Wikimedia Commons.

Hybrid towers combine concrete and steel, with the lower portion constructed from concrete and the upper section from steel. This approach enables greater overall tower height without the need for overly large steel components, making them a practical solution for taller wind turbines.

Finally, there are concrete towers, which are often built in segments onsite. These towers are increasingly used for very tall turbines or in locations where steel is costly or less readily available. Their modular construction also makes them suitable for sites where transportation of large steel sections would be challenging.

Figure 99: Wind turbine tower concrete segment. Cjp24, CC BY-SA 4.0, via Wikimedia Commons.

Loads on the Tower

The wind turbine tower is designed to carry a combination of loads that arise from both the structure itself and the forces generated by the turbine's operation and environment. The most fundamental of these is the dead load, which comes from gravity. This includes the weight of the rotor, nacelle, and the tower structure itself. These forces act axially, compressing the tower from top to bottom and forming the constant baseline stress that must always be supported.

A second major category of loading is the aerodynamic thrust generated by the rotor. This horizontal force is transferred through the nacelle into the tower and is typically the dominant contributor to overturning moments at the foundation. The thrust load varies with wind speed and rotor operation, meaning it is not constant but dynamic in nature, making it critical for tower strength and stability.

The tower itself is also subject to wind drag, which acts directly on the cylindrical structure. While smaller in magnitude compared to rotor thrust, tower wind load still adds both shear forces and overturning moments that the foundation must resist.

Beyond these, dynamic effects also play a significant role. Yawing motions, drivetrain torque reactions, and periodic rotor excitations (1P and 3P frequencies, linked to blade passing and rotor imbalance) introduce fluctuating stresses. Combined with turbulence in the wind, these can cause fatigue loads that accumulate over the turbine's lifetime and must be accounted for in design.

Finally, site-dependent loads such as seismic activity, ice accretion on blades or tower surfaces, and transient loads during installation or maintenance may also affect the tower. While not always present, they are critical in specific geographic regions and must be incorporated into safety checks.

For ultimate limit state (ULS) sizing, the most critical loads are typically the dead load and the aerodynamic thrust, since these govern the maximum stresses and overturning base moments. Once these are addressed, engineers also consider natural frequency and dynamic tuning, ensuring that the tower avoids resonance with rotor excitation frequencies to maintain both structural safety and long-term fatigue resistance.

The following provides worked walk-through of the main tower loads and how to size and-check them with numbers. I'll use a representative modern onshore turbine to keep the math tangible.

1) Loads you always check

A) Dead load (gravity)

This is axial compression from the rotor, nacelle, and tower self-weight.

- Example masses (typical 4–6 MW class):
    - Rotor + nacelle: ~350 t
    - Tower: ~200 t
    - Total mass $m \approx 550 \text{ t} = 5.50 \times 10^5$ kg

Axial force:

$$N = m\,g \approx 5.50 \times 10^5 \text{ kg} \times 9.81 \text{ m/s}^2 = 5.39 \times 10^6 \text{ N} = 5.39 \text{ MN}$$

This is the baseline compression the shell must carry everywhere.

B) Aerodynamic thrust (dominant overturning)

Rotor thrust pushes horizontally at (about) hub height and drives the base bending moment.

Use the thrust formula

$$T = \frac{1}{2}\,\rho\,A\,C_T\,V^2$$

Take:

- Air density $\rho = 1.225\ \mathrm{kg/m^3}$
- Rotor diameter $D = 126\ \mathrm{m} \Rightarrow A = \pi(D/2)^2 = \pi(63)^2 \approx 12{,}469\ \mathrm{m}^2$
- Below rated wind, a representative C_T is 0.8; wind speed $V = 12\ \mathrm{m/s}$

Compute:

$$T = \tfrac{1}{2}(1.225)(12{,}469)(0.8)(12^2) \approx 0.5 \times 1.225 \times 12{,}469 \times 0.8 \times 144 \approx 0.881 \times 10^6\ \mathrm{N} = 0.88\ \mathrm{MN}$$

Overturning moment at the base (lever arm ≈ hub height h):

$$M_T \approx T\,h(\text{take } h = 100\ \text{m}) \Rightarrow M_T \approx 0.88\ \text{MN} \times 100\ \text{m} = 88\ \text{MN·m}$$

This single number often dominates ULS design.

C) Tower wind drag (adds shear and a smaller moment)

Treat the tower as a vertical cylinder with distributed lateral load:

$$w(z) \approx q\,C_D\,D_{\mathrm{tw}} \quad \text{with} \quad q = \tfrac{1}{2}\rho V^2$$

Use:

- $V = 12\ \mathrm{m/s} \Rightarrow q = \frac{1}{2}(1.225)(12^2) \approx 88.2\ \mathrm{N/m^2}$
- Drag coefficient $C_D \approx 1.2$
- Mean tower diameter $D_{\mathrm{tw}} \approx 5\ \mathrm{m}$

Then distributed load:
w ≈ 88.2 × 1.2 × 5 ≈ 529 N/m = 0.529 kN/m

Total tower lateral force over height H = 100 m:
V_{tower} ≈ w · H ≈ 0.529 × 100 = 52.9 kN

Base moment from a uniform cantilever load:
M_{tower} ≈ w · H^2 / 2 ≈ 0.529 × 100^2 / 2 = 2.65 MN·m

Compare base moments: 88 MN·m (rotor thrust) vs 2.65 MN·m (tower drag). Rotor thrust clearly dominates at this wind speed.

D) Base shear and total overturning (ULS snapshot)

$V_{\text{base}} \approx T + V_{\text{tower}} \approx 0.88 \text{ MN} + 0.053 \text{ MN} = 0.93 \text{ MN}$

$M_{\text{base}} \approx M_T + M_{\text{tower}} \approx 88 + 2.65 = 90.7 \text{ MN·m}$

2) Section checks at the tower base (stress from N and M)

Assume a typical thin-walled steel shell at the base:

- Outside diameter $D = 5.5$ m
- Wall thickness $t = 35$ mm $= 0.035$ m

Thin-wall area and inertia (good first approximations):

$$A \approx \pi D t \approx \pi \times 5.5 \times 0.035 \approx 0.604 \text{ m}^2$$

$$I \approx \frac{\pi D^3 t}{8} = \frac{\pi \times 5.5^3 \times 0.035}{8} \approx 2.29 \text{ m}^4$$

$$y = D/2 = 2.75 \text{ m}$$

Axial stress from dead load:

$$\sigma_N = \frac{N}{A} \approx \frac{5.39 \times 10^6}{0.604} \approx 8.9 \text{ MPa} \quad \text{(uniform compression)}$$

Bending stress from overturning:

$$\sigma_M = \frac{M_{\text{base}} y}{I} \approx \frac{90.7 \times 10^6 \times 2.75}{2.29} \approx 109 \text{ MPa}$$

Combined extreme-fiber stresses:

- **Windward (compression):** $\sigma \approx \sigma_N + \sigma_M \approx 8.9 + 109 \approx 118$ MPa
- **Leeward (tension):** $\sigma \approx \sigma_N - \sigma_M \approx 8.9 - 109 \approx -100$ MPa (tension 100 MPa by magnitude)

These values are plausible for ULS checks (before applying factors, corrosion allowance, shell buckling checks, local stiffener design, etc.). In practice, you'd apply partial factors, check shell buckling (axial + bending + ovalization), joint/anchor design, and soil–structure interaction at the foundation.

3) Dynamic "tuning" (natural frequency vs 1P/3P)

To avoid resonance, the tower's first fore–aft natural frequency f_1 is kept between rotor 1P and 3P over the operating range (soft–stiff design). Example:

- Operating rotor speed range: 6–15 rpm
  $\Rightarrow$ 1P $=$ 0.10–0.25 Hz, 3P $=$ 0.30–0.75 Hz

A simple cantilever estimate:

$$f_1 \approx \frac{1.875^2}{2\pi}\sqrt{\frac{EI}{\rho A\,H^4}}$$

Using:

- $E = 210$ GPa, $I \approx 2.29$ m^4
- Steel lineal mass $\rho A \approx 7{,}850 \times 0.604 \approx 4{,}740$ kg/m
- $H = 100$ m

Compute the ratio under the root:

$$\frac{EI}{\rho A\,H^4} \approx \frac{(210 \times 10^9)(2.29)}{4{,}740 \times (100)^4} \approx \frac{4.81 \times 10^{11}}{4.74 \times 10^{11}} \approx 1.02$$

$$f_1 \approx \frac{3.516}{2\pi}\sqrt{1.02} \approx 0.56 \text{ Hz}$$

That sits nicely between 1P (≤0.25 Hz) and 3P (≥0.30 Hz), which is the goal.

(In detailed design you'll refine EI and mass distribution over height, include nacelle/rotor modal participation, soil stiffness, and validate with an aero-elastic model.)

<u>How things scale with wind</u>

- **Thrust** $T \propto V^2$→ base moment $M_T \propto V^2$
- **Power** $P \propto V^3$(below rated)
- **Tower drag** also $\propto V^2$

At higher winds (near rated), the pitch control reduces C_T, so thrust does not grow as V^2indefinitely—this protects the tower and foundation. In storms (approaching cut-out), thrust often drops further due to feathering; ultimate checks then focus on the governing combinations in the design standard (e.g., IEC 61400 load cases).

Beyond dead load and rotor thrust, a wind turbine tower is subject to several other important types of loads that influence both its design and its long-term durability. These loads may not always govern ultimate limit state (ULS) design, but they play a crucial role in fatigue assessment and site-specific safety checks.

One of these categories includes torque and yaw loads. Torque comes from the drivetrain and rotor operation, while yaw loads occur when the nacelle rotates to face the wind. Together, they add torsional stresses and secondary bending moments into the tower shell. Although they are generally smaller than the overturning moments caused by aerodynamic thrust, they are vital when evaluating fatigue because they introduce fluctuating stresses that accumulate over millions of cycles.

Site-dependent loads such as seismic forces and ice accretion also require careful consideration. In earthquake-prone regions, seismic loading can introduce dynamic horizontal forces that rival wind effects, while in colder climates, ice accumulation increases both mass and wind-exposed surface area. These effects can govern the design in specific regions even if they are not globally dominant.

Fatigue loading is one of the most significant long-term concerns. Unlike a one-time extreme load, fatigue results from repeated cyclic stresses caused by rotor imbalances, turbulence in the wind, yaw misalignment, and even normal control actions such as pitching the blades. These fluctuating stresses lead to the gradual degradation of steel structures. Particular attention is paid to areas with stress concentrations, such as door openings, flange connections, and welds. Engineers verify these details using hot-spot stress methods to ensure the tower can withstand decades of cyclic loading without failure.

In short, while the dead load and aerodynamic thrust are the major players for ULS strength checks, yaw, torque, seismic, ice, and especially fatigue loads are equally important for ensuring the tower's safe and reliable operation throughout its design life. These additional loads highlight why wind turbine towers are not just designed for strength but also for resilience against dynamic and environmental challenges.

Below are four short, fully-worked examples showing how "other loads" affect a wind-turbine tower.

1) Drivetrain torque and yaw torque → tower torsion (fatigue-relevant)

Assumptions (rated operating point):

- Rated electrical power: 5 MW
- Rotor speed at rated: 12 rpm
- Tower base: outside diameter D = 5.5 m, wall thickness t = 35 mm → radius R = 2.75 m
- Thin-walled circular tube polar moment of area (good first estimate):
  $J \approx (\pi D^3 t) / 4$

Step 1 — Compute rated rotor torque
Angular speed $\omega = 2\pi \cdot n/60 = 2\pi \cdot 12/60 = 1.257$ rad/s
Torque at rated $T\_rated = P/\omega = 5.0\times10^6 / 1.257 \approx 3.98$ MN·m

Emergency stops or control transients can create torque peaks of ~2× rated. Take T_peak = 8 MN·m transferred through the yaw bearing into the tower shell as torsion.

Step 2 — Torsional shear stress at the base
$J \approx (\pi D^3 t)/4 = (\pi \times 5.5^3 \times 0.035)/4$
$5.5^3 = 166.375 \rightarrow J \approx (\pi \times 166.375 \times 0.035)/4 \approx 4.57\ m^4$

Shear stress $\tau = T \cdot R / J = (8.0\times10^6\ N \cdot m \times 2.75\ m) / 4.57\ m^4$
$\tau \approx 22.0\times10^6 / 4.57 \approx 4.8$ MPa

Interpretation: That shear stress is small compared with bending stresses from thrust (often 80–150 MPa at the extreme fibre). It rarely governs ULS, but it cycles with starts, stops, and power fluctuations, so it matters for fatigue and for details around cut-outs/door openings.

2) Seismic equivalent static load (site-specific)

Assumptions:

- Total mass (tower + nacelle + rotor): $m \approx 550\ t = 5.50\times10^5$ kg
- Peak ground acceleration (PGA): $a = 0.20\ g = 0.20\times9.81 = 1.962\ m/s^2$
- Hub height H = 100 m
- Lumped mass split for a quick estimate: 70% at the top (nacelle/upper tower), 30% distributed over the height

Step 1 — Base shear from inertia
Total horizontal inertial force: $F = m\,a = 5.50\times10^5 \times 1.962 \approx 1.08$ MN

Step 2 — Overturning moment at base

- Top lump (70%): $F\_top = 0.7\times1.08 = 0.756$ MN at 100 m → $M\_top = 0.756\times100 = 75.6$ MN·m
- Distributed lump (30%): $F\_dist = 0.324$ MN acting at mid-height (≈ 50 m) → $M\_dist = 0.324\times50 \approx 16.2$ MN·m

Total M_base (seismic) ≈ 75.6 + 16.2 = 91.8 MN·m

Interpretation: For this PGA, the seismic overturning is on the same order as the wind-thrust example (~90 MN·m at 12 m/s). In high-seismic regions, seismic can govern certain checks (connections, anchor cage, grout, foundation).

3) Ice accretion – added weight and thrust

Assumptions:

- Three blades, each ~60 m long, average chord ~3 m → single-blade wetted area (both sides) ≈ 2 × 60 × 3 = 360 m^2
- Uniform radial ice thickness t_ice = 10 mm = 0.01 m over blade surface (simplified)
- Ice density ρ_ice ≈ 900 kg/m^3
- Baseline rotor thrust at 12 m/s from earlier example: 0.88 MN
- Hub height H = 100 m

Step 1 — Added mass per blade
Volume per blade ≈ Area × thickness ≈ 360 × 0.01 = 3.6 m^3
Mass per blade ≈ 3.6 × 900 ≈ 3.24 t
All blades: ~9.7 t extra → added gravity load ΔN ≈ 9.7×10^3 kg × 9.81 ≈ 95 kN

Step 2 — Effect on thrust and moment
Ice increases drag and roughness. If thrust rises by a modest +10%, new thrust T' ≈ 0.88 × 1.10 = 0.968 MN
Overturning: M_T' ≈ 0.968 × 100 = 96.8 MN·m (vs 88 MN·m baseline)

Interpretation: Even thin ice can add several tonnes and notably raise overturning. It also worsens balance, increasing fatigue at the base and at welded details.

4) Fatigue check with a simple S–N approach (illustrative)

Use a welded detail fatigue class (Eurocode-style) Detail Category 90 MPa at 2×10^6 cycles; slope m = 3 for the high-cycle regime.
Allowable cycles at a given stress range Δσ:

$$N\_allow = 2\times10^6 \times (\Delta\sigma / 90\ \text{MPa})^{-3}$$

Assume the tower-base extreme-fibre stress range from turbulence, yaw misalignment, etc., is approximately constant during operation (simplification). Consider 20-year life with n = 2×10^7 cycles in the governing range.

Case A: Δσ = 60 MPa

N_allow = 2e6 × $(60/90)^{-3}$
= 2e6 × $(0.667)^{-3}$ ≈ 2e6 × 3.375 ≈ 6.75×10^6 cycles

Damage D = n / N_allow = 2e7 / 6.75e6 ≈ 2.96 → Not acceptable (D > 1).

Case B: Δσ = 40 MPa

N_allow = 2e6 × $(40/90)^{-3}$
= 2e6 × $(0.444)^{-3}$ ≈ 2e6 × 11.39 ≈ $2.28×10^7$ cycles

Damage D ≈ 2e7 / 2.28e7 ≈ 0.88 → Borderline/acceptable (depends on partial factors, spectrum).

Case C: Δσ = 30 MPa

N_allow = 2e6 × $(30/90)^{-3}$
= 2e6 × $(0.333)^{-3}$ = 2e6 × 27 ≈ $5.4×10^7$ cycles

Damage D ≈ 2e7 / 5.4e7 ≈ 0.37 → Comfortably acceptable.

Interpretation: Small reductions in stress range produce large gains in fatigue life (m = 3). That's why controlling yaw misalignment, balancing the rotor, and smoothing torque/pitch actions are so important.

Foundations

The foundation of a wind turbine is a critical component that anchors the tower to the ground or seabed while safely distributing the immense loads generated by the structure. The choice of foundation depends on whether the turbine is installed onshore or offshore, as soil stability, terrain, and environmental forces vary greatly between locations.

Figure 100: A turbine foundation at Toddleburn Wind Farm (UK). Walter Baxter, CC BY-SA 2.0, via Wikimedia Commons.

For onshore turbines, the most common type is the gravity foundation, also known as shallow spread footings. This design typically consists of a large circular or octagonal reinforced concrete slab, 15–25 meters in diameter, buried several meters below the surface. Its mass and size prevent the tower from tipping under high wind loads. In areas with weaker or unstable soil, pile foundations are used instead. These involve long reinforced concrete or steel piles driven deep into the ground to transfer loads to bedrock or stronger soil layers. In mountainous or rocky terrain, rock-anchored foundations are preferred. Here, the tower is bolted directly into anchors drilled into the rock, providing excellent stability.

Figure 101: Wind turbine base and access stairs. Patrick Finnegan, CC BY 2.0, via Flickr.

Generally, each wind turbine is secured to the ground using what is known as a gravity foundation. This type of foundation relies on its immense weight to keep the turbine stable, ensuring that the forces generated by the rotating blades and the tower structure are safely transferred into the ground. The engineering design of these foundations is highly specialised, as they must account for site-specific conditions such as soil type, bedrock depth, groundwater levels, and local seismic or climatic risks. A well-constructed foundation is crucial because even the most advanced turbine will not perform effectively or safely if it is not anchored to a stable and durable base.

The construction process begins with a large excavation at the turbine site. Once the excavation is prepared, a flat "blinding" layer of concrete is poured at the bottom. This thin base layer acts as a clean, level platform that ensures the main foundation can be formed on a smooth and stable surface. Without this step, irregularities in the soil could compromise the precision and strength of the foundation.

After the blinding layer sets, a reinforced steel frame is constructed by hand within the excavation. This reinforcement, which uses around 105 tonnes of steel rebar, forms a dense cage that provides the tensile strength needed to complement the compressive strength of the concrete. Because concrete is strong under compression but weak under tension, the rebar is essential to ensure the foundation can resist the bending and torsional stresses caused by turbine operation and wind forces.

Once the steel frame is in place, the foundation is filled with concrete. Approximately 1,700 tonnes of concrete are poured continuously to form a single, solid block. Continuous pouring is critical because it prevents cold joints (weak lines where separate pours would meet), maximising the structural integrity of the foundation. When the pour is complete and the concrete cures, the foundation becomes a monolithic structure capable of withstanding decades of fluctuating loads.

Following the curing process, the excavation is backfilled with aggregate material around the sides of the foundation. This backfilling not only reinstates the ground level but also adds significant weight, further anchoring the turbine against overturning forces. The combined mass of the concrete, steel, and backfill ensures that the structure remains stable even in extreme wind conditions.

The finished foundation is left as a flat, solid platform that serves multiple purposes. It provides a stable base to support the tower once it is erected, while also functioning as a storage area for turbine components before installation. Additionally, the foundation acts as a crane pad, allowing the heavy lift cranes and associated equipment to operate safely and effectively during the erection process. This dual role demonstrates the careful planning behind foundation construction, where both the long-term stability of the turbine and the practical logistics of installation are accounted for in the design.

The choice of foundation type depends heavily on site conditions such as soil strength, rock depth, and groundwater levels. Broadly, foundations fall into two categories: spread foundations and piled foundations. Regardless of type, a carefully designed interface embedded in the concrete is required between the foundation and the tower to ensure strong connectivity, alignment, and load transfer. In some cases, the interface itself dictates which foundation type can be used.

Spread Foundations: Spread foundations distribute the load from the turbine tower over a wide area of soil, much like a slab foundation. By increasing the footprint of the base, the pressure exerted on the soil is reduced, preventing settlement and increasing stability. These foundations are generally made of reinforced concrete and are often circular or square in plan, depending on the site layout and load requirements. The large plate-like structure spreads the tower's forces evenly across the soil, ensuring reliable performance in locations with competent ground conditions.

Gravity Foundation: A gravity foundation is the most common spread foundation type used for turbines. It is constructed by excavating soil to a certain depth, placing reinforced concrete at the base, and then backfilling around the structure with soil or aggregate. The excavation depth and backfilling serve an important role: the overlying soil adds significant weight that enhances resistance against overturning caused by wind loads. This makes gravity foundations both practical and sustainable, as they balance stability with relatively efficient use of concrete. However, the process does require large amounts of excavation and refilling, which can increase construction time and costs. Importantly, gravity foundations must be built on strong soil layers; if the soil is too weak or compressible, they may not provide adequate stability.

Pile Foundations: When soil conditions are weak, soft, or variable, spread foundations may not be sufficient. In these cases, pile foundations are often used. Piles are long, slender columns—made of reinforced concrete or steel—driven or drilled into deeper, stronger soil layers or into bedrock. These piles transfer the loads of the turbine through weaker surface soils to more competent strata below, reducing settlement and increasing resistance to overturning forces. Pile foundations are particularly useful for large turbines in areas with poor ground conditions, as they can sustain both compressive and tensile forces arising from wind-induced bending moments.

The connection between the piles and the turbine's base plate is crucial. Two main connection types exist: clamped and hinged. In clamped connections, the pile head is rigidly fixed to the plate, causing the pile to resist large bending moments. In hinged connections, the pile is free to rotate at the top, meaning bending moments are reduced but vertical and axial loads remain critical. Designers must balance these factors to ensure safe and efficient load transfer.

Placement of Piles on Bedrock: In some sites, bedrock is found at relatively shallow depths. Here, piles can be driven or socketed into the bedrock, effectively eliminating settlement issues that typically arise from soil compression. In this case, settlement is governed by pile deformation rather than soil movement, ensuring high structural stability. If required, piles may also be anchored into the bedrock to resist tension forces generated by wind actions. However, determining the load-carrying capacity of anchored piles is complex, requiring advanced geotechnical expertise, accurate site investigations, and sometimes pull-out testing to validate assumptions.

Piled Raft Foundations: A hybrid system, known as a piled raft foundation, combines the benefits of spread and pile foundations. In this approach, a spread foundation distributes loads across the surface soils while a group of piles transfers excess loads deeper into the ground. The interaction between the raft and piles must be carefully designed to prevent uneven settlement and maximise efficiency. This is usually achieved using advanced finite element modelling to account for soil stiffness, raft stiffness, and pile group capacity. Piled raft foundations are particularly effective in soils of intermediate strength, where neither spread foundations nor pile groups alone provide sufficient stability.

For offshore wind turbines, foundation design is far more complex due to hydrodynamic forces, wave loading, and variable seabed conditions. The most common option for shallow waters (up to about 30 meters) is the monopile foundation, which uses giant steel piles, 4–10 meters in diameter and up to 40 meters long, driven deep into the seabed. Another option is the gravity base foundation, where massive concrete caisson-type structures rest on the seabed, relying on their sheer weight for stability. In deeper waters (30–60 meters), jacket foundations—lattice-style steel frames secured with piles—are often used. For very deep waters beyond 60 meters, floating foundations are employed. These include spar-buoys, semi-submersibles, or tension leg platforms anchored with mooring lines. Though still an emerging technology, floating foundations are rapidly advancing and expanding deployment opportunities in deep offshore regions.

Figure 102: Rambiz with gravity based foundation of wind turbine. Luc Van Braekel, CC BY 2.0, via Flickr.

The foundations used for offshore wind turbine towers are designed to suit a wide range of marine conditions, with the choice of foundation largely determined by water depth, seabed geology, and distance from shore. Offshore foundations are broadly categorized according to sea depth: shallow waters (0–30 m), transitional waters (30–50 m), and deep waters (50–200 m). The cost of installation increases significantly with depth, making the selection of foundation type a critical factor in both economic feasibility and long-term sustainability. While gravity and monopile foundations are common in shallow waters, multipod structures (tripods and jackets), suction caissons, and floating substructures are increasingly used in deeper waters where traditional methods become uneconomical or technically challenging.

Gravity Foundations: Gravity-type foundations were the earliest form used in offshore wind projects. They rely on their massive self-weight for stability, consisting of a large concrete base that sits directly on the seabed. These foundations are particularly suitable for shallow waters near the shore, where construction logistics are simpler and seabed conditions are relatively stable. Their use was widespread in the first generation of offshore wind farms because they leveraged well-established civil engineering techniques and reduced risks associated with new technologies. However, their cost and difficulty of construction increase sharply beyond depths of 10 meters, making them less favourable in deeper waters.

Monopile Foundations: Monopiles are now the most common offshore foundation type, particularly in Europe's North Sea. They consist of a single, large-diameter steel pile driven deep into the seabed. Monopiles are favoured in shallow waters (less than 30 meters), especially where seabeds are composed of sand or gravel, which provide ideal conditions for pile driving. Their popularity comes from being cost-effective, relatively simple to install, and requiring less material than alternative solutions. However, their installation involves jack-up barges and pile-driving equipment, which generate high levels of noise, vibration, and suspended sediment. These disturbances can negatively impact marine ecosystems, making environmental considerations a limitation of this method.

Suction Caisson Foundations: Suction caissons are an innovative, eco-friendly alternative to monopiles. Resembling an upside-down bucket, they are installed by creating suction that draws the caisson into the seabed. This method requires minimal heavy machinery and causes little vibration, noise, or sediment disturbance, greatly reducing environmental impacts. Suction caissons are also cost-efficient and quick to install, with significant savings in steel use compared to monopiles. For example, tripod suction caissons can cost roughly half as much as comparable jacket-pile foundations for the same seabed conditions. These advantages make suction caissons an excellent sustainable option for offshore wind farms, particularly in areas where ecological sensitivity is a priority.

Multipod Foundations (Tripod and Jacket): Multipod foundations, which include tripod and jacket structures, are well-suited for transitional water depths of 30–50 meters. These space frame structures distribute loads efficiently while keeping material usage relatively low. Because they require only shallow penetration into the seabed, they help reduce construction costs in intermediate depths. Multipods are also more resilient to extreme weather events, such as hurricanes or typhoons, compared to monopiles or suction caissons. Their increased stiffness and embedment depth make them especially reliable in challenging conditions, though they are typically more complex to fabricate and install.

Floating Substructures: For deep waters (50–200 meters), floating substructures are increasingly seen as the future of offshore wind energy. These foundations do not rest directly on the seabed but are anchored using mooring lines or ballast systems. They are classified into three types: ballast stabilized, mooring line stabilized, and buoyancy stabilized. Floating foundations allow turbines to be installed far from shore, where wind resources are stronger and more consistent. They are more economical in deep water than traditional seabed foundations, as they avoid costly and technically demanding pile driving or concrete works. A stabilized barge is a common form of buoyancy-stabilized foundation, kept in place with anchors attached to the seabed. While still under active development and commercial demonstration, floating structures promise to expand offshore wind capacity into regions previously unsuitable due to depth constraints.

Offshore wind turbine foundations are selected based on depth, seabed conditions, and cost considerations. Gravity and monopile foundations dominate in shallow waters, while suction caissons and multipod structures provide effective solutions for transitional depths. In deep waters, floating substructures represent the most sustainable and economically viable pathway for expanding offshore wind energy. Each foundation type carries its own balance of cost, environmental impact, and technical feasibility, but all contribute to the overall sustainability of offshore wind power generation.

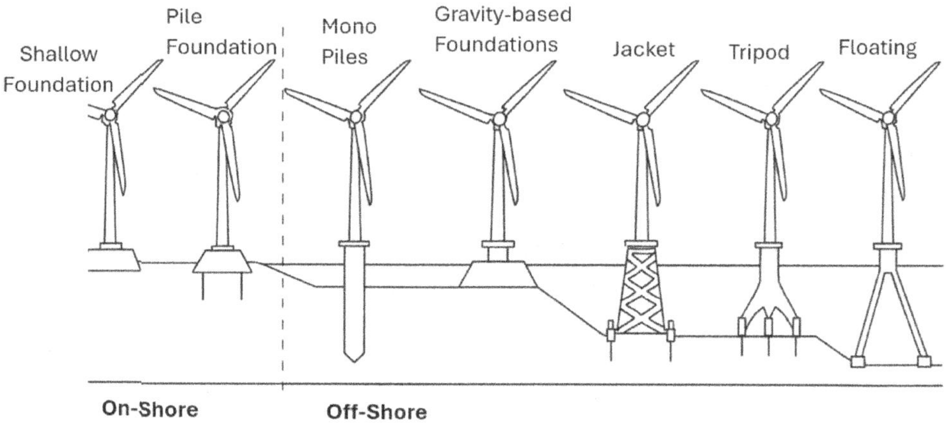

Figure 103: Foundations used for on-shore and off-shore wind turbines.

There are several key engineering challenges associated with wind turbine foundations. One is fatigue resistance, as towers and foundations must withstand millions of load cycles during their service life. Another is corrosion protection, especially offshore, where advanced coatings, cathodic protection, and specialized materials are required to combat saltwater exposure. Transport and installation also pose major challenges, since tower sections and foundation structures are enormous, requiring heavy-duty ships, cranes, and transport vehicles. Finally, dynamic stability must be ensured, meaning both tower and foundation are designed to minimize resonance with rotor frequencies, which could otherwise lead to destructive vibrations.

The design of offshore wind turbine foundations has its roots in the experience gained from offshore oil and gas structures. However, despite this shared origin, offshore wind foundations present unique challenges that require distinct engineering approaches. The primary purpose of the foundation is to safely transfer loads from the turbine's tower and nacelle to the seabed. These loads arise from environmental conditions such as wind, waves, and currents, as well as from the dynamic operation of the turbine itself. Unlike oil and gas platforms, which primarily deal with steady loads and occasional wave-induced forces, offshore wind turbines are highly dynamic structures. The slenderness of their towers and the heavy, rotating mass of the blades and rotor at the top create a system that is very sensitive to resonance effects.

A fundamental distinction lies in the natural frequencies of the two types of offshore structures. For instance, the natural frequency of a typical 3.6 MW wind turbine system is approximately 0.3 Hz, and for larger 8 MW turbines it is closer to 0.22 Hz. This is strikingly close to the excitation frequency of the rotor itself, which operates around 0.2 Hz. This means that offshore wind turbines are dynamically sensitive, as their natural frequencies can overlap with those of operational or environmental loads. In contrast, offshore oil and gas platforms have natural frequencies above 0.6 Hz, with the most significant wave-induced frequencies at about 0.1 Hz. Because their excitation frequencies do not align with their natural frequencies, oil and gas platforms are less susceptible to dynamic amplification. The heavy rotating blades of a turbine—each weighing roughly 30 tonnes—create significant cyclic loads that must be considered in foundation design.

When examining load transfer mechanisms, the differences between monopile foundations and jacket structures become clear. A monopile is a single, large-diameter steel pile driven into the seabed, and it resists loads primarily through lateral soil interaction. The loads are transferred as overturning moments, making monopiles highly dependent on the stiffness of surrounding soils. In contrast, jacket structures distribute loads through multiple piles. Here, the dominant mechanism is push–pull axial action, with piles alternately resisting tension and compression forces. This difference explains why monopiles are generally effective in shallow waters with suitable seabed conditions, while jackets are used in deeper waters where axial load transfer provides greater stability and efficiency.

Floating wind turbine systems introduce yet another layer of complexity. One of the most well-known examples is the spar-supported floating turbine, such as the Hywind project. These structures use catenary mooring systems anchored to the seabed with suction caissons. The design must consider the extreme case where mooring lines are fully stretched, with no section resting on the seabed to absorb friction. In this scenario, loads are transferred directly to the anchor with maximum angles of inclination, making it essential to calculate the ultimate load-bearing capacity of the anchor system. This is a situation not typically encountered in floating oil and gas platforms, which are designed with different load-sharing mechanisms in mind.

Offshore wind foundation design must also account for a uniquely complex set of loadings. Standards such as DNVGL-ST-0437 require designers to evaluate both ultimate limit states (ULS) and fatigue limit states (FLS). Loads include aerodynamic forces from the blades, hydrodynamic forces from waves and currents, hydrostatic forces on submerged parts of the structure, gravitational and inertial loads from the rotating system, and additional considerations such as sea ice, seismic activity, and even boat collisions. Each of these load types must be combined into realistic time series data to simulate the behaviour of the turbine under thousands of possible scenarios.

The computational effort required for this process is immense. A typical fatigue design might require between 5,000 and 10,000 dynamic simulations, while ultimate strength design can involve 10,000 to 15,000 simulations. Each simulation often runs for 600 seconds of model time, with time steps as small as 0.01 seconds. This results in between 12,000 and 60,000 load cases that need to be evaluated for each design. Managing the data, generating input files, and ensuring accuracy across this scale of analysis is both time-consuming and highly susceptible to human error. This highlights one of the central challenges of offshore wind turbine design: the intersection of highly dynamic structural behaviour, complex environmental loads, and the need for robust, error-free computational methods to ensure the safety and sustainability of these vital renewable energy systems.

Access Ladders, Lifts, and Platforms

Access ladders, lifts, and platforms are essential parts of wind turbines, designed to ensure that technicians can safely reach, maintain, and repair critical components located high above the ground or water. Because modern onshore and offshore turbines can exceed 100 meters in hub height, these internal access systems must be robust, reliable, and compliant with international safety standards. They form part of the turbine's Health, Safety, and Environment (HSE) strategy and are crucial for efficient operation and reduced downtime.

Figure 104: Wind turbine access door. Patrick Finnegan, CC BY 2.0, via Flickr.

Access Ladders

Inside the tower of most turbines, vertical ladders provide the primary means of access from the base to the nacelle. These ladders often extend the full height of the tower, segmented with intermediate rest platforms every 10–20 meters for safety and recovery. To reduce fatigue and risk, they are fitted with fall-arrest systems, typically rails or wire systems to which technicians attach harnesses. In some turbines, particularly older or smaller models, ladders are the only means of ascent. Climbing such ladders can take 15–20 minutes, depending on tower height, which makes ergonomics and safety features critical.

Figure 105: Ladder providing access to the nacelle of a wind turbine in Lower Austria. The view is from bottom to top, showing the safety rope for the climb. Überraschungsbilder, CC BY-SA 3.0, via Wikimedia Commons.

Fall arrest systems inside wind turbine towers are essential safety mechanisms that protect technicians during climbs on the vertical ladders. Because the ladders often extend 80–120 meters in modern turbines, the risk of a slip or fall is significant, and a fall arrest system ensures that if a technician loses their footing or grip, they will not fall the full height of the tower. These systems are designed to stop a fall quickly, minimize impact forces on the body, and allow safe recovery.

In many modern turbines, the ladder is equipped with a rigid vertical rail mounted alongside or in the centre of the ladder. Technicians wear a full-body harness fitted with a guided fall arrester device (also called a glider or traveller). This device connects the harness to the rail and allows smooth upward or downward movement as the technician climbs. If the climber slips, the device instantly locks onto the rail through a cam or braking mechanism, halting the fall within a very short distance. Rail systems are considered highly reliable and durable because the rigid track provides consistent engagement and is less affected by wind tower vibrations or environmental factors such as dust or moisture.

Some towers, especially older models, use a vertical wire rope (cable) instead of a rigid rail. The harness is connected to the wire with a cable sleeve or shuttle, which slides freely as the technician climbs. In the event of a fall, the device locks onto the cable, either through centrifugal brakes, a cam mechanism, or friction clamps, stopping the descent. Wire systems are flexible, easier to retrofit, and often cheaper than rigid rails. However, they may require more maintenance to prevent issues such as cable wear, tension loss, or corrosion—particularly in offshore environments where saltwater exposure accelerates degradation.

In both rail and wire systems, technicians wear a certified fall-arrest harness that distributes forces across the thighs, shoulders, and chest, reducing injury risk. The harness is connected to the rail or cable system using a carabiner or locking connector, ensuring that the climber remains continuously attached during ascent and descent. Technicians are trained to check that their fall-arrest device is correctly oriented and locked before beginning a climb.

Figure 106: Climbing Wind Turbine Ladder. Idaho National Laboratory, CC BY 2.0, via Wikimedia Commons.

When activated, the system halts the fall within 30–60 centimetres, preventing free-fall acceleration and reducing the forces transmitted to the body. This is critical because unrestrained falls, even over short distances, can generate forces capable of causing severe injury. By keeping the arrest distance short, the system protects the technician from hitting intermediate rest platforms or structures within the tower.

Intermediate platforms every 10–20 meters allow climbers to rest while still connected to the fall arrest system. These platforms also serve as staging points in case of a rescue: if a technician becomes suspended after a fall, colleagues can reach them from a platform to carry out a controlled descent or recovery. Rescue plans are an integral part of turbine safety procedures, ensuring that a fallen technician is not left suspended, which can cause suspension trauma.

In short, fall arrest systems in wind turbine towers combine a personal harness, a guided fall arrester device, and either a rigid rail or wire system. Together, they provide continuous protection during climbs, ensuring that technicians can safely ascend and descend the long ladders required to reach the nacelle.

Service Lifts

To reduce climbing fatigue and speed up access, many modern wind turbines are equipped with service lifts (also known as personnel hoists). These lifts can carry one or two technicians at a time, along with hand tools, to different levels of the tower. They are designed to withstand the vibrations and movements of the turbine while maintaining safe operation. Service lifts dramatically reduce physical strain, lowering the risk of musculoskeletal injuries, and allow faster response times for maintenance tasks. In offshore wind farms, where technicians may already experience fatigue from transfer by boat or helicopter, lifts are considered vital for health and productivity.

Figure 107: Service Lift.

Platforms and Rest Areas

Intermediate platforms are installed at various heights within the tower. These serve several purposes: they provide rest points during climbs, enable technicians to stabilize themselves in emergencies, and act as workspaces for certain tasks like cable inspections. Platforms are also essential for rescue operations, allowing technicians to recover safely in the event of medical or mechanical emergencies. At the top of the tower, the yaw deck and nacelle platform give access to key components such as the yaw motors, main shaft, gearbox, and generator. External platforms on the nacelle roof are also common, providing technicians with safe access for blade inspections or crane attachment points during major component exchanges.

Safety and Ergonomics

All access systems—ladders, lifts, and platforms—must meet international safety standards such as ISO 13857, EN 50308, or OSHA regulations, depending on the jurisdiction. Safety features include non-slip rungs, self-closing gates on platforms, fall-arrest anchor points, and emergency descent devices in case of lift failure. In offshore turbines, additional safety features include sealed hatches, watertight access doors, and sometimes motion-compensated ladders at the base to handle wave action during technician transfer from service vessels.

As turbines increase in size, with hub heights surpassing 150 meters in next-generation models, the reliance on lifts rather than ladders is growing. Manufacturers are also investing in automation, such as robotic inspection systems, to reduce the need for technicians to climb towers at all. Still, ladders, lifts, and platforms remain essential, as many tasks require direct human presence. The design of these access systems continues to evolve, focusing on minimizing technician fatigue, improving rescue procedures, and ensuring that turbine towers remain safe and accessible workplaces.

Corrosion Protection and Coatings

Corrosion protection and coatings are essential for the design, construction, and maintenance of wind turbines, particularly those installed offshore, which face significant corrosion challenges due to exposure to saltwater, humidity, UV light, and mechanical stresses from winds and waves [197, 198]. The importance of effective corrosion protection cannot be overstated, as failure to implement proper measures can substantially degrade the structural integrity of turbine components, ultimately shortening their serviceable life and increasing operational costs [198, 199].

Wind turbines are engineered for operational lifespans of approximately 20–30 years. However, materials such as steel, which are commonly used in turbine structures, are prone to deterioration if not effectively protected. This concern is exacerbated in offshore installations, where corrosion due to the saline marine environment poses significant threats [199, 200]. Additionally, even onshore turbines located in coastal or humid regions experience significant corrosion risks due to environmental factors, necessitating the application of protective coatings and other mitigation strategies to ensure longevity and efficient maintenance [198, 201].

The multilayer protective coating system is widely accepted as a comprehensive solution for safeguarding steel towers and foundations from corrosion. This system typically comprises a primer layer, often zinc-rich for sacrificial protection, an intermediate epoxy or polyurethane layer to resist moisture and chemicals, and a durable topcoat that provides UV resistance and abrasion protection [200, 202]. For offshore turbines, the conditions are particularly aggressive; thus, specialized coatings are needed to withstand both submersion and exposure in the splash zone, which is prone to cyclic wetting and drying [197, 199].

Moreover, conventional protective measures are often supplemented with cathodic protection systems, which include sacrificial anodes made of zinc or aluminium that corrode preferentially to the steel or Impressed Current Cathodic Protection (ICCP), which employs a current to mitigate corrosion [199]. However, these systems must undergo regular inspections due to their propensity to require maintenance as sacrificial anodes deteriorate over time [198].

Wind turbine blades, constructed predominantly from glass or carbon fibre composites, encounter different degradation mechanisms than steel towers. These materials are vulnerable to UV degradation, erosion from environmental elements, and particulate impacts. To address these issues, coatings such as gelcoats and specialized erosion-resistant tapes are employed to enhance weather resistance and mitigate material loss from impacts [203, 204]. Novel coating technologies, including nanocoatings and hydrophobic treatments, have also gained traction, particularly in cold climates where icing is a risk [203, 204].

Research indicates that the implementation of robust corrosion protection increases the operational efficiency of wind turbine structures while lowering long-term maintenance costs [205]. Therefore, integrating advanced materials and coatings into the design of wind turbines, combined with regular monitoring and maintenance protocols, is vital for extending the service life of these renewable energy assets [205, 206].

The delivery of effective corrosion protection for offshore wind turbines is a collaborative effort involving multiple parties. The steel builder, the paint applicator, and the paint supplier must work closely together to achieve durable and reliable results. Each plays a critical role: the steel builder provides a design and structure suitable for painting, the applicator ensures that the coating is applied correctly under proper conditions, and the paint supplier delivers systems that meet the specific demands of offshore environments. Any failure in coordination can undermine the entire process, leading to premature coating failure and costly repairs [207].

Corrosion protection is always the final step in the production process, which means that the paint shop often works under pressure to compensate for schedule delays. However, rushing the painting process is a dangerous mistake. While painting in a controlled shop environment typically costs between $20–$32 per square meter, repairing failed coatings offshore can be 40–60 times more expensive [207]. Offshore repair not only requires specialized equipment and access but also exposes workers to greater safety risks and turbines to extended downtime. This makes quality in the initial application process critical to long-term economic and structural sustainability.

Environmental exposure also greatly influences the severity of corrosion. Onshore turbines are generally subjected to cyclic dew and condensation combined with moderate sunlight exposure, leading to moderate corrosion risks at weak points in the coating. Offshore turbines, by contrast, face constant high humidity, saltwater spray, ultraviolet radiation, and the highly corrosive splash zone where waves impact the structure. These conditions accelerate corrosion dramatically. For example, unprotected steel in Germany corrodes at a rate of 25–50 µm thickness loss per year onshore (C3 corrosivity according to ISO 12944), but in offshore splash zones, corrosion rates as high as 500 µm per year have been observed [207]. This stark difference underscores why offshore turbines require more robust systems and stricter application practices.

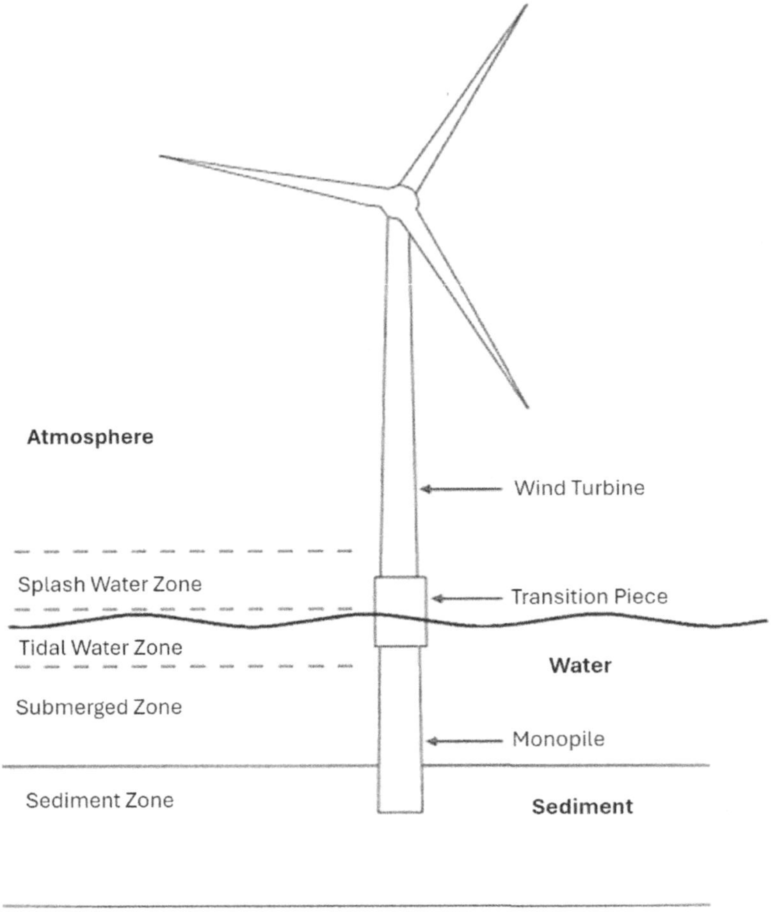

Figure 108: Corrosion Zones.

A well-tested coating system alone does not guarantee success. Structural design plays an equally important role. If the steel builder does not account for accessibility, certain surfaces may be impossible to coat properly. Designers must ensure adequate spacing for tools, appropriate dimensions of openings, and features that prevent water or deposits from collecting. Sharp edges, poor weld seams, and surface irregularities must be avoided, as they create weak points where coatings can fail and corrosion can spread rapidly. Weld seams in particular must be clean and well-profiled, since poor workmanship in this area accelerates degradation [207].

The quality of application is another decisive factor. Studies have shown that roughly half of premature coating failures are linked to application issues rather than the coating system itself. Correct surface preparation, ensuring the right level of cleanliness and roughness, as well as applying coatings to the correct thickness, are essential for success. Frequent quality checks and rigorous adherence to specifications are therefore non-negotiable. The best outcomes are achieved with automated blasting facilities, climate-controlled painting booths, and advanced two-component spray equipment that minimizes mixing errors. This equipment must be operated by skilled and certified applicators to reduce the likelihood of human error [207].

Finally, the growing competition in the market for wind turbine coatings introduces further challenges. There is increasing demand for coatings that dry faster, are solvent-free, and can be applied in fewer coats. While these innovations can improve efficiency and reduce costs, they also heighten the risk of compromising long-term performance if not carefully matched to real-world conditions. Laboratory-tested systems often deliver excellent results under controlled circumstances, but replicating that same performance on thousands of square meters of steel, applied continuously and sometimes under less-than-ideal field conditions, is far more difficult. This gap between theory and practice highlights the importance of coordination, skilled labour, and realistic planning in ensuring that corrosion protection for offshore wind turbines is both technically sound and economically sustainable [207].

Corrosion protection for wind turbines, especially those installed offshore, requires a combination of strategies to safeguard steel and other structural materials from aggressive environmental conditions. The three most common approaches are passive cathodic protection, active cathodic protection, and protective coatings. Each plays a distinct role, and they are often used together to ensure the long service life expected of modern turbines.

Passive Cathodic Protection (Sacrificial Anodes): Passive cathodic protection works on the principle of galvanic corrosion, where a more reactive metal is allowed to corrode in order to protect the steel structure. These more reactive materials, known as sacrificial anodes, are typically made from zinc, aluminium, or magnesium alloys. When attached to parts of the turbine such as the monopile foundation or transition piece, they corrode preferentially and release electrons that flow to the steel, preventing it from oxidizing.

Sacrificial anodes are widely used in offshore foundations, particularly in submerged and splash zones. They offer continuous protection without the need for external power, which makes them a reliable and relatively simple solution. However, they degrade over time and must be replaced periodically, adding to maintenance requirements. Their performance can also be influenced by factors such as salinity, water temperature, and current flow.

Active Cathodic Protection (Impressed Current Cathodic Protection – ICCP): Active cathodic protection uses an external power source to prevent corrosion. In this system, inert anodes—often titanium coated with mixed metal oxides—are connected to a DC power supply. A low-voltage current is impressed into the seawater, which forces the steel structure to behave as a cathode. Since cathodes do not corrode, the steel remains protected for as long as the system is active.

Offshore wind farms often use ICCP alongside coatings, particularly for large monopile and jacket foundations where sacrificial anodes alone would be too heavy or impractical. ICCP offers controlled and adjustable protection that can last decades with proper monitoring and reduces the structural weight compared to passive systems. The downside is that ICCP requires continuous power, monitoring, and maintenance. If the system fails, corrosion protection is lost, which is why redundancy and alarms are essential.

Protective Coatings: Protective coatings form the first line of defence against corrosion by creating a physical barrier that prevents oxygen, moisture, salts, and pollutants from reaching the steel surface. These systems typically consist of multiple layers. Epoxy coatings provide adhesion and chemical resistance, while polyurethane topcoats offer UV and abrasion protection. Zinc-rich primers add both barrier and galvanic protection, and specialized offshore systems often combine these into multilayer arrangements designed for splash-zone exposure and seawater immersion.

In wind turbines, towers and transition pieces are usually coated with thick multilayer systems, with extra reinforcement in the splash zone. Blades are treated with gelcoats and leading-edge erosion-resistant coatings, while internal tower spaces are coated to prevent condensation-related corrosion. Coatings are versatile and can be tailored to different components, but their effectiveness depends heavily on proper surface preparation, environmental conditions during application, and workmanship. Poor application is responsible for nearly half of premature coating failures.

Integrated Protection Strategy: In practice, corrosion protection systems are rarely used alone. Instead, they are combined to create redundancy. Protective coatings form the primary barrier, sacrificial anodes provide backup in submerged areas, and ICCP can be used to reduce the need for anode replacements and extend inspection intervals. This integrated approach ensures that if one system fails, another will take over. For instance, if a coating is damaged and bare steel is exposed, the cathodic system prevents corrosion from initiating at the defect.

The most effective corrosion protection for wind turbines combines barrier systems with electrochemical methods. Passive cathodic protection offers a simple and reliable solution, active systems provide precise and scalable control, and coatings deliver tailored surface protection. Together, these methods ensure that turbine structures remain durable throughout their intended service life, reducing maintenance costs and protecting one of the most valuable assets in the renewable energy industry.

Maintenance

The maintenance of corrosion protection on a wind turbine is a systematic process designed to ensure the tower, blades, nacelle, and offshore foundations remain safe and durable throughout their expected 20–30 year service life. Given the exposure of turbines to harsh environments—particularly offshore with constant salt spray, high humidity, and intense UV radiation—regular inspections, preventive measures, and timely repairs are essential. The overall process involves several interconnected stages.

The first stage is inspection and monitoring, which serves as the foundation of corrosion management. Technicians carry out visual inspections to identify peeling, cracking, rust staining, blistering, or damage to coating systems, paying particular attention to critical areas such as weld seams, joints, splash zones, ladders, and boat landings. Coating thickness is measured with specialized gauges to verify compliance with protective standards, while sacrificial anodes are checked for size and consumption levels to determine their remaining service life. For ICCP systems, electrical readings are logged to ensure protection is functioning correctly. Inspections also extend to blades, where gelcoat erosion, UV degradation, and cracks are identified using rope access or drone technology. These checks are typically annual, though they may also follow storms or unexpected events. Increasingly, drones, ROVs, and corrosion sensors are used to minimize manual risk.

When issues are found, the next stage is surface preparation and cleaning. Surfaces must be free of rust, salt, old paint, and contaminants before repairs can be made. Techniques include abrasive blasting, power tools, and high-pressure water jetting. Offshore, additional containment measures are required to avoid environmental contamination. Sharp edges or poorly finished welds may also be dressed and smoothed to prevent coating breakdown. Proper preparation is vital, as it ensures new coatings bond effectively and last longer.

The third stage is coating repair and reapplication. Repairs are carried out with systems that match or are compatible with the original coating. Small defects may be corrected with localized spot repairs, while larger areas—particularly in high-wear splash zones—often require full recoating. Blade surfaces may need erosion repairs using fillers, tapes, or specialized coatings, sometimes applied in temporary blade shelters to control curing. Offshore repairs are particularly difficult and costly, often 40–60 times more expensive than shop-based applications, which reinforces the importance of preventive maintenance.

Another essential element is cathodic protection maintenance. Sacrificial anodes are routinely inspected, with replacements scheduled as their material is consumed. ICCP systems require continuous monitoring of current output, voltage, and electrode readings, with adjustments made as necessary. Together, coatings and cathodic systems provide layered protection: if a coating is damaged, the cathodic system acts as a backup to prevent immediate corrosion until repairs are completed.

Throughout the process, environmental and safety considerations play a central role. Onshore towers are maintained using fall-arrest systems, internal lifts, or rope access on external surfaces. Offshore, work often requires crew transfer vessels (CTVs), service operation vessels (SOVs), and sometimes ROVs for underwater inspections. Waste materials such as old coatings, blasting media, or corroded debris must be collected and disposed of responsibly to prevent environmental harm, especially in marine settings.

Finally, documentation and life-cycle planning ensure long-term success. Every inspection, repair, and replacement is recorded, creating a detailed history of coating performance, anode consumption, and ICCP system operation. Digital maintenance systems are often used to schedule inspections, track defects, and plan recoating campaigns. These records allow operators to anticipate major interventions, budget effectively, and extend the service life of assets.

In summary, corrosion protection maintenance for wind turbines is a continuous cycle of inspection, preparation, repair, and monitoring. Onshore this process is simpler, but offshore it is considerably more challenging and costly, making preventive care absolutely vital. Protective coatings form the first barrier, while cathodic protection—both sacrificial anodes and ICCP systems—provides backup. Blade coatings further safeguard against erosion and structural damage. When combined, these strategies ensure that wind turbines remain efficient, safe, and operational across their intended lifespan.

Lightening and Surge Protection

Lightning Protection Zones (LPZ) in Wind Turbines

Defining lightning protection zones (LPZ) is a key step in ensuring that surge protection systems within a wind turbine are effective and properly coordinated. Each LPZ represents a boundary where the level of electromagnetic exposure changes, helping engineers determine the correct placement of protective devices such as lightning protection systems (LPS), cable shielding, and surge protective devices (SPDs). The concept is outlined in IEC 62305-1, which provides guidelines for classifying and designing these protection zones based on the potential severity of lightning effects [208].

Tower: In most large wind turbines, the tower is a tubular metal structure that acts almost like a Faraday cage, providing excellent electromagnetic shielding. This means that the electrical field inside the tower is largely isolated from external lightning effects. When the tower is properly grounded and has a secure electrical connection to the nacelle, the interior environment of the tower can be classified as LPZ 1 or even LPZ 2, indicating that it experiences very low electromagnetic interference and minimal risk from lightning-induced surges.

Nacelle: The nacelle, which houses the main mechanical and electrical components (including the gearbox, generator, and control systems), typically incorporates metallic or mesh shielding within its fiberglass structure. This provides partial protection against external electromagnetic fields. The nacelle is generally defined as an LPZ 1 zone, where some lightning current may still flow through the bearings and other conductive paths. To safely direct this current, it is discharged to the ground via conductive brushes or grounding cables. However, because the nacelle's shielding is not perfect, there may still be some residual electromagnetic radiation, which must be considered in surge protection planning.

Power Distribution Room or Cabinet: In onshore wind turbines, transformers and power distribution units are often located outside the tower and are categorized as LPZ 1 zones. In offshore installations, these components are typically placed inside the tower base, providing an additional layer of protection against the environment. In this case, the area can be defined as LPZ 2/3, depending on how effectively it is shielded and grounded.

Hub: The hub — the central component connecting the rotor blades — is typically a hollow cast-iron structure that offers a degree of magnetic shielding. This section is generally considered part of LPZ 1. The hub also contains the pitch control system and various control cables that connect to the nacelle, making it critical to install SPDs and grounding paths to prevent overvoltage damage in this area.

LPZ Division and Surge Protection: The division of a wind turbine into different LPZs allows engineers to evaluate how lightning strikes and surges will affect each subsystem. Components such as blades, generators, and control electronics are designed to withstand the electrical and electromagnetic stresses expected in their designated zones.

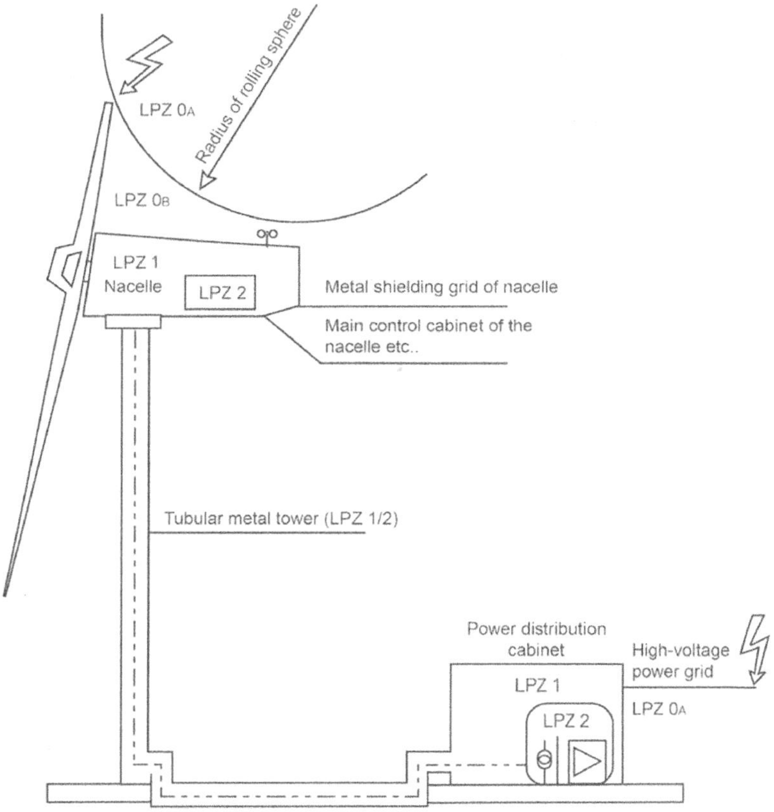

Figure 109: Diagram of division of the typical wind turbine into lightning protection zones [208].

SPDs are then installed at the boundaries between LPZs, where the electromagnetic field intensity changes significantly. This ensures that transient overvoltages are discharged safely and that sensitive electronics within inner zones (such as LPZ 2 and LPZ 3) remain protected. The type, capacity, and installation point of each SPD are determined based on the characteristics of the electrical system at those boundaries [208].

Transient Overvoltages in Wind Turbines

Wind turbines are particularly susceptible to transient overvoltages, which are short-duration spikes in electrical voltage that can occur during normal turbine operation or as a result of external events such as lightning. These overvoltages pose a significant risk to sensitive electrical and electronic components, including converters, generators, and control systems. Two primary types of transients are commonly observed: converter switching transients and cable-induced transients [208].

The first type, converter switching transients, is most commonly found in *Doubly-Fed Induction Generator (DFIG)* systems. These turbines use converters equipped with Insulated Gate Bipolar Transistors (IGBTs), which regulate voltage and current through high-frequency switching of pulse-width modulation (PWM) signals. While efficient, this switching process generates high-frequency voltage spikes, known as *du/dt harmonics*. If these transients are not properly filtered, they can cause stress on insulation materials, degrade cable and motor windings, and interfere with the proper operation of Surge Protective Devices (SPDs). Over time, the repetitive nature of these voltage spikes accelerates insulation aging and increases the risk of premature electrical failure [208].

The second major source of transient overvoltage is cable-induced transients, which are a result of the long electrical cabling typically found in wind turbines. These cables connect the converter—usually located at the base of the tower—to the generator positioned in the nacelle. Due to the inductive and capacitive coupling effects of these long runs, high-frequency overvoltage oscillations can develop along the cable. When the cable's impedance interacts with converter output signals, voltage reflections can occur, forming oscillating waveforms that travel back and forth between components. If these transients exceed the endurance ratings of installed SPDs, they can lead to premature wear or total failure of the protection system. Over time, this can compromise the electrical integrity of the turbine's drivetrain and control systems [208].

In a typical Doubly-Fed Induction Generator (DFIG) configuration, the following voltage parameters are used to define the operating and transient conditions: a maximum operating voltage of 750 V (±10%), transient overvoltage of 1.7 kV line-to-earth (L-PE), and 2.95 kV line-to-line (L-L). The voltage rise rate can reach 1.4 kV per microsecond, and the switching frequency of the converter is approximately 2000 Hz. These parameters highlight the demanding electrical environment within a wind turbine, emphasizing the need for precise surge protection and effective cable management [208].

In summary, transient overvoltages in wind turbines arise from both internal converter dynamics and external cable interactions. If not properly mitigated, these high-energy events can lead to insulation degradation, component failure, and reduced lifespan of SPDs. Proper design measures—such as the use of filters, appropriately rated SPDs, and controlled switching sequences—are critical to ensuring long-term reliability and safety in modern wind turbine electrical systems [208].

High Voltage Ride-Through (HVRT) Requirements

In modern wind power systems, High Voltage Ride-Through (HVRT) capability is a critical requirement for both the turbine and its associated electrical protection systems. HVRT refers to the ability of a wind turbine to remain connected to the grid and continue operating safely during temporary overvoltage (TOV) events, which can occur during grid disturbances or faults. These overvoltages are short-term but intense spikes in grid voltage that can arise from sudden disconnections, switching operations, or faults being cleared elsewhere in the transmission system. If turbines are unable to withstand such events, they may trip offline, potentially destabilizing the power grid and reducing the overall efficiency of wind energy generation [208].

HVRT events are particularly significant in Doubly-Fed Induction Generator (DFIG) and Permanent Magnet Synchronous Generator (PMSG) systems, which are the two most common generator configurations used in modern wind turbines. In DFIG systems, the rotor circuit is connected to the grid through power converters equipped with Insulated Gate Bipolar Transistors (IGBTs). These converters are sensitive to voltage fluctuations, and any temporary overvoltage can cause insulation stress, semiconductor breakdown, or protective tripping if not properly managed. Similarly, in PMSG systems, where the generator is directly coupled to the grid through full-scale converters, HVRT tolerance is essential to maintain electrical stability and protect power electronics from excessive voltage stress [208].

To ensure turbine survival during HVRT events, Surge Protective Devices (SPDs) and converter systems must be specifically rated to handle short-duration overvoltages within defined thresholds and time intervals. Typically, this means that SPDs must endure voltage levels up to 1.3 times the turbine's nominal operating voltage for short durations (e.g., several hundred milliseconds to a few seconds) without losing functionality. For example, if a turbine operates at 400 V nominally, its SPD may need to sustain voltages up to 520 V for up to 500 milliseconds or higher transient voltages for shorter periods, depending on grid conditions. The SPDs act as fast-response devices that divert excess voltage to ground, preventing damaging voltage levels from reaching converters, transformers, and control electronics.

Compliance with HVRT standards ensures that the wind turbine can "ride through" grid disturbances without disconnection. This capability is not only a technical necessity but also a regulatory requirement in many grid codes around the world, including those defined by the IEC 61400-24 and national energy authorities. By meeting HVRT standards, turbines contribute to grid stability by continuing to supply reactive power support during faults, helping restore normal grid conditions more quickly [208].

Environmental Requirements for SPDs in Wind Turbines

Surge Protective Devices (SPDs) used in wind turbines must be engineered to perform reliably under extreme and variable environmental conditions. Wind farms, particularly those located offshore or in high-altitude, coastal, or desert regions, expose electrical systems to severe mechanical, thermal, and chemical stresses. As critical components that protect turbines from transient overvoltages and lightning-induced surges, SPDs must maintain consistent performance despite these challenging surroundings. To ensure durability and operational safety, they are subject to rigorous environmental qualification standards defined by international regulations such as IEC 61400-24 (specific to wind turbine lightning protection) and IEC 61643 (for SPD design and testing) [208].

One of the key environmental requirements for SPDs is temperature resilience. Wind turbines operate in climates that range from sub-arctic to tropical, meaning that SPDs must continue functioning effectively across a broad temperature spectrum, typically from –40°C to +85°C. At very low temperatures, materials can become brittle, and electronic components may exhibit slower response times, while high temperatures can accelerate component aging and degrade insulation or sealing materials. The wide temperature tolerance ensures that SPDs maintain their ability to conduct and dissipate surge currents safely, regardless of ambient or internal nacelle temperatures.

Humidity tolerance is another crucial factor, especially for offshore turbines and coastal installations where moisture levels are persistently high. SPDs must remain functional in relative humidity conditions ranging from 5% to 95% without suffering from internal condensation, corrosion, or electrical leakage. To achieve this, manufacturers design SPDs with sealed housings and employ hydrophobic coatings, ensuring internal circuitry is protected from water vapor ingress. In offshore environments, the constant presence of salt-laden air introduces additional challenges, making salt spray resistance a vital requirement. Prolonged exposure to marine atmospheres can corrode metallic terminals and connectors, reducing SPD effectiveness or causing electrical faults. Therefore, SPDs for offshore use are built with corrosion-resistant materials such as stainless steel or coated aluminium and tested to meet marine-grade salt spray endurance standards outlined in IEC 61400-24 [208].

In addition to thermal and humidity resistance, SPDs must endure the vibrational stresses produced by rotating components and fluctuating wind loads within the turbine structure. These vibrations can cause micro-fractures, loosen electrical connections, or damage internal assemblies if not properly mitigated. To prevent such failures, SPDs undergo vibration testing in accordance with IEC 61400-24, ensuring they can operate continuously in the dynamic mechanical environment of the nacelle and tower without degradation.

Finally, the biological and chemical resilience of SPDs is also important. Offshore and tropical installations can foster the growth of mould, mildew, and bacteria, which may compromise insulation materials or degrade protective housings over time. To combat this, SPD casings and seals are often made from anti-mildew and antibacterial materials designed to resist microbial contamination and ensure long-term reliability in humid, bioactive environments [208].

SPD Key Parameters for Wind Turbine Protection

Surge Protective Devices (SPDs) play a vital role in protecting wind turbine electrical and electronic systems from transient overvoltages and lightning-induced surges. To ensure they operate effectively and provide adequate protection, several key parameters must be considered when selecting and installing SPDs. These include the Voltage Protection Level

(Up), Maximum Continuous Operating Voltage (Uc), and Discharge Current Capacity (In and Iimp). Each of these parameters determines how well the SPD will perform under specific electrical and environmental conditions within a turbine's lightning protection zones (LPZ) [208].

1. Voltage Protection Level (Up)

The Voltage Protection Level (Up) refers to the maximum residual voltage that the SPD allows to pass through to downstream equipment during a surge event. In simple terms, it defines the "let-through" voltage—the highest voltage that sensitive devices will be exposed to even after surge protection is activated. This parameter is critical because if the Up value is too high, it can exceed the rated impulse withstand voltage (Uw) of the protected components, such as converters, transformers, or control systems. Exceeding this threshold can result in insulation breakdown, component degradation, or catastrophic equipment failure.

Therefore, the SPD's Up value must always be lower than the impulse withstand voltage of the devices it protects, ensuring that only a safe residual voltage reaches sensitive circuits. The selection of Up depends on both the system voltage and the location of the SPD within the turbine's electrical architecture. For example, SPDs installed at the tower base or nacelle, where surge energy is highest, require lower Up values for critical components. In wind turbine applications, achieving an optimal balance between effective surge suppression and minimal voltage drop across the system is essential for maintaining electrical reliability [208].

2. Maximum Continuous Operating Voltage (Uc)

The Maximum Continuous Operating Voltage (Uc) defines the highest RMS voltage that an SPD can safely withstand on a continuous basis without degrading or entering a conductive state. This parameter ensures that the SPD remains inactive under normal operating conditions, only responding when voltage exceeds acceptable limits. In wind turbines, Uc must be carefully selected to prevent nuisance triggering—frequent SPD activation due to system harmonics, converter switching transients, or temporary overvoltages caused by grid fluctuations.

If Uc is set too close to the system's nominal voltage, minor variations in voltage can cause the SPD to conduct repeatedly, leading to excessive wear, heat buildup, and premature failure. Conversely, selecting an SPD with too high a Uc value reduces its ability to respond quickly to surges, potentially allowing damaging voltages to reach critical components. Therefore, it is recommended to choose SPDs with a Uc rating slightly above the system's nominal operating voltage but below the threshold of expected overvoltage disturbances.

In the context of wind turbines, where converters and variable-speed drives frequently modulate voltage, harmonic distortion and switching overvoltages are common. This makes proper Uc selection particularly important to ensure long-term SPD reliability and performance, especially in systems with frequent transient activity [208].

3. Discharge Current Capacity (In and Iimp)

The Discharge Current Capacity of an SPD determines how much surge energy it can safely divert to the ground without being damaged. Two main parameters define this capability: the nominal discharge current (In) and the impulse discharge current (limp).

- Nominal discharge current (In) refers to the peak current the SPD can handle during standard surge testing (typically an 8/20 μs waveform) without degradation.
- Impulse discharge current (limp) represents the SPD's capacity to handle the extreme surge energy of a direct lightning strike, simulated using a 10/350 μs waveform.

In wind turbine applications, the required discharge capacity depends on the SPD's installation location within the Lightning Protection Zone (LPZ) framework. SPDs located near potential lightning entry points—such as at the tower base, nacelle, or hub—must withstand higher surge currents compared to those installed in more protected internal circuits like control cabinets.

For this reason, wind turbines typically employ Type 1+2+3 SPDs, which provide comprehensive protection across multiple zones.

- Type 1 SPDs handle direct lightning currents (10/350 μs waveform) and are installed at primary entry points, such as the tower base or main switchboard.
- Type 2 SPDs protect distribution circuits and control systems against residual surges.
- Type 3 SPDs are used near sensitive electronics, offering fine protection against low-energy transient voltages.

Using a coordinated system of Type 1+2+3 SPDs ensures that all levels of surge energy—from high-energy lightning strikes to smaller switching transients—are safely dissipated, protecting the turbine's entire electrical network [208].

Protection Measures for Wind Turbines under IEC 61400-24

Wind turbines are highly exposed to lightning strikes due to their elevation, location, and conductive structures. The *IEC 61400-24 (EN 61400-24)* standard and *Germanischer Lloyd (GL) 2010* guidelines provide detailed recommendations to safeguard wind turbines against both direct lightning strikes and transient overvoltages [208]. These standards recommend that all sub-components of a wind turbine's lightning protection system (LPS) be designed for Lightning Protection Level (LPL) I, which represents the highest protection standard. A lower protection level may only be permitted following a formal risk analysis demonstrating that exposure and potential damage risks are minimal .

According to IEC 61400-24, an effective lightning protection strategy combines external lightning protection systems (LPS) and surge protection measures (SPMs) to safeguard electrical and electronic systems. The LPS intercepts and conducts lightning currents safely to the ground, while the SPM prevents surge-related damage to internal components.

To implement these measures, the wind turbine is subdivided into Lightning Protection Zones (LPZs). Each LPZ defines a specific electromagnetic environment and protection level. For instance:

- LPZ 0A includes parts directly exposed to lightning strikes, such as the blade tips and nacelle roof.
- LPZ 1 and LPZ 2 are internal zones where electromagnetic fields are significantly attenuated through shielding and surge protection devices.

The rolling sphere method—a technique used to simulate the path of a lightning strike—is applied to determine potential strike points and define these protection zones. However, the IEC standard specifies that this method cannot be applied directly to rotor blades, as their aerodynamic shape and internal lightning receptors require specialized testing to confirm lightning current withstand capability.

Shielding is fundamental to reducing electromagnetic interference (EMI) caused by lightning and electrical surges. The nacelle should function as a closed metallic shield, forming a localized Faraday cage that minimizes electromagnetic fields inside. Large tubular steel towers are considered near-perfect Faraday cages due to their continuous conductive structure. For concrete hybrid towers, this shielding effect is achieved by connecting internal reinforcement steel, ensuring full equipotential bonding across all components.

To maintain electromagnetic compatibility (EMC):

- Switchgear and control cabinets should be metallic and grounded.
- Cables must be shielded and connected to the equipotential bonding system at both ends.
- Shielding connections should use 360° contact terminals to minimize resistance and prevent interference loops.
- Cable routing must follow the principles of IEC 62305-4 and IEC/TR 61000-5-2 to ensure electromagnetic field reduction.

Examples of effective shielding techniques include metal braids on GRP-coated nacelles, shielded cable ducts, metal towers, and closed metallic control cabinets.

The external LPS intercepts and safely conducts lightning currents from exposed areas to the ground. Key components include:

- **Air-termination systems**: Metal receptors at blade tips, nacelles, and hubs attract lightning.
- **Down conductors**: Copper or aluminium pathways that channel current from receptors to the tower base.

- **Earth-termination systems**: Networks of electrodes that disperse lightning energy into the soil.

Using the rolling sphere method, a sphere of 20 m radius (for LPS Class I) is "rolled" over the turbine structure to identify strike points. Air-termination systems must be installed wherever the sphere contacts the surface.

For composite nacelles (e.g., fiberglass-reinforced plastic), a metallic cage or braid is integrated to form a conductive pathway. The nacelle and tower's conductive structures are linked, ensuring all lightning currents are discharged without sparking or localized heating that could cause fire or explosion.

Natural conductive components—such as bearings, frames, and tower reinforcements—can be incorporated into the LPS if they meet conductivity and mechanical strength requirements.

The earth-termination system is crucial for dissipating lightning energy into the ground while ensuring personal and operational safety. According to IEC 62305-3, the system serves multiple functions: lightning current dissipation, equipotential bonding, and protection against touch or step voltages.

There are two main types of earthing systems [208]:

- Type A (used in auxiliary buildings): Comprises vertical or horizontal electrodes connected by down conductors.
- Type B (mandatory for turbines): Combines a ring earth electrode around the tower with a foundation earth electrode integrated into the base reinforcement steel.

The ring electrode, typically made of stainless steel (AISI 316 Ti), minimizes step voltage and provides corrosion resistance. All foundation electrodes must be embedded in at least 5 cm of concrete to prevent corrosion and connected to the turbine's main earthing busbar [208].

Internal protection focuses on limiting electromagnetic and conducted interference within the turbine. Key strategies include [208]:

- **Equipotential bonding:** All metallic parts, cable shields, and structural elements are interconnected to maintain a common potential.
- **Surge protection:** Type 1, Type 2, and Type 3 SPDs are installed at LPZ boundaries to discharge surge currents and reduce voltage spikes.
  - Type 1 SPDs (10/350 μs waveform) are installed at LPZ 0A–1 boundaries to handle high-energy lightning currents.
  - Type 2 SPDs (8/20 μs waveform) protect against induced surges and residual voltages.
  - Type 3 SPDs provide fine protection for sensitive electronics in LPZ 2–3.

Each zone transition—such as from the tower base to the nacelle—must have SPDs sized according to the expected discharge currents and system voltage.

Surge arresters must be rated according to IEC 60364-4-44 and IEC 60664-1 to match the turbine's nominal voltage.

- For 400/690 V systems, SPDs with a protection level ≤ 2.5 kV are required.
- For 230/400 V systems, the protection level should be ≤ 1.5 kV.

High-capacity arresters, such as the FLP25-440 or SLP40 series, are installed near transformers, converters, and switchgear. These devices are tested to withstand lightning current waveforms up to 10/350 µs, ensuring resilience against direct and indirect lightning effects.

For aircraft warning lights and other low-voltage systems, Type 2 SPDs protect against surges transmitted through communication and control lines. In DFIG-based turbines, additional SPDs are used on the rotor side to protect the converter and generator interface.

Effective lightning and surge protection in wind turbines is achieved through a combination of external LPS, internal shielding, equipotential bonding, and coordinated SPDs. The IEC 61400-24 standard establishes the design, testing, and verification procedures necessary to ensure the turbine can withstand both direct lightning strikes and transient overvoltages without system failure.

By dividing the turbine into Lightning Protection Zones, ensuring proper earthing and shielding, and installing correctly rated SPDs, operators can significantly reduce risks of fire, equipment damage, and operational downtime. These protection measures form the backbone of safe, reliable, and long-term wind energy generation across both onshore and offshore environments [208].

Surge Arresters for Information Technology Systems

Surge arresters for information technology (IT) systems play a vital role in protecting electronic equipment and communication networks within wind turbines from the damaging effects of lightning strikes and electrical transients. These devices are designed in accordance with the international standard IEC 61643-21 (EN 61643-21), which defines the requirements for surge protective devices (SPDs) used in telecommunication and signalling networks. SPDs are typically installed at the boundaries between lightning protection zones (LPZs), ensuring that induced or conducted surges are safely dissipated before reaching sensitive electronics [208].

Coordinated Protection and Design Considerations

Multi-stage surge arresters are used to achieve layered protection across different voltage and current levels. However, these stages must be carefully coordinated to avoid "blind spots," where one protection stage activates while others remain inactive. Improper coordination can result in unmitigated voltage surges that lead to SPD malfunction or partial protection failure. Coordination ensures that each SPD stage operates sequentially and harmoniously, providing optimal protection from both direct lightning-induced surges and switching transients [208].

Cable Types and Protection Needs

In modern wind turbines, fibre optic cables are frequently used for communication between control systems, such as linking the tower base control cabinets with the nacelle. Because fibre optics do not conduct electricity, they are naturally immune to electromagnetic interference (EMI) and do not require surge protection—unless the cables include a metal sheath. In such cases, the metallic sheath must be integrated into the equipotential bonding system, either directly or through SPDs, to prevent potential differences and induced voltages from causing insulation breakdown or equipment damage.

In contrast, shielded copper cables are commonly used for internal turbine connections, such as between sensors, actuators, and control cabinets. These cables can conduct induced surges caused by lightning electromagnetic pulses (LEMP) or switching transients, so SPDs are essential for protecting connected electronic systems [208].

Signal Lines Requiring Protection

Several critical signal lines within a wind turbine must be protected using SPDs [208]:

- **Weather Station and Aircraft Warning Light Lines:** These connect sensors on the meteorological mast (LPZ 0B) to the turbine's switchgear or control cabinet (LPZ 2). Since they span multiple zones, they are exposed to induced surges. Devices such as FRD2-24 or FRD4-24 surge protectors are typically used to safeguard 4–20 mA analogue signal interfaces.
- **Pitch System Communication Lines:** The pitch control system, located in the hub, relies on real-time communication with the nacelle control unit. Ethernet-based data lines used for this communication must be protected using SPDs like DT-CAT 6A/EA, which are suitable for Industrial Ethernet applications up to 48 V DC and frequencies up to 250 MHz.
- **Yaw and Fire Suppression Systems:** Similar surge protection is applied to the signal lines controlling yaw motors and fire extinguishing units, ensuring continuity of safety-critical operations during transient events.
- **Inverter Signal Lines:** SPDs prevent transient surges from interfering with converter control signals, preserving stability in power regulation systems.

Part 4

Installation and Commissioning

Chapter 9

Pre-Installation Planning

Before a single tower section is lifted or a blade is bolted into place, a wind farm project must pass through one of its most demanding stages — pre-installation planning. This phase determines whether construction proceeds smoothly or descends into costly delays. It links engineering design with the realities of transport logistics, site preparation, and safety compliance. Every kilometre of road, every crane movement, and every checklist entry must be verified long before the first component arrives on site.

Wind turbine components are among the largest structures moved on public infrastructure, so logistics planning begins months or even years in advance. Transport routes are surveyed, bridges assessed for load capacity, and ports evaluated for crane reach and deck strength. The planning process coordinates fleets of specialized vehicles, escorts, and vessels to deliver blades, towers, and nacelles that can stretch or weigh more than most industrial equipment ever handled. Offshore projects add further complexity, requiring synchronized port operations, weather-window forecasting, and heavy-lift vessel scheduling.

At the same time, detailed documentation and checklists establish the foundation for safe and efficient work. These include transport and lift plans, risk assessments, environmental permits, and quality-control records. They provide assurance that every bolt, cable, and crane movement aligns with regulatory standards and engineering specifications.

Finally, pre-installation planning defines the sequencing of construction — ensuring that each delivery, lift, and assembly step happens in the right order, under the right conditions, and with the right people in place. The result is a tightly orchestrated process that transforms logistical complexity into operational precision.

This chapter explores how successful wind farm construction begins long before installation day — through careful coordination, disciplined safety planning, and meticulous attention to every detail of logistics, rigging, and documentation.

Site Logistics and Transport of Components

Transporting Components

The logistics and transport of wind turbine components are among the most complex stages in wind farm construction. Modern wind turbines are massive, with towers exceeding 100 meters in height, blades reaching up to 80 meters in length, and nacelles weighing several hundred tonnes. Coordinating the movement of these oversized and heavy components from the manufacturing facility to often remote or offshore sites requires detailed planning, specialized equipment, and strict compliance with safety and regulatory standards.

Site logistics begins long before components arrive. Detailed surveys are carried out to assess access roads, bridges, ports, and site conditions. Engineers evaluate whether roads are wide enough for long blades, whether bridges can carry the heavy nacelle loads, and whether curves and gradients can accommodate oversized trailers. Where necessary, roads may be reinforced, widened, or temporarily modified to allow transport. In offshore projects, port facilities are examined for load-bearing capacity, quayside space, and crane capability.

Figure 110: Wind turbine towers being unloaded at a port. Dale Frost of the Port of San Diego from San Diego, CA, CC BY 2.0, via Wikimedia Commons.

Wind turbine towers are typically transported in segments due to their size. Each cylindrical section can be 20–30 meters long and weigh up to 70 tonnes. Heavy-duty trucks with extendable trailers are used for onshore transport, often requiring police escorts and road closures for safety. Offshore projects rely on barges or heavy-lift vessels to transport tower sections from port to site. At both ends, large cranes are needed to load and unload the sections safely.

Figure 111: A wind turbine mast transported by truck. J Brew, CC BY-SA 2.0, via Wikimedia Commons.

Turbine blades pose unique logistical challenges due to their extreme length and aerodynamic shape. Blades can exceed 70–80 meters in modern turbines, making them difficult to manoeuvre on winding or narrow roads. Specialized extendable trailers with hydraulic steering axles are used to navigate tight turns, and in some cases, "blade lifter" trailers allow the blade to be tilted upwards to avoid obstacles such as trees, buildings, or power lines.

Figure 112: Unloading wind turbine blades. Jason.nlw, CC BY 2.0, via Wikimedia Commons.

Transport often occurs at night to minimize disruption and traffic hazards. Offshore, blades are transported by specially fitted vessels with supports designed to prevent damage during transit.

Figure 113: Delivery of the UK's longest onshore wind turbine blades (58.7m blades for 145m GE 2.75-120 wind turbines) to Muirhall Wind Farm, South Lanarkshire. ShellAsp, CC BY-SA 4.0, via Wikimedia Commons.

The nacelle, which houses the generator, gearbox, and other key systems, is one of the heaviest components, often weighing between 60 and 120 tonnes for utility-scale turbines. Due to its weight, the nacelle requires multi-axle modular trailers onshore, while offshore transport relies on heavy-lift ships with reinforced decks. The hub, although lighter, is still large and must be carefully supported to avoid stress on its connection points.

For offshore wind farms, transport logistics are even more demanding. Ports must have sufficient storage areas and crane capacity to handle pre-assembly of towers, blades, and nacelles. Installation vessels—jack-up barges or floating heavy-lift ships—are loaded with multiple components at a time to reduce the number of trips to sea. Weather windows are crucial: wave heights, wind speeds, and tides all affect whether transport and installation can proceed safely. Any delay due to weather can significantly impact project costs and schedules.

Figure 114: Arrival of tripods in Wilhelmshaven. DOTI/Jan Oelker, CC BY 3.0, via Wikimedia Commons.

On-Site Storage and Staging

Once delivered, components are stored in laydown areas at the site or port. These areas must be large, level, and reinforced to bear heavy loads without settlement. Blades are stored on cradles to prevent deformation, while tower sections are kept on supports that prevent coating or surface damage. Nacelles and hubs are often stored under protective covers. Efficient site logistics ensures that components are staged in the correct sequence for assembly, reducing unnecessary handling.

Lifting and Installation Logistics

Transport planning is closely tied to installation logistics, as cranes and heavy-lift equipment must be available when components arrive. For onshore turbines, large crawler cranes with boom extensions are brought in to lift tower sections, nacelles, and blades into place. Offshore, jack-up vessels use heavy-lift cranes capable of handling extreme weights and heights. Careful scheduling prevents idle time for these costly machines, which are among the most expensive resources on site.

Key challenges in turbine logistics include narrow rural roads, steep gradients, bridge weight limits, unpredictable weather (particularly offshore), and coordination of multiple contractors and suppliers. Solutions often involve route upgrades, temporary road modifications, traffic management plans, and the use of cutting-edge transport technology. Offshore, accurate weather forecasting and flexible scheduling help reduce risks.

Logistics and Scheduling

Constructing a modern utility-scale wind turbine involves a complex logistical process with various large components, each serving a critical role in the overall structure. The assembly process typically encompasses five primary categories of components: the tower, nacelle, hub, blades, and foundation, along with ancillary systems. Careful planning of the delivery and assembly of these components ensures a streamlined and efficient construction process.

The tower, made of cylindrical steel sections, is fundamental to the turbine's structure. Depending on the design, it typically comprises three to five sections, with hub heights generally ranging from 80 to 160 meters [209, 210]. Transportation of these components is facilitated by extendable trailers for onshore sites or heavy-lift vessels for offshore implementations. The first components to arrive on-site are the tower sections, as their installation establishes the base for the subsequent assembly of the nacelle and other elements [211].

Once the tower is erected, the nacelle is delivered. The nacelle is one of the heaviest components, with weights between 60 and 120 tonnes, and it houses critical systems such as the generator and gearbox [212] . The delivery logistics for the nacelle require heavy multi-axle trailers or specialized vessels for efficient transport [210]. Following the installation of the nacelle, the hub, typically transported separately, is then mounted at the apex of the tower, providing the attachment point for the turbine blades [213].

The blades present another significant challenge due to their considerable size; modern blades can span lengths of 60 to 80 meters and are usually shipped individually, either on extendable trailers with hydraulic steering systems or on specially designed offshore transport vessels

[214, 215]. Arriving toward the end of the construction phase, blades are typically installed using a 'single blade lift' method or by assembling them into a rotor star for lifting in unison [216].

Foundations play a critical role in ensuring the stability of turbines. For onshore installations, massive gravity concrete bases are common, while offshore installations utilize monopile, jacket, or floating foundation systems [209, 217]. The completion of foundations precedes the delivery of turbine components, ensuring a robust base for the heavy structures [218]. Ancillary components, which include electrical systems and transformers, are either delivered during or after the turbine assembly phase, addressing the logistical necessities that arise during construction [219].

Crane Operations and Rigging Safety

Crane operations and rigging are among the most critical and high-risk activities in wind turbine construction, both onshore and offshore. Turbine components such as tower sections, nacelles, hubs, and blades are extremely large and heavy, requiring specialized lifting strategies to install them safely at great heights. Any errors in crane operation or rigging can lead not only to costly delays and equipment damage, but also to serious safety incidents. For this reason, wind farm projects place strong emphasis on careful planning, competent personnel, and strict safety controls.

Crane operations begin with the mobilization of large crawler cranes or heavy-lift cranes to the site. Onshore projects typically use crawler cranes with long boom extensions capable of lifting nacelles weighing 60–120 tonnes and blades up to 80 meters long. Offshore projects use jack-up installation vessels equipped with heavy-lift cranes that can safely operate at sea.

Figure 115: Crawler crane and wind turbine. Cjp24, CC BY-SA 4.0, via Wikimedia Commons.

Because turbine towers can reach 100–160 meters in hub height, cranes must not only have the lifting capacity but also the reach and stability to position components accurately at extreme elevations. Site logistics play an important role—assembly areas must be level and reinforced to bear the weight of crawler cranes, and offshore jack-up barges must be stabilized on the seabed. Weather conditions also have a major impact: operations are typically suspended in high winds, heavy rain, or poor visibility, as even moderate gusts can cause a suspended blade to swing dangerously.

Rigging involves attaching the load (tower sections, nacelle, hub, or blades) securely to the crane for lifting. Due to the size and irregular shape of turbine components, specialized rigging equipment is required.

- Tower sections are typically lifted with lifting lugs or slings connected to designed attachment points.
- Nacelles are rigged using heavy-duty spreader bars to distribute weight evenly.
- Blades are lifted using purpose-built blade yokes or lifting frames that cradle the blade and allow pitch and yaw adjustments mid-air for precise alignment during installation.

Rigging crews must ensure the correct load distribution, sling angles, and centre of gravity control, as improper rigging can cause uncontrolled tilting or rotation during lifts. Clear communication between rigging crews and crane operators is vital, often using hand signals, radios, or standardised lifting commands.

Figure 116: Top section of turbine tower being lifted into position. Paul Anderson, CC BY-SA 2.0, via Wikimedia Commons.

In this context, rigging refers to the process of preparing and securing loads so they can be safely lifted, moved, or positioned by cranes, hoists, or other lifting equipment. In simple terms, it is the art and science of connecting a heavy or oversized object to a lifting device in such a way that the load can be raised and controlled without tipping, swinging, or overstressing the equipment.

In wind turbine construction, rigging is a critical activity because the components—such as blades, nacelles, and tower sections—are extremely heavy, tall, and awkwardly shaped. Specialized rigging gear like slings, shackles, spreader bars, and blade yokes are used to distribute forces evenly and to control the balance of the load. The rigging crew ensures that the centre of gravity is properly accounted for and that attachment points are secure.

Rigging also involves planning how the load will move from one point to another, establishing safe lifting paths, and coordinating with the crane operator and signaller. Safety is central to rigging: poorly executed rigging can lead to dropped loads, equipment damage, or serious injury. For this reason, only trained and certified riggers are permitted to carry out the work.

So, rigging is not just about attaching ropes or chains—it is a highly skilled process that combines knowledge of load dynamics, safety standards, and practical techniques to move massive components safely and precisely.

Wind turbine lifting work is classed as high-risk construction activity under occupational health and safety (OHS/WHS) regulations. Key safety measures include:

1. **Lift Planning and Risk Assessment** – Every lift is planned in detail, including component weight, wind limits, crane configuration, rigging method, and emergency procedures. A lift plan must be approved before work begins.
2. **Competent Personnel** – Only trained and certified crane operators, riggers, and signallers are permitted to conduct lifts. Offshore work may also require additional marine safety training.
3. **Weather Restrictions** – Lifts are halted if wind speeds exceed safe limits (often around 9–12 m/s for blade lifts). Blade lifts are especially sensitive due to their surface area and aerodynamic properties.
4. **Exclusion Zones** – Ground personnel are kept outside exclusion zones beneath suspended loads to prevent injury in case of equipment failure or load drop.
5. **Equipment Inspection** – Rigging gear, such as slings, shackles, hooks, and spreader bars, is inspected before each use for wear, deformation, or defects. Cranes are also regularly serviced and load-tested.
6. **Emergency Procedures** – Rescue plans are in place for situations such as suspended loads in deteriorating weather, equipment failure, or worker injury at height.

The legal requirements for working as a rigger vary between jurisdictions, but they all emphasize the same key principles: safety, competence, and certification. Each region has its own framework to ensure that riggers are properly trained and able to carry out lifting operations safely.

In the United States, rigging is governed by OSHA standards, specifically 29 CFR 1926.1400–1442, which cover cranes and derricks in construction. OSHA requires that a rigger must be "qualified," meaning they must have either formal training, extensive experience, or both, to demonstrate the ability to safely perform rigging tasks. While there is no federally mandated national licence for riggers, employers must verify that their workers are competent. Many riggers choose to obtain certification from the National Commission for the Certification of Crane Operators (NCCCO), which offers two levels of rigger qualification: Level I and Level II. Qualified riggers are mandatory whenever workers are within a crane's fall zone and loads are being hoisted. They are responsible for inspecting rigging equipment, calculating load weight and centre of gravity, and ensuring safe attachment to lifting devices.

In the United Kingdom, rigging is regulated under the Lifting Operations and Lifting Equipment Regulations 1998 (LOLER) and the Provision and Use of Work Equipment Regulations 1998 (PUWER). Employers must ensure that riggers are competent, typically through accredited training schemes such as those offered by the Construction Plant Competence Scheme (CPCS) or the National Plant Operators Registration Scheme (NPORS). A CPCS "Slinger/Signaller" card or equivalent is often required for riggers who attach or detach loads and signal crane movements. Furthermore, all lifting operations must be planned and supervised by a competent person, and riggers are legally obligated to follow established safe systems of work.

In Australia, rigging is classified as a high-risk work (HRW) activity and is regulated by Safe Work Australia's Model WHS Regulations, with enforcement carried out by state and territory authorities. Riggers must hold a High-Risk Work Licence, which is issued by the relevant regulator, and the class of licence depends on the level of rigging work performed. The categories include DG (Dogging), RB (Basic Rigging), RI (Intermediate Rigging), and RA (Advanced Rigging). To obtain a licence, riggers must complete accredited training and assessment with a Registered Training Organisation (RTO). Licensed riggers are also expected to be competent in hazard identification, fall prevention, and the safe use of lifting gear.

In Europe, requirements are guided by EU Directives such as the Machinery Directive 2006/42/EC and the Work at Height Directive 2001/45/EC, although each member state enforces its own national legislation. All riggers must be trained and competent, and in countries such as Germany, the Netherlands, and Denmark, workers must complete certified training in slinging and signalling. Rigging work is required to meet European Norms (EN), such as EN 13155 for lifting accessories, and employers are bound by the EU Framework Directive on Health and Safety (89/391/EEC) to provide safe working conditions. While there is no universal EU-wide rigger licence, most member states require documented training and certification through approved bodies, such as TÜV in Germany.

Overall, while the details differ between regions, the global standard for rigging work is built on the same foundation: ensuring that only competent, trained, and certified personnel are allowed to perform these high-risk tasks.

Offshore wind construction introduces additional hazards. Heavy-lift vessels must maintain stability while installing components, and sea swell can cause crane hooks to heave unpredictably. Specialized motion-compensated lifting systems are sometimes used to counteract wave movements. Crew transfers and rigging at sea require high levels of coordination, and operations are often delayed until a safe weather window is available.

Documentation and Checklists

Pre-installation planning for wind turbines is one of the most critical stages in a wind farm project. Before any heavy components arrive on-site, developers, engineers, and construction teams must prepare detailed documentation and checklists covering logistics, safety, technical compliance, and environmental requirements. These documents ensure that installation can proceed efficiently, safely, and in strict alignment with engineering specifications and regulatory standards.

The purpose of pre-installation documentation is to minimize risks, avoid delays, and confirm that all resources, equipment, and procedures are ready before turbine erection begins. This planning stage provides a structured framework for coordination between multiple stakeholders, including project managers, crane operators, rigging teams, logistics providers, and safety officers. By ensuring everyone has a clear set of responsibilities and procedures, projects can run more smoothly and safely.

Several categories of documentation play an essential role in this process. Site-specific risk assessments identify hazards such as uneven terrain, high winds, limited access, or offshore conditions. These assessments also outline mitigation strategies like exclusion zones, weather restrictions, and emergency procedures while aligning with national and international safety standards such as OSHA in the U.S., WHS in Australia, LOLER/PUWER in the UK, or EU Safety Directives.

Method statements and work procedures provide step-by-step guidance on the sequence of installation, covering foundation readiness, tower erection, nacelle installation, rotor lifts, and cabling. These documents also define team responsibilities, communication protocols between operators and riggers, and contingency measures for equipment failures or weather delays.

Logistics and transport plans detail the delivery schedule of major components including tower sections, blades, nacelles, and hubs. They include approved transport routes, bridge load limits, escort requirements, and port facilities for offshore projects. Laydown areas and staging arrangements are specified to minimize unnecessary re-handling of heavy parts.

Crane and rigging plans are essential for safe lifting operations. They document crane setup requirements, ground-bearing checks, and exclusion zones, while also including rigging diagrams for each component and lifting gear specifications. Large or complex lifts require engineering approval and may involve pre-lift simulations.

Quality assurance and technical documentation ensure that all components meet standards before installation. This includes manufacturer's manuals, torque and alignment specifications, coating and corrosion protection records, and inspections confirming no transport damage has occurred.

Environmental and permitting documents address site-specific environmental concerns. They include impact assessments, wildlife and noise mitigation measures, offshore permits for vessels, sea traffic management, and waste and spill prevention strategies.

Health and safety documentation is central to pre-installation planning. This includes Job Safety Analyses (JSA) or Safe Work Method Statements (SWMS), fall protection plans for tower climbs, rescue and emergency procedures, and first-aid protocols. Toolbox talk templates are also prepared to facilitate daily safety briefings on-site.

Alongside these, detailed checklists are created for every component. Foundation checklists verify curing strength, bolt tension, and leveling. Tower section checklists confirm alignment and coating integrity. Nacelle checklists ensure gearbox, generator, and yaw systems are ready. Blade checklists focus on leading-edge condition, transportation damage, and balance. Electrical checklists confirm readiness of cables, grounding, and SCADA systems.

The use of checklists provides consistency and accountability throughout the process. They reduce the chance of critical steps being overlooked, especially under tight deadlines, and create records that support quality assurance, contractor accountability, and compliance.

Chapter 10

Turbine Assembly

After months of preparation, transport, and groundworks, the most visible and defining phase of a wind farm begins — turbine assembly. This is the point where hundreds of tonnes of steel, composite, and electrical equipment are lifted skyward and transformed into a functioning energy system. Every lift, bolt, and cable must come together with precision. The work combines engineering excellence, logistical mastery, and strict adherence to safety, as crews operate in challenging environments, often at heights exceeding 100 metres.

Turbine assembly is divided into several interdependent stages: tower erection, nacelle installation, rotor and blade assembly, and cable routing and terminations. Each phase demands its own combination of technical skill, specialized machinery, and disciplined coordination between project managers, crane operators, riggers, technicians, and safety personnel. Onshore installations rely on crawler cranes, heavy-lift trucks, and ground crews working in close synchrony. Offshore, jack-up vessels and marine crews face the additional challenges of weather windows, sea states, and vessel stability — where every hour of lost time can have significant financial implications.

The process begins with the tower erection, the backbone of the turbine, where precise alignment and secure bolting ensure structural integrity. Once the tower stands tall, the focus shifts to lifting and securing the nacelle, which houses the generator, gearbox, and control systems — the heart of the turbine. The rotor and blades are then installed, completing the turbine's aerodynamic structure. Each step is governed by rigorous safety procedures, from wind speed limits and exclusion zones to certified rigging practices. Finally, electrical and data cables are routed and connected, linking the turbine's control and power systems to the grid and enabling full operational control.

This chapter explores each of these stages in depth — from planning and workforce coordination to the specialized techniques that make turbine assembly possible both on land and at sea. It explains how teams manage heavy-lift operations, maintain safety under variable conditions, and achieve millimetre-level precision at immense scales. By the end, readers will understand how the towering structures seen on skylines and offshore horizons are meticulously assembled through engineering discipline, teamwork, and adherence to global standards.

Tower Erection

Onshore Wind Turbine Tower Construction

Erecting the tower of an onshore wind turbine is a complex and high-risk task that requires careful planning, specialized equipment, and full compliance with engineering and safety standards. The process is broken down into distinct stages, each of which must be executed with precision to ensure both structural integrity and worker safety.

The first stage is pre-installation planning. Before tower erection can begin, the foundation must be fully cured, surveyed, and approved for use. Access roads, crane pads, and laydown areas are also prepared and stabilized to handle the movement of heavy equipment. Logistics are carefully coordinated so that tower sections, cranes, and rigging gear are delivered in the correct sequence for efficient lifting. At this stage, detailed safety checks are performed, including risk assessments that account for hazards such as high winds, heavy lifting operations, exclusion zones, and emergency response requirements.

The next stage involves cranes and rigging setup. A main crane, often a lattice-boom crawler crane, is positioned on the prepared pad to handle the heaviest lifts, such as the tower sections and nacelle. An assist crane supports the unloading, alignment, and stabilization of sections before final placement. All rigging equipment, including slings, spreader bars, and lifting frames, must be thoroughly inspected and prepared before use.

With the cranes in position, the team begins lifting and erecting the tower sections. The base section is lifted upright and carefully lowered onto the anchor bolts embedded in the foundation. Precision alignment is critical at this stage, and bolts are progressively tensioned to specification. Each middle section is then lifted, aligned, and bolted to the one below, with internal platforms, ladders, or elevators progressively installed as the tower rises. The final top section completes the cylindrical structure of the tower.

Once the sections are assembled, alignment and bolting are finalized. Flange connections are aligned using hydraulic jacks or drift pins, and bolts are tightened in a specific sequence to ensure uniform stress distribution. Torque and tensioning must strictly follow manufacturer specifications. During this stage, internal cabling for power and control systems is often threaded through the tower, either simultaneously with assembly or immediately afterward.

Throughout the process, safety and weather considerations are paramount. Operations are highly weather-dependent, and lifts are suspended when wind speeds exceed safe thresholds—typically around 9–12 m/s. Exclusion zones are enforced to protect personnel from suspended loads, and constant communication is maintained between crane operators, rigging crews, and ground teams to prevent accidents.

Finally, once erection is complete, a series of post-installation checks are performed. Bolts are re-checked and re-tensioned, and the tower's vertical alignment is confirmed through survey. Internal systems such as ladders, lifts, fire safety equipment, lighting, and cabling are also inspected and tested to ensure readiness.

Human Resources

Onshore wind turbine tower construction is a labour-intensive and highly coordinated process that requires a team of skilled personnel. The exact number of human resources can vary depending on the turbine size, site complexity, and contractor practices, but typically 20–30 personnel are directly involved in tower erection activities, supported by additional logistics, engineering, and safety staff. Each individual has a clearly defined role to ensure efficiency, safety, and quality.

1. Project and Site Management

- **Project Manager / Site Manager (1–2 people):** Oversees the overall construction process, ensuring that tower erection is completed on time, within budget, and according to engineering and safety standards. Coordinates between contractors, suppliers, crane companies, and client representatives.

- **Construction Supervisor / Installation Manager (1):** Directly supervises the erection crew, approves lift plans, and ensures work follows the method statements and risk assessments.

2. Crane and Lifting Operations

- **Main Crane Operator (1):** Operates the large crawler or lattice-boom crane responsible for lifting tower sections, the nacelle, hub, and blades. Precision, coordination, and weather awareness are critical.

- **Assist Crane Operator (1):** Operates a smaller mobile crane that helps unload and position components before they are lifted by the main crane.

- **Riggers / Slingers (4–6 people):** Attach lifting gear (slings, spreader bars, lifting frames) to components, ensure correct rigging, guide loads during lifts, and signal to the crane operator. Their work is essential for load stability and safety.

- **Banksman / Signalman (1–2 people):** Acts as the main communication link between crane operators and ground crews, using hand signals or radios to guide safe lifting.

3. Tower Assembly Crew

- **Assembly Technicians / Fitters (6–8 people):** Bolt tower sections together, align flanges, and install internal platforms, ladders, and cabling as the tower rises. They also apply torque and tensioning procedures to manufacturer specifications.

- **Surveyor / Alignment Specialist (1):** Verifies tower verticality, flange alignment, and bolt tensioning using laser equipment or theodolites.
- **Electrical Technicians (2–3 people):** Thread and connect internal cabling (power, grounding, and control) during or after tower assembly. Also prepare connections for the nacelle.

4. Health, Safety, and Quality Assurance

- **Health, Safety, and Environment (HSE) Officer (1–2 people):** Monitors compliance with safety regulations, enforces exclusion zones, checks PPE use, conducts toolbox talks, and prepares emergency response.
- **Quality Assurance / Quality Control Inspectors (1–2 people):** Inspect tower sections, bolts, coatings, and internal systems for compliance with technical standards and manufacturer specifications.

5. Logistics and Support

- **Transport/Haulage Crew (3–5 people):** Deliver tower sections, nacelles, and blades to site using specialized trailers. Escort drivers and spotters may also be required for oversized loads.
- **Ground Crew / General Laborers (2–4 people):** Assist with site preparation, staging of components, moving rigging gear, and supporting other teams.

Typical Human Resource Breakdown (per turbine tower):

- Management & Supervision: 2–3
- Crane Operators: 2
- Riggers & Signalmen: 5–8
- Assembly Technicians: 6–8
- Electrical Technicians: 2–3
- Survey & QA/QC: 2–3
- HSE Officers: 1–2
- Logistics & Ground Crew: 5–7

Total: 20–30 personnel actively engaged in the erection phase, with overlap depending on shift patterns and site conditions.

The duration of onshore wind turbine tower construction depends on several factors, including turbine size, site conditions, weather, and the efficiency of the installation crew. However, with a well-coordinated team of 20–30 skilled personnel, the tower erection phase itself is usually completed in 2–5 days per turbine, once the foundation has already been prepared and cured.

Duration

Typical Timeframe by Phase:

1. **Pre-Installation Preparation (½–1 day):** Before lifting begins, the team prepares the site by staging tower sections in sequence, setting up cranes and rigging, and conducting final safety checks. This includes reviewing lift plans, exclusion zones, and weather conditions.

2. **Base Section Installation (½ day):** The base section is the most critical lift. It must be precisely aligned on the foundation bolts and torqued to specification. Survey checks ensure correct positioning.

3. **Middle Section Installation (1–2 days):** Each subsequent section is lifted, aligned, and bolted to the one below. Depending on tower height (often 80–160 meters), this can involve 2–3 lifts. Internal ladders, platforms, or elevators are also progressively installed, which adds time.

4. **Top Section Installation (½ day):** The final section completes the tower. This is usually faster than the base or middle sections, but precision alignment is still critical.

5. **Bolting, Tensioning, and Cabling (1–2 days, overlaps with above):** Once each section is lifted, bolts are progressively tensioned to the manufacturer's specifications. Internal cabling is threaded through the tower, either during lifting or immediately afterward.

6. **Final Checks and Inspections (½–1 day):** The erection team and QA/QC inspectors check flange connections, verify bolt torque, re-survey tower verticality, and inspect for coating or transport damage. Electrical technicians test internal systems and prepare connections for the nacelle.

Influencing Factors:

- **Weather:** High winds (above ~9–12 m/s), heavy rain, or poor visibility can halt crane operations and extend timelines.

- **Tower Height:** Taller towers with four or five sections take longer to erect than smaller three-section towers.

- **Crew Experience:** Well-practiced crews working with modern cranes and rigging systems can complete installations faster.

- **Parallel Work:** Internal cabling and ladder installations often overlap with lifting, reducing total time.

For a standard onshore turbine, the tower erection phase generally takes 2–5 days, provided weather conditions are favourable. This is part of a larger installation process—foundations may take weeks to prepare, and nacelle and rotor installation follow once the tower is complete. When viewed in the context of a full wind farm, teams often work in parallel across multiple turbines, meaning the construction cycle is optimized for continuous workflow.

Offshore Wind Turbine Tower Construction

Erecting the tower of an offshore wind turbine is one of the most challenging phases of wind farm construction. Unlike onshore projects, the work takes place in a marine environment where vessel stability, weather conditions, and sea state all add complexity. The process requires extensive pre-installation planning, specialized equipment, and strict adherence to engineering and safety standards.

The process begins with foundation readiness. Depending on the design—monopile, jacket, or floating substructure—the foundation must already be installed, level, and surveyed. Transition pieces (for monopiles) are usually pre-installed with embedded bolts, ladders, and boat landing platforms. This ensures that when the tower sections arrive, they can be mounted directly without delays.

Next comes transport and staging. Tower sections, nacelles, hubs, and blades are shipped from port facilities to the offshore site using heavy-lift transport vessels. These vessels are specially designed to carry multiple components while keeping them secured against waves and vibrations. Components are staged on deck in the sequence required for installation, minimizing handling at sea.

For lifting operations, jack-up installation vessels are typically used. These ships lower legs to the seabed to stabilize the platform above wave height, creating a steady base for the heavy-lift crane. The main crane, capable of lifting hundreds of tonnes, is configured to handle tower sections, nacelles, and rotor assemblies. Rigging equipment such as spreader bars and lifting frames is inspected and prepared before lifting begins.

The tower is then assembled in sections, similar to onshore erection but with tighter tolerances due to offshore conditions. The base section is lifted from the vessel deck and placed onto the transition piece. Precision alignment ensures the tower is vertical, and bolts are tightened progressively to specification. Subsequent sections are lifted and stacked one by one until the tower reaches its full height. Throughout this stage, internal cabling, ladders, and platforms may be installed as each section is added.

Safety and weather control are critical factors. Lifting operations can only proceed when wind and wave conditions are within safe limits, typically below 9–12 m/s for lifts. Operations are paused during rough seas, poor visibility, or unexpected weather changes. Exclusion zones are enforced on deck and around the vessel to protect workers from suspended loads. Constant communication between crane operators, riggers, and deck crews is essential.

Once all sections are in place, post-installation checks are carried out. These include re-tensioning bolts, verifying tower verticality by survey, and inspecting coatings for transport or lifting damage. Internal systems such as ladders, fall arrest devices, lifts, and cabling are tested before the nacelle and rotor are installed in the next construction phase.

Human Resources

Offshore wind turbine tower construction is far more complex than onshore because of the marine environment, weather windows, vessel operations, and the need for specialized safety procedures. The workforce required is larger and more specialized, often numbering 40–60 personnel per turbine installation team, supported by marine crews, engineers, and safety officers. Each person plays a well-defined role to ensure that heavy lifts and assembly operations can be carried out safely and efficiently in challenging offshore conditions.

1. Project and Site Management

- **Offshore Installation Manager (OIM) (1):** Overall authority on the installation vessel, responsible for safety and coordination of all activities.
- **Construction/Installation Manager (1):** Directly manages turbine erection operations and liaises with the OIM, marine crew, and contractors.
- **Client/Owner's Representative (1–2):** Ensures compliance with project specifications and contractual obligations.

2. Marine and Vessel Crew

- **Master / Captain (1):** Commands the installation vessel or jack-up barge.
- **Bridge Officers / Navigation Crew (2–4):** Handle positioning, dynamic positioning (DP), and marine safety.
- **Marine Deck Crew / Bosuns (4–6):** Assist with mooring, deck operations, and load transfers.
- **Mooring/Anchor Handling Crew (2–3, as needed):** Manage anchoring systems if not using DP vessels.
- **Jack-Up Technicians / Ballast Control Operators (2–3):** Operate jack-up legs, ballast systems, and maintain vessel stability.

3. Crane and Heavy Lift Operations

- **Main Crane Operator (1):** Operates the heavy-lift crane (up to 1,200 tonnes) for tower sections, nacelle, and blades.

- **Assist Crane Operator (1):** Runs a smaller crane for handling rigging gear and smaller lifts.
- **Rigging Supervisor (1):** Oversees all rigging operations and ensures correct methods are used.
- **Riggers / Slingers (6–8):** Connect tower sections to lifting frames, guide loads during lifts, and secure components once in place.
- **Banksmen / Signalmen (2–3):** Communicate with crane operators and riggers via hand signals and radios to ensure safe lifts.

4. Tower Assembly Crew

- **Mechanical Technicians / Fitters (6–8):** Align and bolt tower flanges, tension bolts, and install internal ladders, platforms, and lifts.
- **Electrical Technicians (3–4):** Install internal cabling, grounding, and prepare connections to the nacelle.
- **Surveyors / Alignment Specialists (1–2):** Use laser or GPS equipment to verify tower verticality and flange alignment.

5. Health, Safety, and Quality

- **HSE Officers (2–3):** Enforce offshore safety protocols, monitor exclusion zones, manage PPE compliance, and coordinate emergency response.
- **Medics / First Aid Officers (1–2):** Provide medical coverage on the vessel.
- **QA/QC Inspectors (2–3):** Inspect coatings, bolts, flange connections, and confirm work meets technical specifications.

6. Logistics and Support

- **Deck Logistics Crew (4–6):** Move components from storage positions on deck to crane hooks, stage equipment, and prepare for lifting.
- **Transport/Haulage Coordination (2–3, onshore and offshore):** Ensure components are loaded, secured, and tracked between port and vessel.
- **Stores/Materials Coordinators (1–2):** Manage rigging gear, spare parts, and consumables.
- **Catering/Hotel Crew (4–6):** Provide meals, cleaning, and welfare support for offshore teams (on longer operations).

Typical Offshore HR Breakdown (per turbine installation team)

- **Management & Supervision:** 3–4

- **Marine & Vessel Crew:** 10–15
- **Crane Operators & Rigging Teams:** 10–14
- **Assembly Technicians (Mechanical & Electrical):** 9–12
- **Survey, QA/QC, HSE & Medics:** 6–8
- **Deck Logistics & Support Crew:** 6–10
- **Catering/Hotel Support:** 4–6

Total: 40–60 personnel per turbine, depending on vessel size and contractor practices.

Offshore wind turbine tower construction requires a large multidisciplinary team. Vessel crews ensure safe marine operations, crane operators and riggers handle the heavy lifts, and assembly technicians complete the bolting, alignment, and cabling of the tower sections. HSE officers and medics manage safety, while QA/QC inspectors guarantee technical compliance. Logistics personnel stage and prepare components, while catering staff support long offshore rotations. The result is a tightly coordinated operation where every role is essential to ensuring safe, efficient, and precise offshore tower erection.

<u>Duration</u>

The time required to erect the tower of an offshore wind turbine is generally longer and more weather-dependent than onshore projects. While the tower assembly itself is fairly similar in sequence, the constraints of working at sea—such as vessel positioning, sea swell, and safe weather windows—mean that operations take more time to plan and execute. On average, it takes between 2 and 4 days to erect the tower sections of a single offshore wind turbine, provided weather conditions remain favourable and all equipment functions as expected.

Typical Offshore Tower Installation Timeline

1. Pre-Installation Preparation (½–1 day)

- The installation vessel or jack-up barge is positioned at the site and jacked up or stabilized.
- Final surveys confirm foundation and transition piece readiness.
- Safety checks, toolbox talks, and equipment inspections are carried out.

2. Base Section Installation (½–1 day)

- The first tower section is lifted from the deck and carefully placed onto the transition piece.

- Precision alignment and bolting are critical at this stage.
- QA inspectors check coatings, flange conditions, and torque specifications.

3. Middle Section Installation (1–2 days)

- Each additional tower section is lifted and stacked in sequence.
- Mechanical crews complete bolting and tensioning, while electrical crews begin threading internal cables.
- Surveyors verify alignment after each section.

4. Top Section Installation (½ day)

- The final tower section is lifted and connected, completing the cylindrical tower.
- Internal ladders, lifts, and platforms are progressively installed.

5. Post-Erection Checks (½–1 day)

- Bolts are re-checked and re-tensioned.
- Tower verticality is verified using GPS or laser survey.
- Internal cabling and systems are tested to ensure readiness for nacelle installation.

Factors Affecting Duration

- **Weather Windows:** Offshore lifts are suspended if wind speeds exceed safe thresholds (usually 9–12 m/s) or if sea swell affects crane stability. Delays of several days are common.
- **Tower Height and Sections:** Taller towers with four or five sections require more time than smaller three-section towers.
- **Vessel and Crew Efficiency:** Modern jack-up vessels with optimized deck layouts and experienced crews can complete installations faster.
- **Parallel Work:** Electrical and internal installations often overlap with lifting operations, reducing total time.

In optimal conditions, offshore tower erection typically takes 2–4 days per turbine, not including mobilization of the vessel or weather-related downtime. In less favourable conditions, the same operation may extend to a week or longer due to waiting on weather. Once the tower is complete, the nacelle and rotor installation follow in subsequent steps, usually adding another 2–4 days depending on turbine size and sea conditions.

Nacelle and Rotor Installation

The installation of the nacelle and rotor is one of the most technically demanding phases of wind turbine construction, both onshore and offshore. By this stage, the tower is complete and surveyed, and the foundation is ready to receive the turbine's power-generating components. These steps require precise heavy lifting, meticulous alignment, and strict safety measures due to the weight and aerodynamic sensitivity of the components.

Nacelle Installation

The nacelle serves as the main housing of a wind turbine, containing key components such as the generator, gearbox (in geared designs), main shaft, yaw system, and various control systems. In modern onshore turbines, nacelles typically weigh between 60 and 120 tonnes, while those used in large offshore turbines can weigh significantly more.

The installation process begins with preparation. The nacelle is transported to the construction site separately, using specialized trailers for onshore projects or heavy-lift vessels for offshore installations. Before lifting can take place, the crane crew inspects the rigging equipment and secures lifting frames designed to manage the nacelle's considerable weight and precise centre of gravity. At the same time, the tower's top flange is thoroughly cleaned, inspected, and prepared, with bolts positioned and ready for the connection.

The next stage involves lifting the nacelle into place. Onshore projects use heavy-lift crawler cranes, while offshore operations rely on cranes mounted on jack-up vessels. The nacelle is hoisted in a single lift, with banksmen and riggers guiding its movement to prevent swinging. This is achieved using tag lines or, in some cases, remote-controlled rigging systems. Because of its size and shape, the nacelle is highly sensitive to wind forces, meaning that lifts can only proceed when weather conditions—especially wind speeds—are within safe operational limits.

Figure 117: Attaching Nacelle to top of Turbine Tower. Paul Anderson, CC BY-SA 2.0, via Wikimedia Commons.

Once lifted, the nacelle is lowered carefully onto the tower in the alignment and bolting phase. Precision is essential at this stage, as the nacelle must be perfectly aligned with the tower's top flange. Bolts are tightened progressively, following the manufacturer's torque specifications to ensure even stress distribution and secure attachment. Finally, electrical technicians connect the internal tower cables to the nacelle systems, ensuring that it is ready for the next major step: rotor installation.

Rotor Installation

The rotor, made up of the hub and blades, is one of the most critical assemblies in wind turbine installation. Its installation can be carried out using two main methods, depending on turbine design, available crane capacity, and prevailing weather conditions.

Figure 118: Wind turbine construction, Pougny, E6-08, blades. Cjp24, CC BY-SA 4.0, via Wikimedia Commons.

The first method is single-blade installation. In this approach, the hub is lifted first and bolted securely to the nacelle. Once in place, each blade is lifted individually and attached to the hub. The main advantage of this method is that it requires lighter lifts, which provides greater flexibility in variable wind conditions and makes blade handling more manageable. However, it does involve multiple high lifts, with each blade requiring careful positioning and precise alignment, which can extend installation time and increase operational complexity.

The second method is rotor-star (pre-assembled) installation. Here, the hub and all three blades are first assembled on the ground or deck, forming a "rotor star." The complete assembly is then lifted in a single large lift and bolted to the nacelle. This method significantly reduces the number of high lifts, as the rotor is installed in one operation. However, it demands an exceptionally large crane, extensive laydown space for pre-assembly, and very calm weather conditions to avoid the risks of swinging or instability during the lift.

Regardless of the method, the installation steps follow a structured sequence. The process begins with lifting the hub, either on its own or as part of a pre-assembled rotor star, and securing it to the nacelle. If blades are installed individually, each one is then lifted using specialized blade yokes or lifting frames that can control blade pitch and yaw during the lift. Tag lines or hydraulic stabilization systems are used to steady the blade in mid-air. Once in position, technicians align the blade root connections and torque the bolts to the manufacturer's specifications, ensuring correct balance and aerodynamic performance. For the rotor-star method, the entire rotor is lifted and mounted directly onto the nacelle's main shaft, before being secured and inspected.

This process requires precision, coordination, and strict adherence to safety and weather limitations, as the rotor assembly is one of the most challenging and high-risk stages in wind turbine construction.

Safety and Weather Considerations

Safety is paramount during rotor and blade installation, as these operations involve some of the largest and most complex lifts in wind turbine construction. Weather is a critical factor, with wind limits strictly enforced; typically, all blade and rotor lifts are suspended when wind speeds exceed 9–10 m/s, as the large surface area of the blades makes them highly susceptible to dangerous swinging. Strict exclusion zones are established beneath suspended loads to prevent personnel from entering hazardous areas during lifts. To ensure smooth and safe execution, crane operators, riggers, and ground crews maintain constant communication, often through radio headsets and standardized hand signals. Additionally, emergency preparedness is essential. Rescue plans must be in place for workers operating at height, and contingency procedures are established for aborted lifts in the event of sudden weather changes.

Post-Installation Checks

Once the rotor is installed, thorough post-installation checks are carried out to confirm the integrity and functionality of all systems. Bolt tensioning is re-verified across nacelle and rotor connections to ensure they meet manufacturer specifications and safety requirements. The blade pitch systems are tested to confirm that they move freely, respond correctly, and can adjust angles as required for aerodynamic performance. Electrical technicians finalize cabling connections, ensuring that power, control, and monitoring systems are fully integrated and ready for commissioning. Finally, a detailed survey is conducted to verify rotor alignment and balance, preventing vibrations or imbalances during operation. Only after these checks are complete can the turbine be energized and prepared for operation.

Onshore Wind Turbine Nacelle and Rotor Installation

Onshore wind turbine nacelle and rotor installation is one of the most resource-intensive phases of construction. While tower erection already involves a large team, lifting and securing the nacelle, hub, and blades demands even more specialized personnel due to the heavy weights, height of operations, and the complexity of rotor balancing. Typically, 25–35 personnel are directly involved in this stage, supported by additional logistics, safety, and inspection staff.

1. Project and Site Management

- **Project Manager / Site Manager (1–2 people):** Oversees the nacelle and rotor installation process, ensuring compliance with engineering standards, safety procedures, and the construction schedule. Coordinates between crane crews, rigging teams, and the client.
- **Installation Supervisor / Lift Supervisor (1):** Directly supervises heavy lifting operations, approves rigging configurations, and ensures method statements and risk assessments are followed.

2. Crane and Heavy Lift Operations

- **Main Crane Operator (1):** Operates the heavy-lift crawler or lattice-boom crane to raise the nacelle and rotor (or blades individually). Requires high precision and constant coordination with ground crews.
- **Assist Crane Operator (1):** Uses a smaller crane to unload, align, and support components before the main lift.
- **Rigging Supervisor (1):** Ensures correct rigging methods, lifting frames, and spreader bars are used, and that load stability is maintained.
- **Riggers / Slingers (5–6 people):** Connect slings, yokes, or lifting frames to components, manage tag lines to control loads, and guide nacelle, hub, and blades during lifts.
- **Banksman / Signalmen (2):** Relay instructions between the rigging team and crane operators using radios or standardized hand signals.

3. Nacelle and Rotor Assembly Crew

- **Mechanical Technicians / Fitters (6–8 people):** Bolt the nacelle to the tower flange, align connections, and torque bolts to specification. For the rotor, they secure the hub and blades, ensuring correct fit and tension.
- **Blade Technicians (3–4 people):** Specialize in lifting and attaching blades, controlling blade yokes or lifting frames, and ensuring correct pitch and root connection.
- **Electrical Technicians (2–3 people):** Connect internal tower cabling to nacelle systems, test yaw and pitch systems, and finalize electrical hookups.
- **Surveyor / Alignment Specialist (1):** Confirms flange alignment, nacelle position, and rotor balance using laser tools and precision instruments.

4. Health, Safety, and Quality Assurance

- **HSE Officer (1–2 people):** Monitors compliance with fall protection, exclusion zones, and weather restrictions. Leads toolbox talks and emergency response planning.

- **QA/QC Inspectors (1–2 people):** Verify bolting torque, flange integrity, coating condition, and rotor alignment before commissioning.

5. Logistics and Ground Crew

- **Transport Crew (3–4 people):** Deliver nacelle, hub, and blades to site using specialized trailers and assist in staging components for lifting.
- **Ground Crew / General Laborers (2–3 people):** Assist with moving rigging gear, handling tools, securing tag lines, and supporting technicians during assembly.

Typical HR Breakdown (Per Turbine, Nacelle & Rotor Stage)

- **Management & Supervision:** 2–3
- **Crane Operators & Rigging Teams:** 9–11
- **Mechanical & Blade Technicians:** 9–12
- **Electrical & Survey Specialists:** 3–4
- **HSE & QA/QC Inspectors:** 2–4
- **Logistics & Ground Crew:** 5–7

Total: ~25–35 personnel, depending on turbine size, rotor lift method (single-blade vs. pre-assembled rotor), and site complexity.

Onshore nacelle and rotor installation is a carefully sequenced operation that depends heavily on weather conditions, site readiness, and the lifting method used. Under normal conditions, this phase of turbine construction typically takes 1.5 to 3 days per turbine, though this can vary based on turbine size and whether blades are lifted individually or as a pre-assembled rotor star.

The nacelle lift itself is usually completed in a matter of hours. Once the heavy-lift crawler crane is set up and the rigging is inspected, the nacelle is hoisted, guided into place, and bolted down. This process—including final torque checks and electrical connections—can take between 6–10 hours. However, crane setup and weather delays can extend this timeline considerably.

The rotor installation is more variable. With single-blade installation, each blade is lifted individually, which requires three separate heavy lifts. Each lift involves rigging, stabilizing the blade, guiding it into position, and securing it with torque and pitch alignment. Depending on conditions, each blade can take 4–6 hours, so the entire rotor assembly might extend over 1.5–2 days. By contrast, a rotor-star installation—where the hub and blades are pre-assembled on the ground—can reduce the number of high lifts to just one. This saves time aloft, but requires much larger crane capacity, more ground space, and very calm wind conditions. If conditions allow, a rotor-star lift can be completed in 6–8 hours, but delays often occur due to stricter weather limits.

Throughout the process, bolt torquing, cable connections, and alignment checks add additional hours. Quality inspections, safety checks, and documentation must be completed before the turbine can move into final commissioning. Weather remains the most critical factor: high winds can halt lifting operations instantly, sometimes delaying work for several days.

In summary, the nacelle and rotor installation stage typically requires 25–35 personnel working over 1.5 to 3 days per turbine, depending on crane setup, lift method, turbine size, and weather conditions. Under ideal conditions, the process is fast and efficient, but in practice, downtime from wind or other environmental factors often dictates the pace of work.

Offshore Wind Turbine Nacelle and Rotor Installation

Offshore nacelle and rotor installation is one of the most demanding stages of wind turbine construction because of the massive component weights, limited weather windows, and reliance on specialized installation vessels and heavy-lift cranes. Compared to onshore work, the number of personnel involved is larger and more diverse, since marine operations, safety, and logistics require additional layers of expertise. A typical offshore installation team for the nacelle and rotor phase involves 50–70 personnel per turbine, spread across marine crew, crane operations, rigging teams, assembly technicians, safety, and support staff. Each role contributes to precise, safe, and efficient offshore lifting and assembly.

1. Project and Site Management

- **Offshore Installation Manager (OIM) (1):** Holds ultimate responsibility for operations and safety aboard the installation vessel or jack-up platform.
- **Installation Supervisor / Lift Supervisor (1):** Directly oversees the nacelle and rotor lifts, ensures compliance with method statements, and coordinates with the crane crew and rigging team.
- **Client/Owner's Representative (1–2):** Monitors work to ensure it meets project specifications and contractual obligations.

2. Marine and Vessel Crew

- **Master / Captain (1):** Commands the installation vessel, overseeing positioning and navigation.
- **Bridge Officers / Navigation Crew (2–3):** Handle vessel manoeuvring and dynamic positioning (DP).
- **Marine Deck Crew / Bosuns (4–6):** Manage mooring, deck operations, and component transfers from supply vessels.
- **Ballast Control / Jack-Up Operators (2–3):** Maintain vessel stability by operating ballast tanks and jack-up legs.

3. Crane and Heavy Lift Operations

- **Main Crane Operator (1):** Operates the heavy-lift crane (often 1,000–1,500 tonnes capacity) to hoist nacelle, hub, and rotor or blades.
- **Assist Crane Operator (1):** Handles smaller lifts such as rigging frames, blade yokes, or tools.
- **Rigging Supervisor (1):** Ensures all rigging configurations are safe, compliant, and suitable for offshore conditions.
- **Riggers / Slingers (6–8):** Attach lifting gear to nacelle, hub, and blades; manage tag lines; stabilize loads during lifting.
- **Banksmen / Signalmen (2–3):** Relay commands between riggers and crane operators using radios and hand signals.

4. Nacelle and Rotor Assembly Crew

- **Mechanical Technicians / Fitters (6–8):** Secure nacelle to the tower top, align flanges, torque bolts, and assemble the hub and blades.
- **Blade Technicians (3–4):** Specialize in controlling blade lifts, attaching blades to the hub, and ensuring correct pitch alignment.
- **Electrical Technicians (3–4):** Connect nacelle power and control systems, integrate cabling from the tower, and test yaw and pitch mechanisms.
- **Surveyors / Alignment Specialists (1–2):** Confirm nacelle alignment, tower verticality, and rotor balance using GPS and laser systems.

5. Health, Safety, and Quality Assurance

- **HSE Officers (2–3):** Monitor weather thresholds, enforce exclusion zones, check PPE, and coordinate emergency response plans.
- **Medics / First Aid Officers (1–2):** Provide medical coverage onboard the installation vessel.
- **QA/QC Inspectors (2–3):** Inspect bolts, coatings, flange connections, and rotor alignment for compliance with specifications.

6. Logistics and Support Crew

- **Deck Logistics Crew (4–6):** Move components from deck storage to crane hook positions, prepare rigging gear, and manage laydown areas.
- **Materials / Stores Coordinators (1–2):** Track consumables, rigging gear, and spare parts needed during installation.

- **Transport/Haulage Coordinators (2–3, split onshore/offshore):** Ensure timely transfer of nacelles, hubs, and blades from port to vessel.
- **Catering and Hotel Crew (4–6):** Provide meals, cleaning, and welfare for the team during offshore rotations.

Typical Offshore HR Breakdown (Per Turbine Installation – Nacelle & Rotor Stage)

- **Management & Supervision:** 3–4
- **Marine & Vessel Crew:** 10–15
- **Crane Operators & Rigging Teams:** 10–14
- **Mechanical, Blade & Electrical Technicians:** 12–16
- **Survey, QA/QC, HSE & Medics:** 6–8
- **Logistics & Support Crew:** 7–10
- **Catering/Hotel Support:** 4–6

Total: ~50–70 personnel per turbine, depending on turbine size, lift method (single-blade vs. rotor-star), and vessel type.

Offshore nacelle and rotor installation requires a multidisciplinary team significantly larger than onshore projects. Marine crew ensure vessel positioning and stability, crane operators and riggers execute heavy lifts, and assembly technicians handle precise bolting, alignment, and electrical integration. Safety officers, medics, and QA/QC inspectors provide oversight, while logistics and support staff keep operations flowing in a challenging offshore environment. Every role is critical because delays offshore are costly and weather windows are short, so the process must be executed efficiently and safely.

The installation of the nacelle and rotor on an offshore wind turbine typically takes 1–3 days per turbine, depending on turbine size, site conditions, and weather windows. This phase is one of the most time-critical elements of offshore construction because the lifts involve some of the heaviest and most complex components, and operations can only proceed under very specific weather limits.

The nacelle, weighing hundreds of tonnes, is lifted and secured onto the tower top using a heavy-lift crane mounted on a jack-up vessel or floating installation ship. The process of lifting, aligning, and bolting the nacelle usually takes 6–12 hours, assuming calm weather and no delays. Electrical technicians then connect the nacelle systems to the tower's internal cabling, which can take an additional 4–6 hours.

Rotor installation time depends on the chosen method:

- **Single-Blade Installation:** Each blade is lifted individually and attached to the hub at height. This method takes longer because it involves three separate high lifts, with alignment and bolting for each blade. A full rotor installation by this method typically takes 1.5–2 days under good conditions.

- **Rotor-Star (Pre-Assembled) Installation:** The hub and blades are assembled into a rotor star on deck, then lifted as one unit. This reduces the number of high lifts but requires more deck space and a very large crane. If conditions are favourable, this method can be completed in 12–18 hours, making it faster than single-blade lifts. However, it is more weather-sensitive.

The biggest factor in offshore installation time is weather. Blade and rotor lifts are typically suspended if wind speeds exceed 9–10 m/s, and wave height restrictions can also prevent safe crane operations. Weather delays may extend nacelle and rotor installation by several days, especially during winter or in rough seas.

After installation, final bolt torquing, alignment checks, and electrical hookups are performed. Surveyors verify nacelle positioning and rotor balance using precision equipment. These post-installation activities usually take half a day to a full day, depending on turbine complexity.

Overall Duration:

- **Best-case scenario (good weather, rotor-star method):** 1–1.5 days
- **Typical scenario (single-blade method with minor weather interruptions):** 2–3 days
- **Delayed scenario (weather holds or technical issues):** 4–5 days or more

In practice, project planners allocate 3 days per turbine for nacelle and rotor installation to allow for weather variability, safety checks, and technical adjustments. This conservative planning ensures deadlines can still be met even with interruptions.

Cable Routing and Terminations

Cable routing and terminations are an essential part of wind turbine construction, ensuring the safe and efficient transfer of both electrical power and communication signals between components and ultimately to the grid. This work requires detailed planning, precise installation, and thorough testing, as it involves high-voltage systems and must withstand harsh environmental conditions. Inside the turbine, cables must be routed through the tower, nacelle, and hub, while in offshore projects they also extend through subsea array and export cables. Long-term reliability depends on correct routing, secure terminations, and effective protection against wear, movement, and corrosion.

Inside the tower, large power cables—typically copper or aluminium conductors—are installed vertically from the nacelle to the base, where they connect to transformers or switchgear. Alongside these are fibre optic and control cables that carry data from sensors and control systems. To support these runs, cables are fixed to trays or ladders with specialized hangers or cleats that prevent strain at terminations. At the top of the tower, additional flexibility is built into the routing to accommodate the yaw movement of the nacelle, often using loops or drag chains to reduce stress. Environmental considerations are also critical; cables are jacketed and routed in ways that minimize exposure to condensation, abrasion, and sharp bends to preserve insulation integrity.

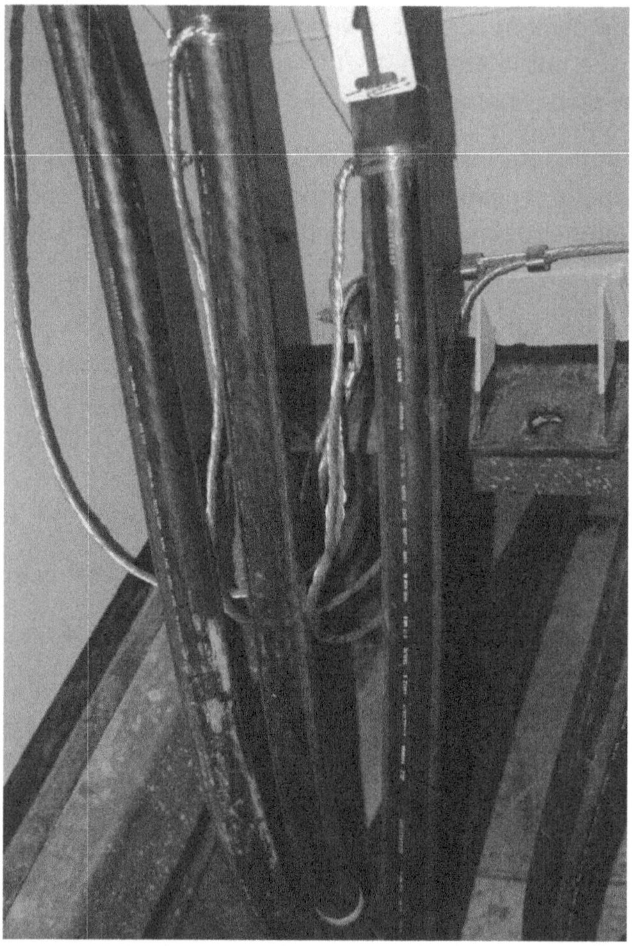

Figure 119: High voltage cables to the transformer. Patrick Finnegan, CC BY 2.0, via Flickr.

In the nacelle and hub, cabling provides connections to the generator, gearbox, yaw drives, pitch systems, cooling systems, and various sensors. High-voltage generator cables link to transformers or switchgear, while smaller cables manage control and instrumentation. Modern turbines also route cabling into the hub to power and control blade pitch systems, using cable loops or slip rings to cope with continuous movement during operation.

For offshore wind farms, cable routing extends beyond the turbine structure itself. Array or inter-array cables connect multiple turbines in a string to offshore substations, typically at medium voltages of 33–66 kV. Export cables then carry electricity from the offshore substation to the mainland grid at much higher voltages, often 132 kV or 220 kV and above. These subsea cables are routed through J-tubes or I-tubes in the turbine foundation, equipped with bend restrictors and protective sleeves to withstand mechanical stress in the splash zone. Burial beneath the seabed provides additional protection from anchors, fishing gear, and strong currents.

Cable terminations are another critical stage, where cables are connected to switchgear, transformers, or other equipment. For high-voltage power cables, terminations use stress control devices, insulation layers, and moisture barriers, often applied with cold-shrink or heat-shrink kits. Fiber optic and data cables require highly accurate splicing or connectors to maintain signal quality. Offshore terminations, in particular, must be watertight and pressure-resistant, often relying on resin-filled joints or specialized connectors designed for subsea conditions.

Before a turbine is energized, extensive testing and commissioning are carried out. Continuity tests confirm correct routing and connection, while insulation resistance tests check for damage or moisture ingress. High-voltage withstand or VLF testing verifies the dielectric strength of power cables. Fiber optic lines are tested with OTDR equipment to ensure signal integrity and detect any breaks or excessive losses. These measures ensure that the system is fully reliable before it enters operation.

Cable routing and terminations in wind turbines combine precise electrical engineering with robust mechanical protection. Inside the tower and nacelle, routing must balance power delivery and data transmission with the movement of components, while offshore systems add the complexity of subsea cable protection. Proper terminations and rigorous testing are vital for ensuring that turbines perform safely and efficiently throughout their 20–30 year service life.

The work of cable routing and terminations in wind turbines is carried out by a mix of electrical specialists, installation technicians, and commissioning engineers, supported by offshore and onshore crews depending on the project location. Because this stage involves both high-voltage power systems and sensitive fibre optic communication lines, it requires a multidisciplinary team with precise training and certifications.

1. Electrical Technicians and Cable Installers

These technicians are responsible for pulling, routing, and fixing the large copper or aluminium power cables that run from the nacelle to the base of the tower. They install the cables onto trays, ladders, or hangers and ensure correct strain relief so that the terminations are not overloaded. They also handle medium-voltage cabling in offshore array systems. Many are qualified electricians with additional training in high-voltage (HV) and medium-voltage (MV) cable handling.

2. High-Voltage (HV) Cable Jointers and Termination Specialists

HV jointers are highly trained professionals who perform the actual cable termination work, including applying stress control devices, insulation, and moisture barriers. Offshore, they also install resin-filled joints and watertight pressure connectors for subsea cables. This role is critical because improper terminations are a leading cause of early failures in wind farm electrical systems. Jointers typically hold specialist certifications for 33 kV, 66 kV, and up to 220 kV cables.

3. Fiber Optic Technicians

Alongside power cables, modern turbines use extensive fibre optic networks for control, monitoring, and communication. Fiber optic technicians are responsible for splicing and terminating these lines with high precision, often using OTDR (Optical Time Domain Reflectometer) testing equipment. Their work ensures low signal loss and reliable data transmission between the turbine, substation, and control centre.

4. Mechanical and Installation Technicians

Cable routing is not purely electrical—mechanical fitters or installation technicians also play a role. They prepare pathways, install trays, ladders, and protective drag chains, and fit bend restrictors or sleeves around cables to prevent abrasion. Offshore, they work closely with marine deck crews to pull subsea cables through J-tubes and secure them with protection systems.

5. Surveyors and Quality Assurance Inspectors

Surveyors verify correct cable routes, bend radii, and strain relief using laser alignment or inspection tools. QA/QC inspectors check coating integrity, verify torque settings on fixtures, and ensure that all terminations meet manufacturer and regulatory standards. Their oversight helps prevent hidden defects that could cause failures years later.

6. Commissioning Engineers

Once the cables are routed and terminated, commissioning engineers test and energize the system. They perform continuity checks, insulation resistance measurements, high-voltage withstand (VLF) testing, and fibre optic loss analysis. They also oversee system integration with SCADA (Supervisory Control and Data Acquisition) and grid connection. These engineers often have a background in electrical power engineering and are responsible for final sign-off before the turbine goes live.

7. Offshore Marine and Deck Crew (for offshore projects)

In offshore wind farms, marine crew, ROV operators, and deck technicians support the subsea cable laying and pull-in process. They assist with burial operations, protect cables with mattresses or rock placement, and guide them into J-tubes on the turbine foundation. Their role is essential to the safe handling of multi-tonne cables in dynamic marine conditions.

Cable routing and terminations are performed by a coordinated team of HV cable jointers, electrical technicians, fibre optic specialists, mechanical installers, surveyors, QA/QC inspectors, and commissioning engineers, supported by offshore deck crews where necessary. Each group brings specialist skills—whether it is securing a 220 kV export cable, splicing fibre optic lines, or testing high-voltage integrity. Together, they ensure the wind turbine's electrical and communication systems are safe, durable, and capable of operating reliably for decades.

Onshore Wind Turbine Cable Routing and Terminations

Typical team size: 10–15 personnel per turbine

- **Electrical Technicians / Cable Installers (3–4):** Pull and route power and control cables inside the tower, nacelle, and hub. Fix cables to trays, ladders, and hangers.
- **HV Cable Jointers / Termination Specialists (2–3):** Perform high-voltage and medium-voltage terminations at the base of the tower and inside the nacelle, using stress control kits and insulation systems.
- **Fiber Optic Technicians (1–2):** Splice and terminate fibre optic cables, ensuring signal quality for control and monitoring systems.
- **Mechanical / Installation Technicians (1–2):** Install cable trays, ladders, and protective systems; fit drag chains or loops at the nacelle interface.
- **Surveyor / QA/QC Inspector (1):** Verify correct cable routes, bend radii, strain relief, and termination quality.
- **Commissioning Engineer (1–2):** Conduct electrical and fibre testing (continuity, insulation resistance, VLF testing, OTDR for optics) and approve readiness for energization.

Onshore teams are smaller because all work is done within the tower structure, without subsea cable installation.

For an onshore turbine, cable routing and terminations typically take 2–4 days per turbine once tower erection is complete.

- **Day 1–2:** Electrical technicians route power, control, and fibre optic cables inside the tower, nacelle, and hub. This includes mounting on trays and ladders, securing with cleats, and accommodating yaw movement at the tower top.
- **Day 2–3:** High-voltage jointers perform cable terminations at the nacelle and base. At the same time, fibre optic technicians splice and terminate data cables.
- **Day 3–4:** Mechanical installers fit protective devices such as drag chains or bend restrictors. Commissioning engineers carry out continuity, insulation resistance, and high-voltage tests, and fibre optic lines are tested for signal quality.
- **Overall:** Smaller teams (10–15 people) make onshore work relatively straightforward, though weather or tower access issues can occasionally cause delays.

Offshore Wind Turbine Cable Routing and Terminations

Typical team size: 20–30 personnel per turbine, plus marine support

- **Electrical Technicians / Cable Installers (4–6):** Route power, control, and data cables inside the tower, nacelle, hub, and transition piece.
- **HV Cable Jointers / Termination Specialists (3–4):** Handle subsea array and export cable terminations inside the transition piece and offshore substation. Install watertight and pressure-resistant joints.
- **Fiber Optic Technicians (2–3):** Splice, terminate, and test fibre optic communication lines within the turbine and subsea connections.
- **Mechanical / Installation Technicians (2–3):** Install J-tubes, bend restrictors, drag chains, and subsea cable protection systems.
- **Marine Deck Crew (4–6):** Support cable pull-in from installation vessels, guide cables through J-tubes, and handle offshore cable protection equipment.
- **ROV Operators (2–3):** Monitor subsea burial, seabed cable routing, and protection in splash zone and seabed environments.
- **Surveyors / QA/QC Inspectors (2–3):** Verify subsea cable burial depth, termination quality, and compliance with standards.
- **Commissioning Engineers (2–3):** Conduct high-voltage testing, system integration, and SCADA communication checks.

Offshore teams are much larger due to subsea cable handling, vessel operations, and safety requirements in marine conditions.

Offshore cabling and terminations are much more complex and usually take 5–10 days per turbine, not including subsea export and array cable laying from the installation vessel.

- **Day 1–3:** Electrical technicians and deck crews pull in subsea cables through J-tubes or I-tubes on the foundation, securing them with bend restrictors and sleeves. Mechanical installers support this process to ensure protection in the splash zone.
- **Day 2–5:** HV jointers perform medium- and high-voltage terminations (33–220 kV) inside the transition piece and connect to offshore substations. Fiber optic specialists splice and test communication lines at the same time.
- **Day 5–7:** ROV operators verify seabed burial depth and protection. Surveyors and QA/QC inspectors confirm compliance with specifications.
- **Day 7–10:** Commissioning engineers carry out high-voltage withstand (VLF) tests, continuity checks, and SCADA integration tests before energisation. Offshore conditions—such as wave height and wind speed—can delay or suspend these activities, often extending the timeline.

- **Overall:** Larger teams (20–30 per turbine, plus marine support crews) are required offshore, and weather windows heavily influence productivity.

Chapter 11

Commissioning Procedures

Once a wind turbine is fully assembled, it is not yet ready to produce power — it must first prove that every component, system, and safety mechanism functions exactly as intended. The commissioning phase represents the bridge between construction and operation, transforming an assembled turbine into a certified, grid-connected energy asset. This stage is one of the most technically demanding and highly regulated in the entire project lifecycle, as it verifies compliance with design specifications, safety standards, and performance guarantees before the turbine is officially handed over to the operator.

Commissioning involves a systematic program of testing, inspection, and verification that ensures the turbine can operate safely, efficiently, and reliably. Electrical, mechanical, hydraulic, and control systems are individually tested, then integrated to confirm that all subsystems work together seamlessly under real-world conditions. Engineers simulate operational and fault scenarios to validate responses from the turbine's automation, braking, yaw, and pitch systems, while electrical specialists confirm grounding integrity, insulation resistance, transformer performance, and protection relay functionality. Every test is meticulously documented to create a verified chain of evidence that supports quality assurance, warranty claims, and regulatory approval.

Equally important is the functional integration phase, where the turbine's automated control logic, power electronics, and grid interface are tested together. This process includes start-up and shutdown sequence validation, grid synchronization, and fault ride-through testing to meet national grid codes and international standards such as IEC 61400. Meanwhile, SCADA and data acquisition systems are checked for accurate monitoring, alarm reporting, and remote operation capabilities — ensuring operators can control and track performance in real time once the turbine enters service.

The final stages of commissioning — power performance verification and handover to operations — mark the turbine's transition into full commercial operation. Independent testing organizations confirm that actual energy output matches the manufacturer's performance curve, and once reliability and safety are demonstrated, ownership formally transfers to the operator.

In this chapter, readers will learn how commissioning procedures turn engineering precision into operational reliability. It explains the testing frameworks, team roles, standards, and documentation processes that collectively ensure every wind turbine operates safely, efficiently, and in compliance with regulatory and contractual obligations.

System Testing and Functional Checks

Commissioning procedures play a critical role in transitioning from construction to operational phases for wind turbines. This transition ensures that all systems within the turbine are safe, functional, and compliant with regulatory and contractual requirements prior to connecting to the electrical grid. The importance of system testing and functional checks is particularly emphasized during this phase, which involves verifying that subsystems and integrated systems function correctly under controlled circumstances. These tests confirm that turbines operate as designed and interact properly with the broader wind farm and grid system [210].

The primary aim of rigorous system testing and functional checks is to validate the correct installation and functionality of electrical, mechanical, hydraulic, and control systems across varied operational scenarios. During this phase, potential defects are identified and rectified before the turbine is energized to commence power production. Such procedures provide documented evidence that the turbine meets the technical specifications, regulatory standards, and contractual obligations established by both the manufacturer and project owners. The systematic approach taken during this testing phase is crucial as it serves to assure quality and mitigate risks associated with operational failures [220].

Electrical system testing is a cornerstone of the commissioning process. This involves conducting continuity and insulation resistance tests to verify that power and control cables are correctly routed and maintain insulation integrity. High-voltage withstand testing (VLF) is utilized to ensure dielectric strength before the turbine's connection to the grid. Grounding systems are assessed for proper bonding and resistance levels, which are critical for ensuring adequate lightning protection and operator safety [221]. Furthermore, transformers and switchgear undergo thorough evaluations, including insulation resistance checks, assessment of turns ratios, and functionality testing of protective relays to ensure reliability [222].

Electrical system testing is a cornerstone of wind turbine commissioning, ensuring that all circuits and equipment are safe, functional, and ready for operation. One of the first checks carried out is continuity and insulation resistance testing. Continuity tests confirm that every conductor in both power and control circuits is correctly routed and free from open circuits, shorts, or cross-connections. A low resistance reading indicates that the pathway is intact. Insulation resistance tests, on the other hand, assess the quality of insulation between conductors and between conductors and ground. High resistance values confirm that insulation is secure, while lower readings may indicate damage, contamination, or moisture ingress. These checks are usually performed with a megohmmeter ("megger") by electrical technicians for low- and medium-voltage circuits, and by high-voltage commissioning engineers for the main power systems. Results are then verified by QA/QC inspectors and logged in commissioning reports.

Another essential procedure is high-voltage withstand testing, often conducted at very low frequency (VLF). In this test, a high alternating voltage is applied to medium- and high-voltage cables at a reduced frequency, typically around 0.1 Hz. The aim is to simulate operational stress on the cable insulation without causing overheating. This test verifies that the insulation can withstand the designed operating voltages plus additional safety margins, and it often reveals weak points or defective terminations that could fail under real operating conditions. Specialist HV test engineers, certified to perform VLF testing, usually carry out this task, while HV jointers or termination specialists remain on hand to correct any defects discovered.

Grounding, or earthing, systems are also rigorously verified during commissioning. Resistance measurements are taken between grounding electrodes, foundation rebar grounding grids, and main electrical components to confirm proper bonding. Methods such as fall-of-potential testing or clamp-on resistance measurements are used to check whether the earthing system will safely dissipate lightning strikes and fault currents into the soil. These tests are typically performed by electrical technicians, but commissioning engineers and safety officers review the results since grounding integrity is directly linked to both equipment safety and personnel protection. Compliance with international standards, such as IEC 61400-24 for wind turbines or IEEE 80 for grounding practices, must also be demonstrated.

Transformers undergo their own set of specialized tests. Insulation resistance checks confirm the condition of windings, while turns ratio tests (TTR) verify that the voltage conversion ratio between primary and secondary windings is correct. Polarity and phase checks ensure that connections are correctly sequenced, preventing phase errors. For oil-filled transformers, dissolved gas analysis (DGA) and oil dielectric strength tests may also be performed to assess internal condition. Finally, protective relays linked to the transformer are tested to confirm that they trip reliably under overcurrent, earth fault, or other abnormal scenarios. These activities are generally performed by transformer test engineers, often from the original equipment manufacturer (OEM) or specialist contractors, with commissioning engineers overseeing integration into the turbine's electrical system. QA/QC inspectors verify that all results meet manufacturer and grid compliance standards.

Switchgear testing forms another vital part of the commissioning process. Mechanical checks confirm that circuit breakers and isolators open and close smoothly, while insulation resistance tests assess the integrity of insulation between phases and to ground. Contact resistance measurements verify that current can flow efficiently across breaker contacts, minimizing losses and heat buildup. Protection relay function tests simulate fault conditions such as short circuits, undervoltage, or overcurrent to ensure breakers respond correctly. Interlock checks, both mechanical and electrical, confirm that unsafe switching sequences cannot occur. These tasks are typically performed by high-voltage commissioning engineers or relay technicians with specialist training, while OEM representatives may be present to validate warranty conditions and certify equipment performance.

In addition to electrical systems, control systems, including Programmable Logic Controllers (PLCs) and SCADA (Supervisory Control and Data Acquisition) systems, are scrutinized for operational integrity. This verification process assures that commands and sensor feedback are accurately processed, enabling effective communication between the turbine, substation, and control centre. Testing SCADA links, including redundancy measures, is essential to confirm these systems can handle faults like overspeed conditions or grid failures seamlessly [223]. Functional checks for alarms and interlocks, as well as validation of remote control capabilities, ensure safe operations and effective management of the turbine systems [224].

Commissioning checks for wind turbine control systems are a vital stage in ensuring safe and reliable operation before grid connection. These checks focus on verifying the turbine's automation, communication, and safety functions so that it responds correctly to both operator commands and fault conditions. Several specialist roles are involved in carrying out this work, each with defined responsibilities.

One of the first steps is PLC (Programmable Logic Controller) verification. The PLC serves as the turbine's core automation controller, managing startup, shutdown, and emergency procedures. During commissioning, engineers simulate a range of operational and fault scenarios—such as overspeed, overheating, or excessive vibration—to verify that the PLC correctly interprets sensor inputs and triggers appropriate outputs. This ensures that systems like yaw drives, pitch motors, and brakes operate as programmed. These checks are performed by PLC commissioning engineers or automation specialists, often with the support of turbine manufacturer representatives. QA/QC inspectors then review and sign off on the results before the turbine is prepared for synchronization.

Another essential step is SCADA (Supervisory Control and Data Acquisition) testing. SCADA links each turbine to the substation and the central control room, enabling remote monitoring and control. Commissioning involves verifying that operational data—such as wind speed, power output, temperatures, and vibration levels—is transmitted accurately, and that operator commands (start, stop, reset) are reliably executed. Redundancy features are also tested to ensure that backup communication paths activate if a fibre optic line or network switch fails. In some cases, cybersecurity and signal integrity are assessed to confirm compliance with grid requirements. SCADA engineers or IT/communications specialists carry out these checks, with oversight from commissioning engineers, and often involve control centre operators to validate remote communication.

Equally important are alarm and interlock functional checks. Each critical turbine subsystem, including yaw motors, pitch systems, generator cooling, lubrication pumps, and braking mechanisms, has alarms and interlocks programmed into the control system. During commissioning, engineers simulate fault conditions—for example, triggering overspeed or simulating a loss of lubrication flow—to confirm that alarms activate, events are logged, and signals are sent to SCADA. Interlocks, such as preventing turbine restart after an emergency stop until safety resets are confirmed, are also validated. Control system engineers and commissioning engineers typically perform these checks, supported by electrical technicians simulating inputs. QA/QC inspectors confirm that alarms occur at the correct thresholds and are properly documented.

Remote control functionality is verified to ensure that turbines can be operated reliably from the wind farm control centre. This includes testing remote start, stop, and monitoring functions, as well as verifying that operational status, alarms, and performance data are correctly displayed in real time. SCADA engineers and commissioning engineers lead this process, with input from grid operators or control centre staff. In many cases, OEM representatives are present to validate compliance with project specifications and warranty conditions.

Mechanical and hydraulic systems are subjected to rigorous testing to confirm operational reliability. The yaw system, responsible for the nacelle's orientation, is checked for responsive rotation and braking capabilities. Pitch systems are cycled fully to validate actuator performance, sensor accuracy, and emergency fail-safe operations. Brake systems are tested under both standard and emergency conditions to ensure readiness, and lubrication and cooling systems are verified for proper fluid management and component functionality [225].

Mechanical and hydraulic system checks during wind turbine commissioning are designed to prove that all moving and fluid-based subsystems operate safely, reliably, and as specified. Because these components control nacelle orientation, blade pitch, braking, and cooling, their performance directly affects turbine safety and power generation. The checks are carried out in several stages, each led by specialist personnel.

Yaw System Testing involves rotating the nacelle to verify that yaw drives, motors, and gearboxes respond smoothly to PLC commands. Braking functions are also validated to ensure the nacelle can stop at precise orientations and hold position against wind loads. These checks are performed by mechanical technicians and commissioning engineers, with surveyors confirming yaw alignment accuracy. Electrical and control engineers may also be involved to validate feedback sensors and PLC integration.

Pitch System Verification requires cycling each blade through its full range of motion. Hydraulic or electric pitch actuators are tested for speed, accuracy, and smoothness of movement. Sensors and encoders are checked for proper feedback, and fail-safe mechanisms are verified by simulating power loss to confirm that blades feather automatically. Blade technicians, hydraulic specialists, and control system engineers typically carry out these checks, while QA/QC inspectors ensure the results match manufacturer specifications.

Brake System Testing is critical for safety. Mechanical and hydraulic brakes on the main shaft are tested under both normal operating conditions and emergency stop scenarios. The response time, braking force, and system redundancy are confirmed to guarantee that the turbine can be safely stopped in high winds or fault situations. Mechanical fitters and commissioning engineers oversee these procedures, supported by electrical technicians who simulate trigger signals.

Lubrication and Cooling System Checks focus on ensuring correct fluid circulation and temperature control for gearboxes, bearings, and power electronics. Pumps, filters, flow meters, and oil levels are inspected, while cooling fans, radiators, or heat exchangers are tested for airflow and heat dissipation. These checks are performed by mechanical technicians and lubrication specialists, often under the supervision of commissioning engineers.

In each case, QA/QC inspectors review test records, confirm compliance with OEM standards, and document results for warranty and regulatory requirements. HSE officers also monitor safety during tests, especially for emergency braking and hydraulic system simulations.

Together, these checks validate that yaw, pitch, braking, and fluid systems are ready for service. Mechanical technicians, hydraulic engineers, blade specialists, control engineers, and commissioning supervisors work as a coordinated team to ensure that every subsystem responds correctly before the turbine is allowed to operate under load.

Functional integration testing brings together these subsystems under simulated operational scenarios, ensuring that start-up sequences occur correctly and that manual and automated shutdown procedures are reliable under fault conditions. Testing aligns power electronics with grid voltage and frequency while also assessing fault ride-through capabilities, which must meet grid code requirements [226]. Verification of redundant systems—such as emergency lighting and fire suppression—further enhances operational resilience.

Functional integration testing is the stage in wind turbine commissioning where all subsystems—electrical, mechanical, hydraulic, and control—are tested together under simulated operational conditions. The goal is to confirm that the turbine behaves as an integrated system, starting up and shutting down in the correct sequence, aligning with grid requirements, and responding safely to fault conditions. Because this phase involves both system-level logic and hardware, it requires a multidisciplinary team of engineers and technicians.

Start-Up and Shutdown Sequences are tested first. Engineers simulate turbine start-up, verifying that each system (yaw, pitch, lubrication, generator excitation, and power electronics) engages in the correct order. Similarly, shutdown procedures—both normal and emergency—are triggered to confirm the turbine responds safely, such as feathering blades, applying brakes, and disconnecting from the grid. These tests are typically performed by commissioning engineers and control system engineers, with PLC and SCADA specialists monitoring logic execution.

Grid Synchronization Checks confirm that the turbine's power electronics align with the grid's voltage, frequency, and phase. Engineers simulate synchronization and confirm that the turbine exports power without causing disturbances. Fault ride-through tests may also be conducted, where grid faults (voltage dips or frequency changes) are simulated to ensure the turbine remains connected in compliance with grid codes. High-voltage commissioning engineers, electrical power engineers, and grid integration specialists usually lead these tests, often with representatives from the grid operator or utility present.

Redundancy and Backup Systems are also tested. Emergency lighting, fire suppression systems, and backup batteries are verified to ensure they function during power loss or fault events. Safety engineers, commissioning engineers, and HSE officers are responsible for these checks, while QA/QC inspectors document results for compliance.

Who Performs It:

- Commissioning Engineers oversee the integration process, verifying that subsystems work together.
- Control System/Automation Engineers validate PLC and SCADA logic during start-up, shutdown, and fault simulations.
- Electrical Power Engineers and HV Test Specialists perform synchronization and fault ride-through tests.
- Safety and HSE Officers confirm redundancy systems (lighting, fire suppression, backup power) are functional.
- QA/QC Inspectors review results, ensuring compliance with manufacturer, client, and grid code requirements.
- Grid Operator Representatives may also participate in final synchronization and grid compliance testing.

Together, these checks ensure that the turbine not only functions as designed in isolation but also operates safely and reliably as part of the wind farm and wider electrical grid.

Data logging and initial performance verification are vital components of the commissioning process. Sensors measuring wind speed, temperature, direction, and vibration are calibrated and assessed for accuracy, with data acquisition systems being tested to ensure effective logging and transmission of operational metrics. Initial performance evaluations establish baseline values for key operational parameters, which are essential for ongoing monitoring and maintenance strategies [227].

Data logging and initial performance verification represent some of the final and most important stages in the commissioning of a wind turbine. These processes ensure that the turbine not only functions as intended but also provides accurate and reliable operational data that can be used throughout its service life. The checks focus on sensor calibration, validation of data acquisition systems, and the establishment of baseline performance values. This work requires technical accuracy and the combined expertise of electrical specialists, instrumentation technicians, and commissioning engineers.

One of the first steps is sensor calibration and verification. Wind turbines are equipped with numerous sensors, including anemometers and wind vanes to measure wind speed and direction, thermocouples and RTDs for temperature monitoring, and accelerometers for vibration detection. Each sensor must be calibrated against a known standard to ensure accuracy. For instance, wind sensors may be compared against meteorological mast data or handheld reference devices, while vibration sensors are validated using test signals or shaker devices. These calibrations ensure that the control system receives precise input for safe and efficient turbine operation. Instrumentation technicians, turbine manufacturer engineers, and commissioning engineers typically perform this work, with QA/QC inspectors reviewing calibration certificates to ensure compliance.

The next stage involves data acquisition and logging checks. All sensor outputs are fed into the turbine's SCADA (Supervisory Control and Data Acquisition) system or local data loggers. Commissioning checks confirm that data is being captured without corruption, logged continuously, and transmitted reliably to the wind farm's control centre. These tests also assess timestamp accuracy, signal integrity, and redundancy, ensuring that backup systems maintain data capture even in the event of a communication pathway failure. SCADA engineers, IT specialists, and commissioning engineers—often supported by OEM software engineers—are responsible for these tasks.

Once data logging is validated, baseline performance verification is carried out. The turbine is run under controlled conditions to record key operational parameters such as power curves, drivetrain vibration levels, nacelle and generator temperatures, and hydraulic pressures in the yaw and pitch systems. These baseline values provide a reference point for future condition monitoring and predictive maintenance, enabling operators to detect early signs of deviation or wear. Commissioning and performance engineers usually handle this stage, while results are reviewed by the owner's representatives and QA/QC inspectors to confirm compliance with project requirements.

The successful execution of these steps relies on a range of specialists. Instrumentation technicians handle calibration, SCADA engineers and IT teams ensure reliable data transmission, and commissioning engineers oversee performance validation. QA/QC inspectors provide oversight by verifying compliance and documentation, while owner's representatives and performance engineers confirm that results meet contractual and regulatory standards.

Finally, documented commissioning reports encapsulate the results of system testing and functional checks. These reports detail any non-conformities identified during the testing phase, along with subsequent corrective actions taken and re-testing performed. Such documentation is essential for proving compliance with warranty stipulations, facilitating project handovers, and securing regulatory approvals [228].

Table 3: Commissioning and testing areas, checks, purposes, and responsible roles.

Commissioning Area	Checks / Activities	Purpose	Who Performs It
Electrical Systems	**Continuity & Insulation Resistance Tests**	Verify correct routing of power/control cables, ensure insulation integrity	Electrical technicians (LV/MV), HV commissioning

Commissioning Area	Checks / Activities	Purpose	Who Performs It
			engineers, QA/QC inspectors
	High-Voltage Withstand Testing (VLF)	Apply AC voltage at low frequency to confirm dielectric strength; detect weak insulation/terminations	Specialist HV test engineers, HV jointers/termination specialists
	Grounding / Earthing Verification	Measure resistance between grounding systems and components; ensure safe dissipation of fault currents/lightning	Electrical technicians, commissioning engineers, safety officers
	Transformer Testing	Insulation resistance, turns ratio (TTR), polarity/phase checks, oil analysis (if applicable), protective relay tests	Transformer test engineers (OEM or specialist contractors), commissioning engineers, QA/QC inspectors
	Switchgear Testing	Mechanical operation, insulation resistance, contact resistance, protection relay simulation, interlock verification	HV commissioning engineers, relay technicians, OEM reps
Control Systems	**PLC Verification**	Simulate scenarios (startup, shutdown, emergency stops, overspeed, overheating) to check logic and outputs	PLC commissioning engineers, automation specialists, OEM reps, QA/QC inspectors
	SCADA Connectivity Testing	Verify data transmission (wind, power, temperature, vibration), operator command execution, redundancy, cybersecurity	SCADA engineers, IT specialists, commissioning engineers, control centre operators
	Alarm & Interlock Functional Checks	Trigger simulated faults to confirm alarms activate, interlocks prevent unsafe restart, events are logged	Control engineers, commissioning engineers, electrical technicians, QA/QC inspectors
	Remote Control Functionality	Validate remote start/stop, monitoring, alarm reporting to control centre	SCADA engineers, commissioning engineers, control centre/grid operators, OEM reps
Mechanical & Hydraulic Systems	**Yaw System Tests**	Verify nacelle rotation, braking functions, and sensor feedback	Mechanical technicians, commissioning

Commissioning Area	Checks / Activities	Purpose	Who Performs It
			engineers, surveyors, control engineers
	Pitch System Tests	Cycle blades fully, test actuator performance, sensor feedback, and fail-safe feathering	Blade technicians, hydraulic specialists, control engineers, QA/QC inspectors
	Brake System Tests	Test mechanical/hydraulic brakes under normal and emergency conditions	Mechanical fitters, commissioning engineers, electrical technicians
	Lubrication & Cooling Systems	Check pumps, filters, oil flow, cooling fans, radiators, and heat exchangers	Mechanical technicians, lubrication specialists, commissioning engineers
Functional Integration	**Start-Up & Shutdown Sequences**	Verify correct order of subsystem activation and shutdown (normal and emergency)	Commissioning engineers, control engineers, PLC/SCADA specialists
	Grid Synchronization & Fault Ride-Through	Align turbine output with grid voltage/frequency/phase; test response to dips/faults	HV commissioning engineers, electrical power engineers, grid operator reps
	Redundancy & Backup Systems	Test emergency lighting, fire suppression, backup power	Safety/HSE officers, commissioning engineers, QA/QC inspectors
Data Logging & Performance Verification	**Sensor Calibration**	Calibrate anemometers, wind vanes, temperature and vibration sensors against known standards	Instrumentation technicians, OEM engineers, commissioning engineers, QA/QC inspectors
	Data Acquisition & Logging Checks	Confirm continuous logging, timestamp accuracy, redundancy, reliable data transmission	SCADA engineers, IT specialists, commissioning engineers, OEM software engineers
	Baseline Performance Tests	Collect power curves, vibration, temperature, and hydraulic pressures as reference for future monitoring	Commissioning engineers, performance engineers, QA/QC inspectors, owner's representatives
Documentation & Handover	**Commissioning Reports**	Record results, non-conformities, corrective	Commissioning engineers, QA/QC inspectors,

Commissioning Area	Checks / Activities	Purpose	Who Performs It
		actions, compliance for warranty/regulatory approval	client/owner representatives

Power Performance Verification

Power Performance Verification (PPV) is a critical process in the commissioning and certification of wind turbines, which aims to confirm that each turbine achieves the expected performance levels based on the manufacturer's specifications. This verification is fundamental for ensuring efficient and safe operation prior to commercial deployment and for instilling confidence among both manufacturers and operators. By evaluating the turbine's power curve—the relationship between wind speed and electrical output—manufacturers can identify any deviations that may affect operational efficiency, thereby fulfilling contractual obligations and addressing potential issues before full service.

The primary goal of PPV is to ensure that the turbine's operational output aligns with the predicted power curve provided by the manufacturer. Deviations are analysed using statistical methods to assess their significance and explore impacts on energy production and contract compliance [229, 230]. This process enables the detection of inefficiencies that could impede optimal turbine performance, confirming the turbine's effectiveness under anticipated wind conditions [231, 232].

PPV is standardized under international guidelines such as IEC 61400-12-1 for onshore turbines and IEC 61400-12-2 for offshore installations. These standards dictate the methodology for measuring wind speed, air density, and power output in a structured manner, ensuring data collection occurs under controlled conditions [233, 234]. Accurate results depend on the test sites' terrain characteristics and low turbulence levels; these factors are essential for eliminating inconsistencies in data that may arise from external variances [235, 236].

A typical PPV setup includes the installation of a meteorological mast (met mast) near the turbine to obtain reference wind conditions. This mast, positioned at a recommended distance of two to four rotor diameters, is equipped with anemometers and sensors to capture relevant atmospheric data such as wind speed, direction, temperature, and pressure, all crucial for correcting air density calculations [229, 234]. Concurrently, internal systems within the turbine collect operational data via SCADA systems, which monitor key performance metrics including power output and rotational speed. This dual data collection method ensures a comprehensive understanding of turbine performance by synchronizing external wind conditions with internal turbine responses [232, 237].

Testing typically spans several weeks, allowing for a wide range of wind speed data, generally between 4 to 25 m/s, to be accumulated. This data is segmented into 0.5 m/s intervals for constructing the power curve, which is then compared against the manufacturer's reference curve [229, 237]. Key performance indicators such as power curve agreement and annual energy production (AEP) are vital for this assessment, providing insights into turbine efficiency and reliability [230, 238].

The verification process must be conducted by an Independent Testing Organization (ITO) accredited under ISO/IEC 17025, providing impartial validation of results. The wind turbine manufacturer typically assists with technical support, including access to SCADA data and reference curves, while developers handle logistical elements such as met mast installations, ensuring comprehensive adherence to safety protocols [239]. Data analysts employ statistical filtering to enhance data integrity by eliminating anomalies caused by icing or malfunctions [231, 234].

The culmination of the power performance verification process results in a certified report confirming that the turbine's power curve meets established design specifications and contract requirements. This certification ensures that the turbine is ready for commercial operation and qualifies for warranty coverage [240, 241]. PPV is a thorough, standardized evaluation method that effectively aligns turbine performance with contractual commitments, providing a systematic approach for validating wind turbine operational readiness.

Handover to Operations

Handover to operations marks the final stage in the wind turbine commissioning process, signifying the transition from construction and testing to full operational control by the wind farm operator or owner. This phase ensures that all systems are fully functional, safety-certified, and compliant with performance guarantees, and that the operational team has the required documentation, training, and authority to manage the turbine independently.

The purpose of this phase is to verify that every component of the turbine and its supporting infrastructure—electrical, mechanical, hydraulic, and control systems—has passed all inspections and functional tests. The commissioning engineers prepare a comprehensive set of handover documents, which typically include test certificates, calibration reports, performance data, safety verifications, and maintenance records. These records serve as official evidence that the turbine meets all design, safety, and contractual obligations.

Before handover, a final inspection and acceptance test is conducted. This includes reviewing the results of electrical testing (such as insulation resistance and high-voltage withstand tests), control system verifications, mechanical integrity checks, and power performance verification. Any non-conformities identified during testing must be rectified and verified through re-testing. Once the turbine demonstrates stable operation over a defined period—often called the *reliability run* or *performance demonstration period*—it is deemed ready for commercial operation.

During the handover meeting, the commissioning team provides the operations personnel with detailed documentation and training. This may include turbine operating manuals, emergency shutdown procedures, maintenance schedules, and SCADA system access protocols. The operations team is briefed on site-specific conditions such as grid interface requirements, safety zones, and environmental constraints. Control of the turbine is formally transferred from the commissioning engineers to the operator, often documented through a *Provisional Acceptance Certificate (PAC)* or *Final Acceptance Certificate (FAC)*.

The key roles in this phase include commissioning engineers, quality assurance and control (QA/QC) inspectors, health and safety officers, and representatives from the turbine manufacturer (OEM) and site owner. The commissioning engineers verify all systems and prepare final reports; QA/QC inspectors ensure compliance with standards; and OEM representatives validate that warranty conditions have been satisfied. The operator's representatives then sign off on the acceptance documents, confirming readiness for operational use.

Part 5

Operations and Maintenance

Chapter 12

Routine Inspections

Once a wind turbine has been commissioned and handed over for operation, its long-term performance and safety depend on one critical factor — routine inspection and maintenance. Wind turbines operate under intense mechanical, electrical, and environmental stresses. Over time, these stresses can lead to component fatigue, wear, and degradation if not regularly monitored and addressed. Routine inspections form the backbone of preventive maintenance, ensuring that turbines remain reliable, efficient, and compliant with safety and warranty requirements throughout their service life.

This chapter provides a comprehensive guide to the inspection processes that underpin safe and sustainable turbine operation. It begins with daily, weekly, and scheduled inspection tasks, outlining the purpose, scope, and procedures for each level of maintenance. Daily checks focus on early fault detection — such as oil leaks, abnormal noises, or SCADA system alerts — enabling technicians to address minor issues before they escalate. Weekly inspections delve deeper into mechanical and electrical systems, verifying lubrication, fasteners, and brake functionality. Scheduled inspections, conducted monthly, quarterly, or annually, involve detailed structural, electrical, and hydraulic assessments designed to maintain turbine integrity and performance standards.

The chapter also explores the methods and technologies used in turbine inspections, from traditional rope-access techniques to advanced drone-based and AI-driven systems that have revolutionized modern wind turbine maintenance. These innovations allow operators to detect micro-defects, assess internal blade structures, and perform predictive maintenance with unprecedented accuracy, reducing downtime and improving safety.

Further sections detail specific maintenance activities, including torque checks, lubrication, oil changes, and blade repairs — all essential for sustaining mechanical reliability and preventing catastrophic failures. The importance of remote monitoring and diagnostic systems is also highlighted, showing how real-time data analysis enables predictive maintenance and performance optimization across entire wind farms.

By integrating manual inspections with digital diagnostics, operators can transition from reactive to proactive maintenance models — identifying faults before they cause breakdowns. In doing so, routine inspections not only extend turbine lifespan and maximize energy output but also uphold the high safety and environmental standards expected of modern renewable energy operations.

This chapter serves as both a technical reference and a practical field guide for wind turbine technicians, site managers, and engineers committed to keeping wind energy assets running safely, efficiently, and sustainably.

Daily, Weekly, and Scheduled Inspection Tasks

Wind turbines require a structured inspection and maintenance regime to ensure continuous safe operation, maximum availability, and long-term reliability. These inspections are typically divided into daily, weekly, and scheduled (monthly, quarterly, or annual) intervals. The scope and frequency of these checks are defined by the turbine manufacturer's maintenance manual, industry standards such as IEC 61400-1 and ISO 10816, and regulatory requirements specific to the country or operator.

Daily Inspections

Purpose: Daily checks are basic operational and visual inspections performed to detect early signs of malfunction or wear. These are typically conducted by on-site technicians or remotely monitored through the SCADA system.

Typical Tasks:

- **Visual inspection from ground level:** Check for oil leaks on the tower or around the nacelle, unusual noises, or visible damage to the blades.
- **SCADA monitoring:** Review turbine operational data such as wind speed, power output, rotor speed, vibration levels, and temperatures. Identify alarms, fault codes, or anomalies.
- **Safety system check:** Ensure warning lights, access locks, and communication systems are functional.
- **Environmental monitoring:** Inspect for obstructions (ice, bird debris, or vegetation) around the tower base that might affect access or safety.
- **Housekeeping:** Check that the turbine interior (if entered) remains clean, with no oil spills, loose tools, or trip hazards.

Performed by: Wind turbine technicians or site operators; remote engineers may assist with SCADA-based monitoring.

Common Faults identified in Daily Inspections

Daily maintenance of wind turbines plays a vital role in ensuring the long-term reliability, efficiency, and safety of wind energy systems. Through regular inspection and prompt diagnosis, maintenance teams can identify and rectify issues before they lead to significant failures or downtime. The most common areas requiring daily monitoring and fault diagnosis include the main control system, gearbox, pitch system, generator, and blades—each of which is essential to turbine performance and energy conversion efficiency.

1. Main Control System: The main control system serves as the "brain" of the wind turbine, responsible for managing and coordinating all operating functions. Most modern wind turbines employ Programmable Logic Controllers (PLCs) that use modular designs and backplane bus connections. This configuration allows efficient data exchange and logical control but also introduces potential vulnerabilities. Common issues include abnormal digital or analogue input/output signals, unresponsive modules, or inactive indicator lights on control boards.

To address these faults, maintenance personnel first check the system's diagnostic logs and SCADA (Supervisory Control and Data Acquisition) alarms to pinpoint the fault location. Software-related faults are often resolved by refreshing or reloading the PLC program, while hardware failures typically require replacement of faulty modules or components. When external faults occur—such as sensor disconnections or actuator failures—the SCADA system provides real-time alarms, helping technicians to quickly isolate the issue and restore normal operation. Regular functional tests of sensors, relays, and control signals are essential to prevent unexpected turbine stoppages.

2. Gearbox: The gearbox is a crucial mechanical component that transmits torque from the main rotor shaft to the generator, increasing rotational speed for efficient power generation. Because it operates under high mechanical stress, gearbox health directly affects turbine availability and output. Daily maintenance involves checking seals, pipelines, end covers, and coolers for oil leaks or damage. Any sign of leakage can indicate seal degradation or internal wear, both of which require immediate attention to prevent contamination and mechanical failure.

Technicians also inspect the oil flow rate and lubrication system through observation ports or borescopes (endoscopes) to ensure adequate oil circulation and correct injection volumes. Abnormalities such as metallic particles in the oil, unusual noise, or elevated vibration levels may indicate bearing or gear wear. Early detection through these checks helps avoid catastrophic gearbox failures, which are among the most costly issues in wind turbine maintenance.

3. Pitch System: The pitch system controls the angle of the blades to regulate rotor speed, optimize energy capture, and protect the turbine under high wind conditions. As one of the most critical actuating systems, it relies on precise mechanical and hydraulic performance. Daily inspection focuses on the torque of hub bolts and the tightness of all mechanical connections, since loose fasteners can lead to severe operational instability. Specialized hydraulic wrenches and tensioners, such as those designed for wind turbine applications, are used to maintain precise bolt tension.

Technicians also check for hydraulic fluid leaks, actuator responsiveness, and sensor feedback accuracy. Any irregularities in the pitch angle response or hydraulic pressure readings are logged and investigated immediately. Routine inspection and torque verification significantly reduce the likelihood of pitch system failure, which could otherwise result in loss of control and turbine overspeed incidents.

4. Generator: The generator converts mechanical energy from the drivetrain into electrical power and operates continuously under varying loads and temperatures. With the increasing capacity and size of modern turbines, generator maintenance has become more challenging due to greater thermal and mechanical stresses. Common issues include bearing overheating, rotor bar fractures, insulation degradation, and excessive vibration. These faults often arise from prolonged operation under fluctuating loads, misalignment, or inadequate cooling.

During daily inspections, maintenance teams monitor temperature and vibration data through the SCADA system and perform physical checks for oil leaks, abnormal noises, or excessive heat at bearing housings. Any deviation from normal operating parameters triggers further diagnostic tests using thermal imaging cameras or vibration analysers. Early intervention prevents costly generator rewinding or replacement and maintains consistent electrical output.

5. Blades: Blades are the primary components for capturing wind energy and are typically constructed from fibre-reinforced composite materials. Because of their exposure to environmental stress, blades are vulnerable to cracks, delamination, erosion, and lightning strike damage. Daily inspections focus on detecting these surface defects, which can compromise aerodynamic efficiency and structural integrity.

One of the most effective diagnostic methods is infrared thermography, which identifies thermal anomalies caused by cracks or material separation. Areas with inconsistent heat distribution suggest internal defects not visible to the naked eye. When small cracks or peeling are detected, technicians apply adhesive repairs and surface polishing to restore the blade's aerodynamic shape. In cases of significant damage, blades are taken out of service for professional composite repair. Regular blade condition monitoring ensures optimal energy capture and minimizes the risk of catastrophic blade failure during high winds.

Daily maintenance of wind turbines involves a combination of visual inspection, sensor monitoring, and targeted diagnostic testing across critical systems. The main control system and SCADA diagnostics ensure smooth coordination and fault detection, while gearbox and pitch inspections focus on mechanical integrity and lubrication. Generator checks prevent electrical failures, and blade inspections maintain aerodynamic performance. Together, these activities enable early fault detection, reduce downtime, and extend the overall service life of the wind turbine.

Inspection Methods

The global wind energy sector has grown immensely, with over 341,000 wind turbines now operating across the world [242]. However, maintaining such a vast fleet comes with substantial logistical and financial challenges—especially when it comes to inspections. On average, a manual inspection—whether performed through rope access or ground-based monitoring—takes around six hours per turbine and is generally conducted only once or twice per year. When multiplied across all turbines globally, this equates to more than 4.1 million labour hours, or roughly 467 years of cumulative downtime [242]. This staggering figure doesn't even account for weather delays, change orders, or other unforeseen interruptions, which could extend the total duration considerably. Clearly, relying solely on manual inspection methods is both time-consuming and inefficient.

Despite their towering presence—often exceeding 466 feet in height with 418-foot blade spans—wind turbines are not as indestructible as they appear. Their size can be misleading, as turbines are mechanically and structurally sensitive systems, especially in their rotating components and composite materials [242]. For instance, rotor blades made of fiberglass are vulnerable to erosion, cracks, or minor surface rust, particularly on their leading edges. Such small imperfections can cause a significant reduction in aerodynamic efficiency and energy output. This fragility underscores the importance of regular and accurate inspections, which help prevent minor wear from developing into serious, costly faults that could shorten turbine lifespan or compromise safety [242].

Traditionally, turbine inspections have been performed using two primary manual approaches: rope access and ground-based monitoring. In a typical rope-access inspection, technicians climb the turbine tower, secure themselves with harnesses and ropes, and descend along the blades to visually inspect for cracks, corrosion, or delamination. Although this method allows for close-up examination, it is dangerous, labour-intensive, and costly, requiring highly skilled teams and extensive safety precautions. A sudden gust of wind or rope entanglement with the blade can result in severe accidents. The inherent risks are reflected in high insurance premiums and operational costs [242].

Figure 120: Inspection of a rotor blade and the tower of a wind turbine, via rope access. Markus Bergmann, CC BY-SA 3.0, via Wikimedia Commons.

Ground-based inspection, on the other hand, involves using high-resolution cameras and telescopic lenses to photograph turbine blades from below. This approach is far safer but considerably less accurate. Even with advanced optics, it is difficult to detect millimetre-scale surface defects on blades hundreds of feet above ground. As a result, ground-based inspections often yield poor-quality or inconclusive data, limiting their value in preventive maintenance planning [242].

The limitations of manual inspection methods have paved the way for a new era of drone-based turbine inspections. Today, up to 90% of U.S. wind farms use uncrewed aerial vehicles (UAVs) to conduct inspections. Drones dramatically reduce inspection times and operational risks while maintaining—or even improving—accuracy [242]. Compared to manual techniques, UAV inspections can cut inspection costs by up to 70% and downtime by up to 90%, allowing turbines to remain productive for longer periods. Autonomous drones equipped with high-definition cameras and AI-based analytics can rapidly capture and process thousands of blade images in minutes, eliminating the need for rope-access crews or extensive setup time [242].

This technological shift also enables more frequent inspections. Previously, wind farms limited inspections to one or two sessions per year due to the cost and downtime involved. With drones, inspections can be conducted quarterly or even monthly, providing continuous insight into turbine condition without compromising profitability. By catching early signs of wear, erosion, or lightning damage, operators can plan predictive maintenance instead of reacting to breakdowns, significantly improving efficiency and asset longevity [242].

In addition to external inspection, drones have begun transforming internal diagnostics as well. Using infrared cameras, ultrasonic sensors, and fibre-optic probes, UAVs can detect hidden subsurface defects that are invisible to the naked eye. This includes delamination in blade materials, internal corrosion, and micro-fractures caused by fatigue loading. Regulatory limits, such as restrictions on human access beyond 91 feet within a blade, mean that manual methods can only inspect about 40% of its internal surface. Drones bypass these limitations entirely, enabling full coverage of internal blade structures, especially critical for offshore turbines exposed to saltwater corrosion and harsh weather [242].

Beyond data collection, the integration of AI-powered analytics platforms now allows real-time processing and interpretation of inspection data. These platforms automatically identify anomalies, classify defects, and generate maintenance reports, turning raw data into actionable insights. This digital workflow eliminates the need for technicians to manually review thousands of images, accelerating decision-making and reducing human error. The result is a smarter, safer, and more cost-effective maintenance ecosystem [242].

Looking ahead, the future of turbine inspection lies in autonomous, data-driven maintenance systems. As global investment in renewable energy accelerates, the demand for scalable, reliable inspection solutions will only grow. Manual inspection methods—once the norm—are gradually being phased out in favour of UAVs and AI-based predictive maintenance tools. This evolution not only enhances safety and efficiency but also ensures that wind farms can sustain continuous, clean power generation with minimal interruption. In short, modern challenges in wind turbine maintenance demand modern, automated solutions—and drones are at the forefront of this transformation [242].

<u>Inspection Processes</u>

These activities are part of a wind turbine's routine preventive maintenance and operational safety checks.

1. Visual Inspection from Ground Level

Purpose: To identify visible abnormalities that may indicate developing mechanical or electrical faults.

Procedure:

1. Stand at a safe distance from the turbine (at least the total height of the tower plus 10 m) to ensure full visibility of the structure.

2. **Inspect the tower:** Look for oil streaks or stains that could indicate hydraulic or gearbox oil leaks. Pay close attention to the base, access doors, and joints between tower sections.
3. **Observe the nacelle:** Use binoculars to check for oil leaks near the nacelle base, yaw area, or cooling vents.
4. **Examine the blades:** Rotate the turbine slowly in maintenance mode if possible, and visually inspect for cracks, erosion, lightning strikes, or discoloration.
5. **Listen for unusual noises:** From the ground, note any abnormal sounds such as grinding, knocking, or irregular pitch changes that may suggest mechanical wear.
6. Record all observations in the daily inspection log and report anomalies to the maintenance supervisor.

Tools: Binoculars, handheld camera (optional), maintenance logbook.

2. SCADA Monitoring

Purpose: To evaluate turbine performance and identify abnormal trends through digital data.

Procedure:

1. Log in to the Supervisory Control and Data Acquisition (SCADA) system or the wind farm monitoring platform.
2. Review real-time operational parameters such as:
    - Wind speed and direction
    - Power output
    - Rotor and generator speed
    - Vibration levels (main bearing, gearbox, generator)
    - Temperatures (gearbox oil, generator, transformer)
3. Check for active alarms or fault codes and note any deviations from normal ranges.
4. Compare today's data with historical performance to identify trends (e.g., decreasing output at given wind speeds).
5. Document any alerts or irregular readings and assign follow-up actions to field technicians if necessary.

Tools: SCADA terminal or remote monitoring software, alarm log, trend analysis reports.

3. Safety System Check

Purpose: To confirm that all safety and access control systems are functional and compliant.

Procedure:

1. Verify that warning lights (aviation and obstruction lights) are operational, especially for turbines near air routes.
2. Check access locks on tower doors and maintenance hatches to ensure they secure properly and prevent unauthorized entry.
3. Test emergency stop buttons and confirm that control panels indicate proper reset status.
4. Verify that two-way communication systems (radio, intercom, or GSM link) between tower and control centre are active.
5. Ensure that fire extinguishers, first aid kits, and emergency signage are present and up to date.

Tools: Multimeter or test device (for circuit checks), radio unit, inspection checklist.

4. Environmental Monitoring

Purpose: To ensure safe access and protect the turbine from environmental hazards.

Procedure:

1. Inspect the area surrounding the turbine's base (within at least a 10 m radius).
2. Clear any obstructions such as ice accumulation, fallen branches, or vegetation that could block access paths or affect tower ventilation.
3. Remove bird debris or nests from near nacelle vents or access ladders (following environmental regulations).
4. In winter, check for ice buildup on blades and tower surfaces; if present, activate de-icing systems or suspend operation until cleared.
5. Confirm that drainage systems around the tower base are free of blockages.

Tools: Safety gear (helmet, gloves), ice scraper (if applicable), maintenance checklist.

5. Housekeeping (Interior Inspection)

Purpose: To maintain a clean, safe, and hazard-free work environment inside the tower and nacelle.

Procedure:

1. Before entry, follow lock-out/tag-out (LOTO) procedures and obtain access authorization.
2. Inspect all interior floors and platforms for oil spills, water leaks, or slippery residues; clean immediately using approved absorbents.
3. Check for loose tools, spare parts, or debris that could cause trip hazards.
4. Ensure that all lighting, ventilation, and emergency exits inside the tower are operational.
5. Verify that all storage cabinets, fire extinguishers, and first aid kits are in their designated places.
6. Record inspection results and corrective actions in the turbine maintenance log.

Tools: Flashlight, cleaning materials, spill kit, tower entry permit.

Weekly Inspections

Purpose: Weekly checks are more detailed than daily observations and focus on verifying that the turbine's mechanical and electrical systems are stable and performing within expected parameters. These often include limited hands-on inspection within the tower.

Typical Tasks:

- **Operational verification:** Confirm that turbine start/stop cycles, yaw, and pitch systems are functioning correctly.
- **Lubrication and hydraulic checks:** Inspect oil and hydraulic fluid levels in yaw motors, brakes, and gearboxes; top up if required.
- **Fastener and joint inspection:** Check access platform bolts, tower bolts (visible ones), and nacelle hatches for tightness and signs of corrosion.
- **Cooling and ventilation systems:** Clean or replace filters, check coolant flow, and verify that fans are operational.
- **Brake and safety systems:** Test emergency stop buttons and ensure mechanical and hydraulic brakes are working correctly.
- **Data verification:** Review SCADA logs for repeated fault codes or performance drift.

Performed by: Qualified wind turbine service technicians under supervision of site supervisors; maintenance engineers may analyse SCADA trends for developing issues.

These procedures would typically be conducted by certified wind turbine technicians or commissioning engineers during scheduled maintenance or post-commissioning checks.

1. Operational Verification

Objective: Confirm that turbine control and mechanical systems (start/stop, yaw, and pitch) are functioning correctly.

Procedure:

1. Access the turbine control panel or SCADA interface.
2. Initiate a manual start-up sequence and observe the turbine as it transitions to operational speed. Confirm that yaw and pitch systems respond automatically to wind direction and speed.
3. Perform a manual stop command to verify smooth deceleration and braking.
4. Check yaw drive operation by commanding small directional changes and confirming correct rotation and braking action.
5. Test pitch system functionality by cycling each blade through its full pitch range; verify synchronization, actuator response, and feathering during emergency stop mode.
6. Record response times, alarm activations, and performance data in the maintenance log.

Personnel: Commissioning engineers, control system engineers, and SCADA technicians.

Safety Note: Always conduct these tests under safe wind conditions (<10 m/s) and ensure exclusion zones are maintained around the rotor.

2. Lubrication and Hydraulic Checks

Objective: Ensure adequate fluid levels and prevent wear or overheating in mechanical systems.

Procedure:

1. Power down the turbine and isolate it according to lockout/tagout procedures.
2. Inspect gearbox oil level using the sight glass or dipstick; top up with OEM-specified lubricant if below the minimum mark.
3. Check yaw and pitch hydraulic reservoirs for correct fluid level and contamination.
4. Inspect brake fluid and cooling oil levels.
5. Examine hydraulic hoses and connectors for leaks, cracks, or wear.
6. If necessary, filter or replace fluids following manufacturer guidelines.

7. Record maintenance actions and oil batch numbers for traceability.

Personnel: Mechanical technicians and lubrication specialists.

Safety Note: Use spill kits and drip trays to prevent environmental contamination.

3. Fastener and Joint Inspection

Objective: Verify that structural and mechanical fasteners remain tight and corrosion-free.

Procedure:

1. Conduct a visual inspection of external tower bolts, access platform bolts, and nacelle hatches.
2. Use a torque wrench to check critical bolt tightness as specified in the manufacturer's manual.
3. Look for signs of corrosion, rust streaks, or fretting at joints.
4. Apply anti-corrosion coating or grease where necessary.
5. Replace any damaged fasteners immediately and record torque values in the inspection report.

Personnel: Mechanical fitters, structural technicians, and QA/QC inspectors.

Safety Note: Use approved climbing equipment and fall arrest systems when accessing elevated areas.

4. Cooling and Ventilation Systems

Objective: Maintain optimal operating temperature for nacelle electronics, generator, and power converters.

Procedure:

1. Isolate power before inspection.
2. Open cooling compartments and check air filters; clean or replace them as per the maintenance schedule.
3. Inspect fan blades and housings for dust buildup or damage.
4. Confirm coolant levels in liquid-cooled systems; inspect hoses for leaks or blockages.
5. Power on and verify that all fans and pumps start automatically during normal turbine operation.
6. Record filter replacement dates and coolant measurements.

Personnel: Electrical technicians and mechanical maintenance engineers.

Safety Note: Ensure system depressurization before opening any cooling circuit.

5. Brake and Safety Systems

Objective: Verify the readiness and functionality of all braking and safety mechanisms.

Procedure:

1. Test all emergency stop buttons (nacelle, tower base, and control panel) to confirm they halt turbine rotation immediately.
2. Engage mechanical brakes manually and observe rotor deceleration time.
3. Activate hydraulic brake system and monitor pressure readings and leakage.
4. Check backup brake systems or spring-loaded actuators for proper engagement.
5. Confirm that limit switches and interlocks are functioning correctly and that fault resets require manual confirmation.
6. Record response times and system pressures for QA documentation.

Personnel: Electrical engineers, hydraulic specialists, and safety officers.

Safety Note: Inform the control centre before initiating brake tests to avoid false alarms.

6. Data Verification

Objective: Identify developing faults or performance deviations through SCADA and data analysis.

Procedure:

1. Access the SCADA system and retrieve operational data for the last 24–72 hours.
2. Review key parameters — wind speed, power output, rotor speed, generator temperature, and vibration levels.
3. Check for recurrent fault codes or alarms, such as high bearing temperature, over-speed warnings, or converter faults.
4. Compare current trends with historical performance baselines.
5. Generate a report summarizing anomalies and recommend corrective actions if performance drift or pattern faults are detected.

Personnel: SCADA engineers, data analysts, and commissioning engineers.

Safety Note: Only authorized personnel should access control data to avoid inadvertent system changes.

Scheduled (Planned) Inspections

Purpose: Scheduled inspections—conducted monthly, quarterly, semi-annually, or annually—are comprehensive maintenance and diagnostic tasks designed to ensure the turbine meets performance, safety, and warranty standards. These follow the manufacturer's prescribed Preventive Maintenance Schedule (PMS).

Typical Tasks:

- **Full mechanical inspection:** Inspect and torque-check tower and nacelle bolts, yaw ring, gearbox mounts, and coupling components.
- **Blade inspection:** Conduct close visual or drone inspection for cracks, erosion, lightning strikes, or delamination. Clean and repair as required.
- **Gearbox and drivetrain checks:** Sample oil for contamination, inspect for leaks, and analyse vibration data for bearing or gear faults.
- **Electrical system tests:** Perform insulation resistance tests, verify grounding systems, inspect transformers and switchgear, and confirm protection relays are functional.
- **Hydraulic and lubrication systems:** Replace filters, check actuator seals, and flush systems if needed.
- **Calibration and safety verification:** Recalibrate sensors (temperature, vibration, pressure, pitch), test alarms, and validate fire suppression systems.
- **Performance analysis:** Compare actual power output with expected power curve to detect underperformance.
- **Documentation and reporting:** Record all findings, corrective actions, and component replacements in the maintenance management system (CMMS).

Performed by: Multidisciplinary maintenance teams comprising electrical, mechanical, and hydraulic technicians; condition monitoring specialists; and OEM representatives during warranty periods. QA/QC inspectors and site managers review documentation for compliance and performance reporting.

These procedures are typically carried out by qualified mechanical, electrical, and commissioning technicians during quarterly, semi-annual, or annual maintenance cycles.

1. Full Mechanical Inspection

Objective: Verify structural integrity and mechanical stability across all major joints and assemblies.

Procedure:

1. Lockout/tagout the turbine to ensure zero-energy conditions before mechanical work begins.

2. Inspect tower bolts, nacelle frame bolts, and yaw ring fasteners for tightness, corrosion, or elongation.

3. Use a calibrated torque wrench or hydraulic tensioning system to confirm bolt preload matches OEM specifications.

4. Examine gearbox mounts, couplings, and main bearing supports for looseness or visible wear.

5. Inspect weld seams and structural joints for cracks or fatigue signs.

6. Record torque values and any component replacements in the maintenance report.

Performed by: Mechanical fitters, structural technicians, and QA/QC inspectors.

2. Blade Inspection

Objective: Detect surface and structural defects that could affect aerodynamic performance or safety.

Procedure:

1. Conduct close visual inspection using rope access, drone-based imaging, or telescopic cameras.

2. Examine blade surfaces for cracks, erosion, lightning strike marks, or delamination.

3. Pay particular attention to the leading and trailing edges, blade roots, and bonding lines.

4. Clean dirt, insects, or oil residues from blade surfaces.

5. If damage is found:

    - Small chips or erosion: Sand, fill, and recoat using OEM repair kits.

    - Cracks or delamination: Tag and report for structural repair.

6. Document all findings with high-resolution photos and location markings.

Performed by: Blade technicians, rope-access specialists, or drone inspection teams.

3. Gearbox and Drivetrain Checks

Objective: Assess gearbox condition, lubrication quality, and alignment for reliable power transmission.

Procedure:

1. Inspect for oil leaks around the gearbox, couplings, and seals.

2. Take oil samples from the gearbox and send for laboratory analysis to check for metal particles, water contamination, or oxidation.

3. Connect vibration sensors or portable analysers to monitor bearing and gear vibration signatures.
4. Compare vibration spectra with baseline readings to detect early-stage wear or imbalance.
5. Check gearbox temperature sensors and oil flow through the cooler.

Performed by: Mechanical technicians, vibration analysts, and oil monitoring specialists.

4. Electrical System Tests

Objective: Ensure electrical integrity, safety, and grounding reliability.

Procedure:

1. Conduct insulation resistance testing on power and control cables using a megohmmeter.
2. Verify earthing system continuity and resistance per IEC 61400-24 or IEEE 80 standards.
3. Inspect transformers and switchgear for insulation condition, mechanical operation, and overheating signs.
4. Perform protective relay tests, simulating fault conditions (overcurrent, earth fault, undervoltage).
5. Confirm lightning protection system continuity from the blades to the grounding system.

Performed by: Electrical engineers, HV commissioning technicians, and QA/QC inspectors.

5. Hydraulic and Lubrication Systems

Objective: Maintain optimal system pressure, cleanliness, and sealing to prevent wear or contamination.

Procedure:

1. Replace hydraulic and oil filters as per maintenance schedule.
2. Check actuator seals, hoses, and fittings for leaks or cracks.
3. Inspect reservoir levels and top up fluids with OEM-specified lubricants.
4. If contamination or sludge is detected, flush the system and refill with new fluid.
5. Verify that yaw, pitch, and brake hydraulic systems operate at correct pressure and response time.

Performed by: Hydraulic technicians and mechanical maintenance engineers.

6. Calibration and Safety Verification

Objective: Ensure accuracy of monitoring sensors and functionality of safety systems.

Procedure:

1. Recalibrate temperature, vibration, pressure, and pitch sensors using certified reference equipment.
2. Test all alarm functions and emergency stop circuits from nacelle and tower control panels.
3. Verify operation of fire detection and suppression systems, including gas, foam, or water mist units.
4. Inspect backup power supplies and emergency lighting for correct operation.
5. Document calibration results and system responses in maintenance records.

Performed by: Instrumentation technicians, safety engineers, and commissioning engineers.

7. Performance Analysis

Objective: Evaluate the turbine's operational efficiency and detect underperformance trends.

Procedure:

1. Extract operational data from the SCADA system, including wind speed, power output, and rotor speed.
2. Plot the measured power curve against the OEM reference curve.
3. Identify any deviations in power output, particularly in the mid-range wind speeds (6–12 m/s).
4. Review availability and downtime logs to assess reliability and utilization.
5. If performance drift is detected, investigate aerodynamic, mechanical, or control system causes.

Performed by: Performance engineers, data analysts, and commissioning engineers.

8. Documentation and Reporting

Objective: Ensure traceability, compliance, and communication of maintenance outcomes.

Procedure:

1. Record all inspection findings, torque values, oil samples, and test results in the Computerized Maintenance Management System (CMMS).
2. Attach photos, sensor calibration certificates, and vibration reports.

3. List all corrective actions and component replacements with part numbers and technician details.
4. Submit a final report to the site manager and operations team for review and scheduling of follow-up tasks.

Performed by: Maintenance coordinator, QA/QC inspector, and site manager.

Torque Checks and Lubrication

Wind Turbine Lubrication and Oil Changes

Regular lubrication and oil maintenance are essential for ensuring the smooth and efficient operation of wind turbines. These machines consist of numerous moving components—gearboxes, bearings, yaw drives, and pitch mechanisms—that are subject to high loads and friction. Without proper lubrication, excessive wear, overheating, and premature component failure can occur, leading to costly repairs and extended downtime.

<u>Lubrication Points</u>

The gearbox is one of the most critical components in a wind turbine. It transfers rotational energy from the rotor to the generator and operates under high torque and pressure. Lubrication minimizes metal-to-metal contact between gears and bearings, reducing friction and dissipating heat. Regular oil sampling and changes help maintain viscosity and prevent the build-up of contaminants that can cause wear.

The pitch system, which adjusts the angle of each blade to capture the optimal amount of wind energy, also depends on lubrication. Hydraulic pitch actuators, bearings, and gears must move smoothly to respond quickly to changes in wind speed. Adequate lubrication ensures consistent operation and prevents jerky or delayed movement that can affect efficiency.

The yaw system, responsible for rotating the nacelle so that the turbine faces into the wind, contains yaw gears and bearings that are under constant stress. Lubricating these components prevents binding and reduces wear caused by friction and environmental exposure. Proper lubrication here is crucial for maintaining directional control and minimizing energy losses.

<u>Types of Lubricants</u>

Two main types of lubricants are used in wind turbines:

- **Synthetic Oils:** These are preferred in modern turbines due to their superior thermal stability, resistance to oxidation, and ability to perform well under extreme temperatures. Synthetic oils provide longer service intervals and maintain consistent viscosity, which is vital for components such as gearboxes and hydraulic systems.

- **Grease:** Grease is used for components that experience less frequent movement, such as main bearings and yaw bearings. It adheres well to surfaces, providing long-term protection and sealing against contaminants like water, dust, and debris.

Routine oil changes and lubrication schedules are established based on the manufacturer's recommendations, typically every 12 to 24 months for the gearbox and annually for bearings and pitch systems.

Oil Volume

The amount of oil required in wind turbines is influenced by several design and operational parameters, making a universal standard challenging to establish. Key factors include the turbine's size, gearbox design, drivetrain configuration, and various manufacturer specifications. For example, utility-scale wind turbines typically require between 250 to 400 litres of gearbox oil, while smaller turbines rated under 1 MW generally consume significantly less, often around 50 to 200 litres [243]. This variation in oil capacity underscores the importance of referencing the original equipment manufacturer (OEM) specifications for precise oil requirements.

It has been found that larger turbines, particularly those with power ratings of 5 MW or higher, necessitate more oil due to their complex gearboxes designed for increased torque and operational reliability [244]. These gearboxes, particularly multi-stage or planetary gear systems, comprise more components and require a greater volume of lubricant for adequate cooling and efficiency under high-load operations [245]. Additionally, the oil plays a crucial role in both lubrication and thermal management, absorbing and dissipating heat generated during operation, thereby preventing overheating and maintaining optimal gearbox performance [246].

Typical examples illustrate these general findings. For instance, the GE 1.5 MW turbine is documented to require approximately 380 litres of oil, indicating that larger gearboxes used in higher-rated turbines can present a more substantial oil requirement to maintain performance standards [243, 247]. Many onshore turbine systems have reported oil capacities near this upper limit, reinforcing the need for robust oil management systems within their design [248].

Thermal management supports the need for specified oil volumes, emphasizing the system geometry, which involves internal fluid channels, sumps, reservoirs, and filters that collectively augment the total oil volume necessary [249]. Furthermore, many turbine designs consider safety margins and operational adaptability by incorporating specifications for expandable oil fill levels to accommodate circulation and foam control, ensuring continuous lubrication amidst fluctuating load conditions [250].

As such, the oil capacity of modern utility-scale wind turbines generally falls within the range of 250 to 400 litres. Smaller turbines often require significantly less oil. However, it is critical to refer to OEM specifications for precise capacity requirements, as various factors directly affect these numbers, including gearbox complexity, turbine size, and the architecture of the lubrication system.

Transporting new oil up to the nacelle of a wind turbine and removing used oil is a highly controlled process designed to ensure safety, cleanliness, and environmental protection. This is especially important because the nacelle is often situated more than 100 metres above the ground. Although the procedure may vary slightly depending on the turbine model, site layout, and maintenance protocols, the core steps remain consistent across the wind energy industry.

The process begins with preparation and safety setup. Before any oil exchange, the turbine must be fully shut down and locked out under Lockout/Tagout (LOTO) procedures to prevent accidental startup. Technicians confirm that the gearbox and oil system have cooled to below 40 °C to avoid burns or pressure issues from thermal expansion. All necessary equipment, such as portable pumps, hoses, and filtration units, is prepared in advance. To prevent environmental contamination, spill trays or absorbent mats are positioned beneath the tower to capture any leaks or drips during hose connections or disconnections.

When transporting new oil to the nacelle, there are two primary methods used. The pumping system is preferred for large turbines. A portable electric or hydraulic oil pump is positioned at the base of the tower and connected to a high-pressure hose system (rated around 200 bar or higher). The hose is routed up through the tower's internal duct or service ladder shaft and connected to a quick-coupling port in the nacelle or gearbox service panel. The pump then transfers clean oil from a sealed storage drum or IBC (Intermediate Bulk Container) to the nacelle's reservoir or gearbox. Filtration systems built into the pump remove contaminants, ensuring oil cleanliness meets ISO 4406 14/12/9 standards before entering the gearbox.

For smaller turbines or remote sites, oil may be delivered manually. In such cases, technicians hoist 20–25 L pre-filled containers using the tower's internal lift or hoist system. The oil is then poured or pumped into the gearbox using portable transfer pumps or funnels equipped with fine filtration.

Removing used oil follows a similar but reversed process. The preferred method is pumping, where the same hose system is used to extract oil from the gearbox and transfer it down to a sealed waste container at the tower base. Modern gearboxes often include dedicated drain ports or quick-connect manifolds to minimize spillage. The pump may also have a filtration and monitoring unit to allow for oil sampling and analysis, providing insights into wear metals, contamination, and oil degradation. In smaller or older turbines, oil may be drained by gravity through a controlled line into a drum, though this method is less common due to limited control and higher spill risk.

Before refilling, technicians typically perform an oil flush by circulating a small amount of clean oil to remove any remaining debris or residues. Samples of the used oil are sent for laboratory analysis to detect wear, oxidation, or water ingress, which supports predictive maintenance. Once flushing is complete, new oil is pumped into the gearbox and other lubrication circuits. Technicians verify oil levels via sight glasses or dipsticks, bleed air from the lines, and briefly run the turbine in service mode to stabilize levels before final checks.

Used oil is classified as hazardous waste and handled according to strict environmental regulations, such as those under EPA or DEFRA guidelines. All waste containers are labelled, sealed, and transported off-site by licensed contractors for recycling or disposal, accompanied by the necessary waste transfer documentation.

Common equipment used in the process includes portable oil change pumps, industrial-rated hoses, quick-connect fittings, filtration units (10–25 µm), oil storage drums or IBCs, and oil sample bottles for testing.

In summary, oil exchange in wind turbines is a carefully managed hydraulic process using pumps and hoses to transfer oil between the ground and nacelle—typically involving 250–400 litres per turbine. The used oil is extracted simultaneously to prevent spillage, and samples are analysed for condition monitoring. The full operation, including setup, flushing, refilling, verification, and cleanup, usually takes 4–6 hours per turbine, depending on size and accessibility. All work is carried out by certified wind turbine technicians following rigorous work health, safety, and environmental protection standards.

Wind Turbine Monitoring and Diagnostic Systems

Modern wind turbines incorporate advanced monitoring and diagnostic systems that continuously collect and analyse operational data. These systems are part of predictive maintenance strategies that help operators detect faults early, plan maintenance activities efficiently, and reduce unplanned downtime.

<u>Key Performance Indicators to Monitor</u>

- **Rotor Speed:** The turbine's control system monitors rotor speed to ensure it remains within safe limits. Deviations may indicate generator issues, gearbox faults, or control system errors.
- **Vibration Levels:** Excessive vibration is one of the most critical warning signs of mechanical problems, such as bearing degradation, rotor imbalance, or drivetrain misalignment. Continuous vibration monitoring allows early fault detection before damage escalates.
- **Wind Speed:** Monitoring wind speed ensures the turbine operates within its designed range. If winds exceed safe thresholds, the control system automatically shuts down the turbine to prevent damage.
- **Power Output:** Comparing actual power generation against expected values helps identify efficiency losses or mechanical/electrical faults.
- **Ice Build-up:** In cold climates, ice formation on blades can create dangerous imbalances. De-icing sensors detect ice accumulation, prompting the turbine to stop operation until conditions are safe.

Predictive maintenance software uses this real-time data to detect irregularities and trigger maintenance alerts. This proactive approach minimizes breakdowns, reduces costs, and extends turbine lifespan.

Wind Turbine Blade Maintenance and Repair

Blades are exposed to continuous environmental stress from wind, rain, dirt, UV radiation, and debris. Over time, these conditions can cause erosion, cracking, or delamination, reducing aerodynamic efficiency and energy production. Regular inspection and maintenance are critical for blade performance and safety.

Blade Maintenance Steps

1. **Visual Inspection:** Technicians perform close-up inspections or drone-based surveys to identify cracks, chips, erosion, or lightning strike damage. The leading edge of the blade is particularly vulnerable and should receive extra attention.
2. **Cleaning:** Dirt, dust, and organic matter such as bird droppings can accumulate on blades, altering airflow and reducing power output. Cleaning restores aerodynamic efficiency and improves performance.
3. **Repairs:** Minor surface damage can be repaired with resin or composite patching materials, while significant structural damage may require professional composite repair teams or blade replacement.

Many operators now rely on UAV (drone) inspections equipped with high-resolution cameras to detect damage quickly and safely. This technology eliminates the need for rope-access inspections except when repair work is required.

Wind Turbine Cleaning and Dirt Removal

Cleanliness is essential to maintaining efficient operation. Dirt and debris can affect not only blades but also cooling systems and electrical enclosures. Regular cleaning ensures components operate at peak performance and reduces the risk of overheating or corrosion.

Cleaning Tasks Include:

- **Blades:** Remove dirt and residue using specialized cleaning tools, rope-access methods, or drones with spray attachments. Regular cleaning prevents aerodynamic losses and improves energy production.
- **Cooling Systems:** Clean filters, vents, and fans in cooling circuits to maintain proper airflow and prevent overheating in generators and gearboxes.

- **Electrical Components:** Dust and contaminants can cause overheating and short circuits. Wipe and blow clean electrical panels and cabinets during scheduled maintenance.

Wind Turbine Testing and Calibration

Testing and calibration are vital to ensuring that turbine control systems operate precisely as intended. Over time, sensors and actuators can drift or lose calibration, which affects turbine performance and safety.

Areas to Test and Calibrate

- **Pitch Control:** Verify that blades adjust their angle correctly according to wind conditions. Miscalibration can cause inefficient energy capture or stress on components.
- **Yaw Control:** Test the yaw drive to ensure the nacelle aligns with the wind direction accurately. Misalignment can lead to reduced power generation and mechanical wear.
- **Brake System:** Test both mechanical and hydraulic brakes to ensure they engage properly during shutdown or emergency stops. The brake system must respond instantly to safety commands to prevent overspeed incidents.

Performing a Wind Turbine Torque Check

A torque check is one of the most critical maintenance procedures in wind turbine servicing. It ensures that all bolted connections—especially in the tower, nacelle, yaw system, and blades—remain tight and within the manufacturer's specified torque values. Loose bolts can lead to misalignment, vibration, component fatigue, and, in severe cases, catastrophic mechanical failure. The procedure requires precision tools, adherence to safety protocols, and accurate documentation.

Preparation and Safety Precautions

Before any torque check is performed on a wind turbine, strict safety and preparation measures must be followed to protect personnel and ensure the accuracy of results. The first step is to implement Lockout/Tagout (LOTO) procedures. This involves de-energizing and isolating the turbine so that no power can reach moving components. All rotating parts must be immobilized, and lockout tags applied according to the site's safety protocols to prevent accidental startup.

Environmental conditions must also be suitable. Torque checks should only be conducted when wind speeds are below 9 m/s to minimise tower sway and ensure technician safety.

Proper personal protective equipment (PPE) is mandatory, including a full-body fall protection harness, gloves, helmet, safety glasses, and any site-specific protective gear.

Technicians must also ensure they have the correct tools and documentation before beginning. This includes a calibrated torque wrench (manual, hydraulic, or electronic), a torque multiplier or hydraulic tensioner for high-torque applications, manufacturer torque charts for each bolt size and location, and torque paint or markers to identify checked bolts. An inspection checklist or CMMS (Computerised Maintenance Management System) form must also be prepared for accurate documentation.

Identifying Components and Torque Locations

Wind turbines contain thousands of fasteners, but torque checks focus primarily on critical load-bearing connections specified in the manufacturer's maintenance manual. These include:

- **Tower sections:** Flange bolts connecting each tower segment.
- **Nacelle frame:** Bolts securing the yaw bearing, generator mounts, gearbox housing, and main shaft couplings.
- **Blade hub and pitch system:** Blade root bolts, pitch actuator mounts, and hub covers.
- **Yaw system:** Bolts fastening the yaw ring, yaw drives, and bearing segments.
- **Drive train and couplings:** Coupling bolts between the main shaft, gearbox, and generator.

Each of these components has specific torque values and tightening sequences defined by the manufacturer, typically ranging from 300 Nm to over 1,000 Nm, depending on the bolt size, grade, and application.

Step-by-Step Torque Checking Procedure

The torque check must be performed methodically to maintain safety and accuracy.

1. **Access the Component:** Safely enter the nacelle, hub, or tower section using fall protection. Ensure that your position is stable and that the working area is secure.
2. **Clean the Bolt Area:** Remove dirt, grease, corrosion, or any debris using a clean cloth and solvent. Contamination can lead to inaccurate torque readings and must be eliminated before testing.
3. **Check Bolt Identification:** Verify the bolt's size, grade, and lubrication condition. Some bolts may be pre-lubricated or treated with anti-seize compounds, which can alter torque readings if not accounted for.

4. **Set the Torque Tool:** Adjust the torque wrench or hydraulic tensioner according to the torque values specified in the manufacturer's manual. Always check that the tool's calibration is current—ideally within the last six months.
5. **Apply Torque:** Apply smooth, steady pressure to tighten the bolt until the wrench clicks or the hydraulic gauge reaches the required value. For tensioned bolts, monitor the hydraulic pressure to ensure it matches the torque equivalent. Follow the correct tightening sequence, typically a crisscross or star pattern, to ensure even load distribution.
6. **Mark the Bolt:** Apply torque paint or a visible marker on each completed bolt. This prevents duplication and provides a clear record of progress during inspection.
7. **Record Results:** Log each bolt location, torque value, and any deviations in the inspection report or CMMS. If a bolt rotates significantly before reaching its torque limit, note this as a potential sign of loosening, thread wear, or corrosion.
8. **Re-Torque if Required:** If a bolt moves or fails to hold its specified torque, retighten it to the full value and recheck after several turbine rotations to confirm stability.

Post-Torque Verification

Once torque checks are completed, perform a visual inspection of surrounding structures to detect signs of misalignment, corrosion, or vibration-related wear. Review and sign off on all inspection records. Quality assurance (QA/QC) personnel should verify that all bolts conform to torque specifications. Any anomalies, such as stripped threads, cracked flanges, or corroded fasteners, must be reported immediately for corrective maintenance.

Documentation and Follow-Up

All inspection results and corrective actions must be recorded in the site's maintenance management system (CMMS). Each record should include the inspection date, turbine location, specific torque values, technician and supervisor signatures, and details of any bolts replaced or re-torqued.

Torque checks are conducted at manufacturer-specified intervals—typically every six months to two years, or following major component replacements. Accurate documentation ensures traceability, supports warranty compliance, and provides valuable data for trend analysis in predictive maintenance planning

Remote Monitoring and Diagnostics

Remote monitoring and diagnostics form the backbone of modern wind turbine operations, enabling continuous, data-driven oversight of turbine health, performance, and safety. Instead of relying solely on scheduled physical inspections, operators can now monitor thousands of turbines in real time from centralized control centres. These systems provide early warnings of faults, optimize maintenance scheduling, and maximize energy production while minimizing downtime and operational costs.

The purpose of remote monitoring and diagnostics (RMD) is to detect potential issues before they escalate into failures. By continuously analysing data from sensors throughout the turbine—located in the nacelle, tower, gearbox, generator, and blades—operators can identify abnormal operating trends such as overheating, vibration, or power deviations. RMD also allows predictive maintenance, enabling faults to be addressed before they result in costly breakdowns. This proactive approach enhances turbine availability, extends component life, and reduces the need for manual inspections, especially in offshore or remote installations.

A typical wind turbine remote monitoring system integrates hardware, software, and communication technologies. It begins with sensors and transducers, which measure critical parameters like temperature, vibration, oil pressure, and rotor speed. The SCADA system (Supervisory Control and Data Acquisition) functions as the turbine's central nervous system, collecting sensor data and transmitting it to the operations centre [251, 252]. A Condition Monitoring System (CMS) complements SCADA by continuously analysing vibration and acoustic data to detect wear, imbalance, or bearing damage. This information is sent via communication infrastructure, such as fibre optics or satellite links, to centralized control platforms. Data analytics and diagnostic software, often powered by artificial intelligence and machine learning, interpret the incoming data to identify anomalies and recommend corrective actions.

Remote systems monitor hundreds of variables, but certain key parameters are particularly important. These include vibration levels that reveal drivetrain misalignment or bearing wear, temperature readings across bearings and generators to detect overheating, and oil quality indicators that warn of contamination or metal wear. Electrical performance data—such as voltage, current, and harmonics—helps diagnose generator or converter faults, while rotor speed, pitch, and yaw data are analysed to ensure aerodynamic efficiency. Environmental sensors for humidity, icing, and lightning protection provide additional context for operational reliability.

Diagnostic systems use a combination of analytical techniques. Trend analysis detects gradual degradation over time, while threshold alarms automatically flag values outside normal limits. Spectral analysis identifies specific fault signatures in vibration data, and oil debris analysis tracks the presence of metal particles in lubricants. Increasingly, machine learning models are trained on historical data to predict failures before they happen, allowing maintenance teams to prioritize repairs. When anomalies are detected, automated alerts are sent to engineers with information about likely fault causes and severity levels.

Beyond monitoring, many systems enable remote control and response. Operators can start or stop turbines, adjust pitch and yaw settings, reset alarms, or limit power output in response to grid conditions or extreme weather. These capabilities reduce response times significantly—an essential advantage for offshore wind farms where physical access is limited.

RMD also plays a vital role in predictive maintenance. By combining SCADA and CMS data with oil analysis and environmental information, engineers can estimate a component's remaining useful life (RUL). For example, rising vibration levels may signal bearing fatigue, while increased gearbox temperature and oil contamination could indicate internal wear. Detecting these patterns early allows maintenance to be planned efficiently, reducing unplanned downtime and extending turbine lifespan.

Most operators manage their fleets from Remote Operations Centres (ROCs) or Network Control Centres (NCCs). These facilities operate 24/7, staffed with engineers and analysts who monitor hundreds of turbines across multiple sites. Real-time dashboards and automated alerts allow them to track turbine status, coordinate maintenance, and ensure compliance with grid requirements. Many centres also integrate meteorological forecasting to predict production levels and manage turbine operation during severe weather.

The benefits of remote monitoring and diagnostics are substantial. They increase availability through early fault detection, reduce maintenance costs by enabling targeted interventions, and improve safety by minimizing tower climbs. Continuous optimization of performance ensures turbines operate near their efficiency peak, and the accumulation of long-term operational data supports better design and lower lifecycle costs.

Chapter 13

Troubleshooting and Fault Finding

Even with the most advanced technology and rigorous maintenance regimes, wind turbines remain highly complex mechanical and electrical systems operating under continuous stress. Over time, exposure to wind loading, vibration, temperature variation, and environmental conditions inevitably leads to wear, degradation, and occasional failure. Troubleshooting and fault finding are therefore essential skills for wind turbine technicians and engineers, enabling them to diagnose problems quickly, restore operation efficiently, and prevent small issues from escalating into costly breakdowns.

This chapter explores the systematic approach to identifying, analysing, and correcting faults across the turbine's major subsystems — from mechanical components such as gearboxes, bearings, blades, yaw and pitch systems, to electrical systems including generators, converters, and control circuits. It begins by examining common mechanical failures, explaining the causes, symptoms, and consequences of wear, misalignment, and fatigue in critical components. Gearbox and bearing faults, for example, are among the most common sources of turbine downtime, while blade erosion and structural fatigue can significantly reduce aerodynamic efficiency and energy yield. Understanding these issues — and the vibration, temperature, or performance patterns that signal their onset — is central to predictive maintenance.

The chapter then addresses electrical and control system faults, which are equally vital to turbine reliability. Electrical failures such as insulation breakdown, short circuits, or generator imbalance can trigger automatic shutdowns and, if undiagnosed, cause extensive secondary damage. Power electronics faults in converters or inverters are particularly costly, as they directly affect grid synchronization and energy output. Readers will learn about the use of diagnostic tools — including insulation resistance testers, thermal imaging cameras, vibration sensors, and power quality analysers — to detect faults before they become catastrophic.

An in-depth section is devoted to interpreting SCADA (Supervisory Control and Data Acquisition) data, which provides technicians with real-time insight into turbine performance.

The chapter explains how to read alarms, analyse time-series data, and correlate parameters such as temperature, vibration, and power output to pinpoint the root causes of faults. Integrating SCADA with condition monitoring systems (CMS) allows maintenance teams to cross-verify electrical and mechanical data, enhancing diagnostic accuracy and reducing unplanned downtime.

Finally, the chapter highlights the importance of structured troubleshooting procedures — from initial fault identification and system isolation to root cause analysis and corrective action documentation. Through case-based explanations and reference to industry best practices, readers will understand how effective troubleshooting not only restores functionality but also improves long-term reliability and operational efficiency across the entire wind farm.

In short, this chapter provides a practical and analytical framework for maintaining high turbine availability. By mastering the principles of fault detection, diagnosis, and resolution, wind technicians ensure that modern wind farms continue to deliver safe, sustainable, and uninterrupted renewable energy production.

Common Mechanical Failures

Common mechanical failures in wind turbines are a major concern for operators and maintenance teams because they directly affect energy output, safety, and the lifespan of equipment. Although modern turbines are designed for reliability and endurance, the constant exposure to dynamic loads, vibrations, and environmental stresses means that mechanical wear and fatigue are inevitable over time. Understanding these failure modes — and their early indicators — is essential for preventive and predictive maintenance strategies.

Gearbox Failures

The gearbox is one of the most complex and heavily loaded components in a wind turbine. It converts the slow rotational speed of the rotor (around 10–20 rpm) into a much higher speed (around 1,000–1,800 rpm) suitable for the generator. This function places extreme stress on bearings, gears, and lubrication systems.

Common failure modes include:

- **Bearing wear and pitting:** Caused by misalignment, lubrication degradation, or contamination with dirt or metal particles.
- **Gear tooth damage:** Fatigue cracking, micro-pitting, and scuffing occur due to high torque fluctuations and surface wear.
- **Oil contamination or degradation:** Leads to poor lubrication and overheating, accelerating wear.

- **Misalignment of shafts:** Caused by foundation settling, improper installation, or thermal expansion.

Detection methods: Vibration analysis, oil particle monitoring, thermographic imaging, and SCADA-based temperature and torque trend analysis help identify faults before catastrophic damage occurs.

Gearbox failure in a wind turbine refers to the malfunction or breakdown of the mechanical transmission system that converts the slow rotation of the turbine blades into the high-speed rotation required by the generator. The gearbox is one of the most complex and heavily loaded components within a wind turbine, consisting of multiple gears, shafts, and bearings that operate under high torque and variable loads. Its primary function is to efficiently transfer mechanical energy from the rotor hub to the generator while maintaining precise alignment and minimizing energy loss through friction.

The most common cause of gearbox failure is mechanical wear, which occurs naturally over time due to the continuous transfer of large loads and torque fluctuations. The internal components—particularly bearings and gear teeth—experience cyclic stresses that can lead to fatigue cracks, pitting, or spalling. Inadequate maintenance or extended operation beyond service intervals accelerates this deterioration.

Lubrication issues are another major factor. The gearbox relies on a constant supply of high-quality oil to reduce friction, dissipate heat, and prevent metal-to-metal contact. When lubrication is insufficient—due to contamination, incorrect viscosity, or degraded oil—friction and temperature rise, leading to accelerated wear and possible seizure of bearings or gears. Poor filtration or delayed oil changes often exacerbate these problems.

Misalignment within the gearbox or between the gearbox and the main shaft is also a significant cause of failure. Misalignment creates uneven load distribution on gear teeth and bearings, increasing localized stress and vibration. Over time, this can result in premature wear, cracks, or even catastrophic gear failure. In some cases, structural deformation in the nacelle or improper installation contributes to ongoing misalignment issues.

Gearbox failure has a direct and severe impact on wind turbine performance. Since the gearbox is central to transferring energy from the rotor to the generator, any malfunction typically forces the turbine to shut down entirely. A failed gearbox means zero power output for the duration of the repair or replacement process, which can span several weeks or even months depending on part availability and crane access. In addition to lost energy production, secondary damage to nearby components—such as bearings, couplings, or generators—may occur due to vibration or overheating, compounding the overall downtime.

Gearbox repair or replacement is among the most expensive maintenance events in the wind industry. Depending on the turbine size and location, costs can easily reach hundreds of thousands of euros, factoring in spare parts, labour, transportation, and crane rental. Offshore repairs can be even more costly due to vessel logistics and limited weather windows. Beyond direct repair costs, the financial impact of lost generation time can be substantial, especially for large wind farms under performance-based contracts.

Bearing Failures

Bearings are used throughout the turbine — in the main shaft, gearbox, generator, yaw system, and pitch mechanisms. They are subject to constant cyclic loads and variable stresses.

Common causes of bearing failure:

- **Fatigue spalling:** Surface cracks develop from repetitive stress cycles.
- **Electrical pitting:** Electrical currents passing through bearings can erode surfaces (especially in generator bearings).
- **Lubrication failure:** Inadequate lubrication or contamination leads to increased friction and overheating.
- **Corrosion:** Moisture ingress causes rusting and surface degradation.

Prevention: Regular lubrication, use of insulated bearings (to prevent current leakage), and monitoring of vibration and temperature trends can extend bearing life.

Bearing failure in wind turbines refers to the breakdown or malfunction of the bearings that support the rotating components — primarily in the main shaft, gearbox, or generator. These bearings are critical for ensuring smooth rotation, efficient power transmission, and structural stability. When they fail, the entire turbine's operation can be severely compromised, often leading to prolonged downtime and expensive repairs.

Several factors contribute to bearing failure, often interacting in complex ways. One of the most common causes is mechanical stress, which occurs when excessive loads, vibrations, or misalignments exceed the bearing's design limits. This can happen due to unbalanced rotor loads, tower deflection, or poor installation practices. Over time, repetitive mechanical stresses lead to fatigue cracks, spalling, or deformation of the bearing surfaces.

Another major cause is inadequate lubrication. Bearings depend on a continuous, clean film of lubricant — either grease or oil — to minimize friction and wear. When lubrication levels drop or oil degrades, metal-to-metal contact occurs, generating heat and accelerating wear. This often results in scoring, pitting, or complete seizure of the bearing. Contamination also plays a critical role; debris accumulation from dust, moisture, or metal particles in the lubrication system can cause abrasive wear, leading to premature bearing degradation.

Bearing failures have a severe impact on wind turbine performance. They typically result in excessive vibration, noise, and increased frictional losses, which reduce mechanical efficiency and energy output. In many cases, bearing damage progresses rapidly, forcing operators to shut down the turbine to prevent secondary damage to other drivetrain components. Because accessing and replacing bearings — especially those in the main shaft or gearbox — requires significant disassembly, bearing failures are among the leading causes of turbine downtime in the wind industry.

The financial consequences of bearing failure can be substantial. Repair or replacement costs vary depending on the turbine's size and design, but even minor bearing replacements can cost several thousand euros. For large turbines, the expense can escalate into tens of thousands of euros, particularly when factoring in crane hire, specialized labour, and lost energy production during downtime. Moreover, if bearing failure leads to collateral damage in the gearbox or generator, total repair costs can multiply.

Blade and Rotor Failures

Blades are large composite structures designed to endure constant aerodynamic and gravitational forces. They are also exposed to environmental wear such as rain, hail, UV radiation, and lightning.

Typical blade failure modes include:

- **Leading-edge erosion:** Caused by repeated impact of rain or particulates, reducing aerodynamic efficiency.
- **Cracks and delamination:** Result from cyclic stress, lightning strikes, or manufacturing defects.
- **Surface wear and coating failure:** Leads to roughness that lowers energy capture efficiency.
- **Structural failure:** In extreme cases, undetected cracks or fatigue can lead to blade separation or collapse.

Detection and maintenance: Regular drone or rope-access inspections, thermographic and ultrasonic testing, and lightning protection checks are essential for preventing catastrophic blade failure.

Blade failure is one of the most significant and costly mechanical issues that can occur in a wind turbine. It refers to the damage, deterioration, or structural failure of one or more turbine blades, which are responsible for converting wind energy into rotational motion. Because the blades are continuously exposed to environmental forces, they are subjected to high levels of stress, fatigue, and wear over their service life. Even minor surface defects can escalate into severe structural problems if left unchecked, affecting not only performance but also turbine safety.

There are several primary causes of blade failure, the most common being material fatigue. Over time, the continuous flexing and cyclic loading caused by wind forces weaken the composite materials (such as fiberglass or carbon fibre) used to construct the blades. This can lead to cracking, delamination, or even full structural collapse if not detected early.

Another frequent cause is impact with foreign objects, such as birds, ice, or airborne debris. These impacts can chip or puncture the blade surface, altering the aerodynamic profile and leading to further stress concentrations. In addition, adverse weather conditions—including high winds, lightning strikes, and extreme temperatures—pose serious risks. Lightning, in particular, can burn through blade surfaces or internal structures, while high winds can cause bending or twisting beyond design limits, resulting in catastrophic failure.

Blade damage has an immediate and significant impact on turbine performance. Since the blades are the primary interface for wind energy capture, even small defects can reduce aerodynamic efficiency and lower power output. Surface erosion, cracks, or delamination can disrupt airflow, leading to turbulence and imbalance. In severe cases, damaged blades may cause dangerous vibrations or imbalance in the rotor, forcing operators to shut down the turbine to prevent further damage. Compared to some electrical or control failures, blade issues often cause larger and more immediate losses in energy production, making early detection and repair crucial.

Repairing or replacing turbine blades is one of the most expensive maintenance tasks in the wind industry. Minor surface repairs or composite patching can cost tens of thousands of euros, but full blade replacements—especially for large, utility-scale turbines—can easily reach hundreds of thousands of euros per blade. These costs include the fabrication or procurement of new blades, specialized labour for installation, and the use of heavy-lift cranes or offshore vessels in marine environments. Additionally, downtime during repairs can result in substantial lost revenue from halted energy generation.

Yaw System Failures

The yaw system aligns the nacelle with the wind direction. It consists of yaw motors, drives, brakes, and a large yaw bearing that carries the nacelle's weight.

Common issues include:

- **Yaw drive gear wear:** Due to continuous small adjustments and backlash.
- **Brake or clutch malfunction:** Causes uncontrolled nacelle rotation ("free yawing").
- **Bearing wear:** Excessive friction or uneven load distribution can lead to uneven movement.
- **Sensor failure:** Faulty wind vane or yaw encoder readings result in poor wind alignment.

Detection: Abnormal yaw motor current readings, uneven wear on the yaw ring, and vibration data often signal developing faults.

Yaw system failures can result from several mechanical, electrical, or control-related issues. One of the most frequent causes is gear or bearing wear in the yaw drive or yaw bearing. Continuous oscillations as the turbine adjusts to wind direction gradually wear down the teeth of yaw gears or degrade the grease lubrication in yaw bearings, leading to increased friction or slippage.

Another major cause is motor or brake malfunction. Yaw motors are subjected to repetitive start-stop cycles, which can lead to overheating or winding damage, while yaw brakes—used to hold the nacelle in position—may seize or fail if not regularly serviced. Inadequate or contaminated lubrication in the yaw gear ring can also lead to premature wear or corrosion, particularly in offshore environments with high humidity and salt exposure.

Electrical faults and control system errors may also prevent the yaw system from responding to wind direction changes. These can include PLC or sensor failures, broken limit switches, or faulty wind vane readings, resulting in misalignment between the nacelle and the wind direction.

Yaw system failure directly impacts turbine performance and longevity. When the nacelle cannot align with the wind, the rotor blades no longer face the optimal direction, leading to significant energy loss — in some cases, reducing power output by 10–20%. Persistent yaw misalignment also increases asymmetric loads on the rotor and tower, accelerating fatigue and raising the risk of blade or bearing damage.

In severe cases, a failed yaw brake or gear system may allow the nacelle to rotate uncontrollably ("free yawing") during high winds, posing serious safety risks and potential structural damage. This condition can also trigger automatic shutdowns, leading to extended downtime until repairs are made.

The financial impact of yaw system failures depends on the extent of the damage. Minor issues such as motor replacement or sensor repair may cost a few thousand euros. However, severe damage to the yaw bearing or gear ring can require major component replacement, often costing tens of thousands of euros in parts and labour. Offshore repairs are especially expensive due to the logistics of accessing the turbine and the need for specialized vessels.

Indirect costs, such as lost production during downtime, can be equally significant. A turbine offline for several days or weeks due to yaw malfunction can lose substantial revenue, particularly in high-wind seasons. Preventive maintenance — including regular lubrication, torque checks on yaw bolts, and SCADA-based monitoring of yaw motor activity — is therefore essential to minimize repair frequency and extend component lifespan.

Pitch System Failures

The pitch system adjusts blade angles to optimize aerodynamic efficiency and control rotor speed. It is critical for power regulation and turbine safety.

Common failure modes include:

- **Hydraulic leaks:** Common in older or fluid-driven pitch systems, leading to pressure loss.
- **Actuator or motor failure:** Causes the blades to stick in an incorrect pitch angle.
- **Sensor faults:** Misreads pitch angle, resulting in poor control or imbalance.
- **Bolt loosening or fatigue:** Causes play in the hub-to-blade interface, leading to vibration and stress.

Preventive measures: Regular torque checks, hydraulic fluid monitoring, and redundancy testing (especially for fail-safe feathering) are vital.

Pitch system failures can arise from a combination of mechanical, electrical, and hydraulic issues.

Hydraulic or Mechanical Wear: In turbines that use hydraulic pitch systems, components such as actuators, pumps, seals, and hoses are subject to continuous pressure cycles and high mechanical loads. Over time, seals may deteriorate, causing internal or external leaks that reduce system pressure and responsiveness. Mechanical linkages or bearings within the hub can also experience wear, leading to uneven or delayed blade movement.

Electrical and Control System Faults: In electrically driven pitch systems, faults often stem from malfunctioning pitch motors, gearboxes, or position encoders. Electrical failures may occur due to insulation breakdown, power surges, or wiring degradation. Sensor or encoder errors can lead to incorrect blade angle readings, causing the control system to command inaccurate pitch adjustments.

Battery or Backup Power Issues: Each blade's pitch control system typically has a backup battery or capacitor bank to move the blades to a feathered (safe) position during power loss. If these batteries degrade or fail, the blades may remain in a power-generating position during an emergency, greatly increasing the risk of overspeed and mechanical damage.

Software or Communication Errors: PLC programming faults, software bugs, or loss of communication between the nacelle controller and blade pitch units can prevent coordinated blade movement. In offshore environments, moisture ingress and corrosion in connectors can further disrupt signal transmission.

Pitch system failures can have immediate and severe consequences for turbine performance and safety. A malfunctioning pitch system can prevent blades from adjusting to the optimal aerodynamic angle, resulting in reduced energy capture and efficiency losses. Even small deviations in pitch angle can lead to a drop in power output, typically by 5–10%. If one or more blades fail to pitch correctly, the rotor becomes unbalanced, increasing vibration and stress on the drivetrain and tower structure.

In more severe cases, if the blades cannot feather during high winds or emergency shutdowns, the turbine risks overspeed conditions. Overspeed can cause catastrophic damage to blades, hub components, and bearings, posing serious safety hazards to personnel and surrounding assets. Consequently, most turbines automatically shut down when pitch faults are detected, leading to immediate downtime and lost revenue.

Long-term effects include accelerated wear on the main shaft, gearbox, and bearings due to unbalanced aerodynamic loads. This not only reduces component lifespan but also increases the frequency of costly maintenance interventions.

The financial impact of pitch system failures varies depending on the type and severity of the malfunction. Minor issues such as sensor replacements, seal leaks, or system recalibration generally involve relatively low repair costs and minimal downtime. In contrast, major failures—such as actuator or pitch motor replacement, or full hydraulic system rebuilds—can result in significant expenses when factoring in component costs, labour, and equipment access.

If the malfunction leads to severe outcomes such as turbine overspeed or blade damage, overall costs can escalate dramatically, particularly when heavy-lift cranes or offshore service vessels are required. Furthermore, extended downtime during periods of strong wind resource can lead to considerable losses in energy production, amplifying the total financial impact of the failure.

Preventive measures such as regular hydraulic fluid checks, pitch bearing lubrication, sensor calibration, and monitoring of SCADA pitch error trends are critical for avoiding costly failures. Modern condition monitoring systems can detect early signs of pressure drops, actuator delays, or abnormal motor currents, allowing maintenance teams to act before failures occur.

Generator and Coupling Failures

The generator converts mechanical energy into electricity, while couplings transfer torque from the gearbox.

Typical problems include:

- **Rotor imbalance:** Causes vibration and electrical inefficiency.
- **Bearing damage:** Due to electrical discharge or poor lubrication.
- **Coupling misalignment:** Leads to torsional stress and bearing overload.
- **Cooling failure:** Results in overheating and insulation breakdown.

Detection: Thermal imaging, current and voltage analysis, and vibration monitoring can detect imbalance and bearing issues early.

Brake System and Hydraulic Failures

Braking systems are vital for stopping the turbine during high winds, maintenance, or emergency shutdowns.

Common failures:

- **Hydraulic fluid leaks:** Reduce braking efficiency.
- **Worn pads or discs:** Increase stopping distance and noise.
- **Sensor malfunctions:** Prevent brake activation during emergencies.
- **Contaminated fluid:** Causes sluggish or uneven braking.

Maintenance methods: Regular fluid checks, hydraulic filter replacement, and emergency stop testing under controlled conditions.

Brake failure in a wind turbine occurs when the mechanical systems designed to slow or stop the rotor malfunction or lose effectiveness. These braking systems are crucial for ensuring safe operation—particularly during high winds, maintenance procedures, or emergency shutdowns. Modern turbines typically use a combination of mechanical disc brakes and aerodynamic braking (by pitching the blades to reduce lift), but mechanical brakes remain the final safety measure. When these systems fail, the turbine can spin out of control, posing severe safety risks and the potential for catastrophic mechanical damage.

One of the most common causes of brake failure is overheating. Continuous or excessive braking during turbulent wind conditions can generate significant frictional heat, leading to brake pad glazing or material degradation. As the pads overheat, they lose their ability to generate the friction necessary to slow the rotor effectively.

Another frequent issue is wear and tear. Like any mechanical system exposed to constant use and environmental stress, brake pads, discs, and hydraulic seals gradually deteriorate. Without regular replacement or adjustment, these components may lose their effectiveness or fail completely.

Improper maintenance also contributes significantly to brake problems. Failure to conduct periodic inspections, replace worn components, or check hydraulic fluid levels can lead to partial or total brake malfunction. Inadequate torque checks on bolts, leaks in the hydraulic circuit, or contamination of brake fluid can all cause the system to perform below design standards.

The consequences of brake failure are among the most serious in turbine operations. A failed braking system can lead to turbine overspeed, where the rotor exceeds its safe design limits. Overspeeding not only places enormous stress on the gearbox, generator, and blades but can also cause catastrophic mechanical failure, including blade detachment or nacelle damage. Even if the turbine's emergency pitch control activates, a faulty brake system can prolong deceleration, increasing the likelihood of secondary failures. As a result, brake integrity is directly tied to turbine safety, reliability, and regulatory compliance.

Repairing or replacing a turbine's brake system can be costly, depending on the extent of damage and accessibility of the site. Simple pad replacements and hydraulic fluid changes may cost a few thousand euros, but full brake system replacements—including discs, actuators, and hydraulic lines—can range from tens to hundreds of thousands of euros. If overspeed damage occurs, secondary repairs to the gearbox, bearings, or generator can multiply costs dramatically, especially for offshore installations where crane mobilization and downtime are more expensive.

8. Structural and Fastener Failures

Tower sections, nacelle frames, and flanges are joined by thousands of high-torque bolts. These are subject to vibration, temperature changes, and fatigue.

Common issues:

- **Bolt loosening:** From vibration or improper torque application.
- **Cracked welds or corrosion:** Often found around tower flanges or access platforms.
- **Foundation settlement:** Leads to misalignment and uneven load distribution.

Detection and prevention: Scheduled torque verification, corrosion inspections, and ultrasonic weld testing are crucial for structural safety.

Common Electrical Faults and Diagnostic Tools

Electrical systems play a vital role in the operation and safety of wind turbines, as they manage every aspect of power generation, control, and communication between subsystems. When electrical faults occur, they can lead to reduced efficiency, unplanned shutdowns, or even catastrophic equipment damage. A strong understanding of common electrical issues and the diagnostic tools used to identify them is therefore essential for maintaining reliable turbine performance and ensuring cost-effective maintenance.

One of the most common electrical problems in wind turbines is insulation breakdown, which occurs when the insulation in cables, generators, or transformers deteriorates over time. Factors such as excessive heat, moisture ingress, and mechanical vibration can weaken insulation materials, leading to short circuits or ground faults. When this happens, protective relays may trip, or dangerous arc flashes can occur, posing serious safety risks and damaging high-voltage components.

Another frequent issue involves ground faults and short circuits. Ground faults happen when electrical current unintentionally flows to the turbine's metallic structure or the ground, often due to damaged cables, moisture infiltration in junction boxes, or degraded insulation. Similarly, short circuits between conductors can result from worn cables or faulty connectors. Both types of faults typically cause an immediate turbine shutdown and must be quickly isolated and tested to prevent further damage.

Overvoltage and undervoltage conditions also present challenges within turbine electrical systems. Fluctuations in voltage can lead to overheating of power electronics, malfunctioning of control systems, or activation of protective relays. Overvoltage events are often caused by lightning strikes or switching surges, while undervoltage conditions may arise during startup or periods of grid instability. Sustained voltage irregularities can cause serious wear on sensitive electrical equipment.

Generator faults are another significant source of downtime. These can include stator winding faults, rotor imbalances, or bearing insulation failures — typically resulting from overloading, inadequate cooling, or mechanical vibration. If such problems go undetected, they may lead to the complete replacement of the generator, one of the most costly electrical repairs in a turbine's lifecycle.

Converter and inverter failures are common in modern turbines that rely on power electronics to regulate voltage and frequency for grid compliance. Overheating, capacitor degradation, or failure of insulated-gate bipolar transistors (IGBTs) can cause system alarms and force the turbine offline. Since converters are critical to energy output and grid synchronization, any malfunction can result in significant operational downtime until parts are replaced.

Wind turbines are also highly exposed to environmental stress, making lightning and surge damage a major concern. Although turbines are equipped with grounding and surge protection systems, improper installation or lack of maintenance can allow overvoltages to reach sensitive components such as sensors, transformers, and control systems. Such damage may not always be visible but can severely impair turbine performance and reliability.

Connector and terminal failures occur when electrical connections loosen, corrode, or degrade over time. Poorly crimped or oxidized connectors can cause intermittent faults, excessive heat generation, and voltage drops. These problems are especially common in offshore wind turbines, where humidity, salt, and constant vibration accelerate corrosion.

Together, these electrical faults underscore the importance of routine diagnostics, preventive maintenance, and environmental protection measures to ensure the safe and efficient operation of modern wind turbines.

Diagnostic Tools

Wind turbine technicians rely on a range of specialized diagnostic tools and methods to detect, analyse, and prevent electrical faults before they lead to equipment failure or unplanned downtime. These tools help verify the condition of insulation, connections, and electrical components, allowing maintenance teams to identify issues early and maintain turbine reliability.

One of the most fundamental tools is the insulation resistance tester, commonly known as a *megger*. This device applies a high DC voltage to electrical circuits, cables, transformers, or motors to measure insulation resistance. A declining resistance reading over time indicates the presence of moisture, contamination, or insulation degradation, which can eventually lead to ground faults or short circuits if left unchecked.

Multimeters and clamp meters are essential for field diagnostics. Multimeters allow technicians to measure voltage, current, and resistance within circuits, while clamp meters provide a non-intrusive way to measure current flow on live conductors without disconnecting wires. These portable instruments are used for quick verification of electrical health during both routine maintenance and fault investigations.

Thermal imaging cameras, which use infrared thermography, detect abnormal heat patterns in electrical panels, generators, and converters. Hotspots revealed by thermal images can point to overloaded circuits, loose terminals, unbalanced loads, or failing electrical components. Regular thermal inspections are an effective way to catch developing issues before they escalate into major failures.

Partial discharge (PD) testers are used for identifying early-stage insulation defects. By detecting small electrical discharges that occur inside deteriorating insulation materials, PD testing provides valuable insight into hidden faults that could lead to catastrophic insulation breakdown if not addressed.

A power quality analyser monitors voltage, current, frequency, harmonics, and transients in real time. It helps engineers assess power stability and detect issues such as harmonic distortion or voltage fluctuations that can harm sensitive power electronics and reduce turbine efficiency. These measurements are especially important for grid-connected turbines, where compliance with power quality standards is critical.

Vibration and acoustic sensors, although primarily used for mechanical condition monitoring, can also reveal electrical faults indirectly. Abnormal vibration patterns may indicate rotor imbalance, generator bearing wear, or electromagnetic interference caused by current leakage or irregular magnetic fields. Integrating vibration analysis with electrical diagnostics provides a more complete picture of turbine health.

The SCADA system (Supervisory Control and Data Acquisition) serves as the central diagnostic and monitoring platform for wind turbines. It continuously collects and analyses performance data, including generator voltage, current, power factor, and converter temperatures. When a fault occurs, SCADA generates alarms and fault codes that help technicians pinpoint the affected subsystem quickly and accurately.

For more advanced troubleshooting, oscilloscopes and signal analysers are used to visualize voltage and current waveforms. These tools allow engineers to examine the behaviour of control and converter systems in detail, detecting irregular switching patterns, timing errors, or signal distortions that may cause instability or energy losses.

Finally, insulation oil analysis is critical for transformers and other components that use dielectric fluids. Laboratory testing of oil samples can reveal moisture content, oxidation levels, and dissolved gases—early warning signs of insulation breakdown or internal component wear.

Diagnosis of Specific Electrical Faults

Diagnosing electrical faults in wind turbines requires a systematic approach using both on-site and remote tools. Each type of fault presents specific symptoms that can be identified through testing, measurement, and data analysis. Below is a detailed explanation of how each fault type is typically diagnosed by technicians and engineers.

Insulation Breakdown

To diagnose insulation breakdown, technicians use an insulation resistance tester (megger) to apply a controlled DC voltage across cables, generator windings, or transformer circuits. The instrument measures resistance between conductors and ground — a healthy insulation system will show very high resistance, often in the megaohm range. A low or declining resistance value indicates moisture ingress, contamination, or thermal degradation. In more advanced diagnostics, partial discharge (PD) testing or tan delta testing is used to detect early-stage insulation deterioration before complete failure occurs. Thermal imaging may also reveal localized heating where current is leaking through degraded insulation. Data from the turbine's SCADA system can further confirm the presence of ground faults or arc events triggered by protection systems.

Ground Faults and Short Circuits

Ground faults are diagnosed through ground fault detectors, continuity tests, and insulation resistance measurements between live conductors and the turbine's metal frame. Technicians isolate circuits systematically to identify where current is leaking to earth. For short circuits between conductors, low-resistance ohmmeters and circuit isolation tests are used to locate the faulted section. The SCADA system often logs an instantaneous trip event when a short or ground fault occurs, allowing engineers to narrow down the location. Visual inspections of junction boxes, cable terminations, and conduits are then performed to confirm physical damage or moisture intrusion.

Overvoltage and Undervoltage Conditions

These voltage anomalies are identified using power quality analysers or data logs from the SCADA system. The analyser records voltage fluctuations, harmonics, and transient surges in real time, helping engineers pinpoint whether the problem originates from the turbine, grid, or control system.

Lightning and surge counters can provide evidence of overvoltage events caused by lightning strikes or switching transients. For undervoltage, diagnostic efforts focus on the grid connection, transformer taps, or converter control system, as low voltage can indicate grid instability or converter faults.

Generator Faults

Generator faults are diagnosed through a combination of electrical, mechanical, and vibration analysis. Electrical tests include stator winding insulation resistance, winding resistance balancing, and polarization index (PI) measurements to detect degradation. Vibration analysis and bearing temperature monitoring can reveal mechanical issues such as imbalance, misalignment, or bearing wear.

Current signature analysis is another method used to identify rotor bar damage or eccentricity in the magnetic field. Data from the condition monitoring system (CMS) and SCADA alarms (e.g., overtemperature or phase imbalance) are cross-checked to confirm the fault.

Converter and Inverter Failures

Diagnosing converter and inverter faults involves reviewing SCADA fault logs for IGBT failure, overtemperature, or DC-link voltage deviations. Technicians use oscilloscopes or signal analysers to examine the switching patterns and waveform integrity of the converter circuit. Thermal imaging helps identify overheating in capacitors or power modules, while continuity and diode tests with a multimeter are used to verify the health of semiconductor components. If internal faults are suspected, the converter cabinet may be opened for inspection of burnt or swollen capacitors and damaged fuses.

Lightning and Surge Damage

When a lightning or surge event occurs, technicians start diagnostics by visually inspecting lightning receptors, down conductors, and grounding points for burn marks or fractures. Continuity tests verify that the lightning protection path remains intact from the blade tip to the ground grid.

Surge arresters and filter units are tested using insulation resistance or leakage current measurements. The SCADA system and lightning counters also record overvoltage events, helping correlate damage timing with storm activity. If power electronics or sensors fail immediately after a strike, surge damage is likely the root cause.

Connector and Terminal Failures

Connector and terminal issues are diagnosed through visual inspection, thermal imaging, and voltage drop testing. Loose or corroded connections often show heat buildup, which appears as bright spots in infrared thermography.

Technicians may also perform contact resistance tests using a micro-ohmmeter to identify high-resistance joints that cause intermittent faults or overheating. In offshore turbines, corrosion checks and re-termination of connectors are common preventive actions due to the high humidity and salt exposure.

Reading SCADA Data for Fault Analysis

Reading SCADA (Supervisory Control and Data Acquisition) data for fault analysis is one of the most important diagnostic skills in wind turbine operation and maintenance. The SCADA system continuously collects, processes, and displays operational data from sensors and control modules across the turbine, allowing operators and engineers to identify, interpret, and resolve faults before they escalate into major failures. Understanding how to interpret SCADA data helps pinpoint the cause of abnormalities, optimize turbine performance, and support predictive maintenance strategies.

Understanding the structure of SCADA (Supervisory Control and Data Acquisition) data is essential for effective monitoring and analysis of wind turbine performance. A modern wind turbine SCADA system continuously tracks hundreds of parameters in real time, with data typically recorded at intervals ranging from one second to ten minutes. This constant flow of information provides operators with a detailed picture of how the turbine is performing under varying environmental and operational conditions.

The data collected by the SCADA system can be grouped into several key categories. Operational data includes parameters such as rotor speed, power output, pitch angle, yaw position, wind speed, and wind direction. These values are used to assess whether the turbine is operating efficiently and responding correctly to wind conditions. Electrical data covers measurements of voltage, current, frequency, power factor, and grid connection status, ensuring that power generation and delivery remain stable and compliant with grid requirements.

Mechanical data focuses on parameters such as vibration, temperature, hydraulic pressure, and gearbox oil condition. These indicators help identify potential wear, imbalance, or lubrication issues before they escalate into major mechanical failures. Environmental data — including ambient temperature, humidity, and air density — provides essential context, as changes in environmental conditions directly influence turbine performance and energy yield.

Alarm and event logs record fault codes and time-stamped alerts for each subsystem, such as the gearbox, generator, or converter. These logs are vital for diagnosing problems and tracing the sequence of events leading to a fault. All turbines within a wind farm transmit their SCADA data to a central database, where it is visualized through dashboards, trend graphs, and event logs. This centralized system allows operators and engineers to analyse data across multiple turbines, identify patterns, and make informed decisions that enhance reliability, safety, and efficiency across the entire wind farm.

In modern wind turbine operations, SCADA systems are essential for fault detection and management. They generate alarms that encapsulate various critical details whenever a fault occurs, including a timestamp that documents when the fault happened, the subsystem responsible, the severity level, and a description of the fault type. This structured information allows technicians to conduct thorough analyses of the alarms to understand the underlying issues more clearly.

The timestamp is crucial for correlating the timing of a fault with other operational events that might have occurred immediately prior. For instance, if a "Generator Overtemperature" alarm follows a "Cooling Fan Fault," it may suggest a causal relationship between the two issues, guiding technicians toward a more accurate diagnosis of the root cause of the fault [253, 254]. The subsystem indicated in the alarm helps pinpoint which component triggered the alert, enabling focused inspections [254].

Severity levels classified as warning, fault, or trip play an essential role in determining operational responses. These classifications indicate whether the turbine remains operational or if it requires immediate shutdown for safety reasons [255, 256]. Clear communication of such information is vital for reducing downtime and operational risks [255, 257].

Furthermore, the description embedded within the alarm is instrumental for technicians, offering concise information about the nature of the fault, such as "Phase imbalance detected" or "Hydraulic pressure low" [258]. This descriptive element enhances the understanding of potential operational impacts and informs any necessary remedial actions [182].

As noted in various studies, the capabilities of SCADA systems extend beyond merely generating alarms; they archive comprehensive signal information, historical alarms, and fault logs that include environmental and operational conditions relevant to the fault [254, 255]. This archival capability is significant since it allows for historical analysis that can reveal patterns over time, triggering proactive maintenance strategies [259, 260]. The integration of such data facilitates advanced diagnostic capabilities that improve overall reliability and maintenance practices in wind turbine operations [253, 255].

When a fault occurs, the SCADA system generates an alarm or fault code that includes:

- Timestamp – When the fault occurred.
- Subsystem – Which component triggered the alarm (e.g., “Yaw System,” “Generator Bearing,” “Converter Overtemperature”).

- Severity Level – Warning, fault, or trip (indicating whether the turbine is still operational or has shut down).
- Description – A short text explaining the fault type (e.g., "Phase imbalance detected" or "Hydraulic pressure low").

To analyse an alarm, technicians review the exact time and sequence of events surrounding the fault. For example, if a "Generator Overtemperature" alarm occurs after a "Cooling Fan Fault," it suggests a secondary fault caused by cooling failure rather than an isolated thermal issue.

Analysing trends and time-series data is one of the most powerful diagnostic capabilities within a SCADA system. The trend or graphing function enables operators to visualise and compare multiple parameters over time, revealing relationships between operating conditions and developing faults. By examining how different data points—such as temperature, vibration, or power output—change together, technicians can identify the root causes of abnormal behaviour and distinguish between mechanical, electrical, or control-related issues.

For instance, a rise in gearbox temperature followed by a drop in power output may indicate a problem with lubrication, suggesting that the gearbox oil is deteriorating or circulation is restricted. Similarly, high vibration levels that occur alongside fluctuating rotor speed can point to drivetrain misalignment or bearing wear. Voltage drops that coincide with converter alarms may highlight grid instability or internal electrical faults.

A key part of this analysis involves zooming in on the time window surrounding an alarm or event. Doing so helps operators determine whether the issue appeared suddenly—implying a component failure—or developed gradually over days or weeks, which often signals progressive degradation or maintenance neglect. Through this type of time-series analysis, SCADA data transforms from raw numbers into meaningful insights, supporting faster, more accurate fault diagnosis and proactive maintenance planning.

Each event or fault in SCADA is logged with a unique status code, which corresponds to predefined messages in the turbine manufacturer's diagnostic manual. Reading these codes accurately allows maintenance engineers to isolate the subsystem involved.

Typical examples include:

- Code 304 – "Yaw position error."
- Code 511 – "Pitch communication fault."
- Code 812 – "Generator bearing temperature high."

After identifying the code, the technician consults the fault tree or diagnostic guide to determine likely causes and recommended actions. Many modern SCADA systems also include built-in fault tracing tools that automatically suggest root causes based on historical data.

Cross-referencing parameters is an essential technique for conducting deeper fault analysis in wind turbine operations. By comparing related data points, operators can differentiate between electrical, mechanical, and control-related issues, ensuring that maintenance teams address the true cause of a fault rather than just its surface symptom. This method relies on analysing how one parameter affects or correlates with another within the same subsystem or across multiple subsystems.

For example, in the case of pitch faults, technicians compare the pitch angle commands issued by the programmable logic controller (PLC) with the actual blade pitch angles recorded by sensors. Any discrepancy between the two values suggests a potential sensor calibration issue, actuator malfunction, or control lag. For yaw faults, operators examine the nacelle's position relative to the wind direction. Persistent misalignment or slow correction indicates possible problems with the yaw motors, gear drives, or encoder feedback. Similarly, generator faults can be diagnosed by correlating electrical current imbalances with vibration levels or bearing temperature increases, revealing a coupling between electrical irregularities and mechanical stress.

By systematically cross-referencing these parameters, operators can verify the underlying cause of anomalies with greater accuracy. This approach not only improves fault diagnosis but also reduces unnecessary component replacements, optimises maintenance efficiency, and enhances the overall reliability of wind turbine operations.

Raw SCADA data can be noisy due to sensor errors, communication glitches, or transient conditions. Before analysis, data must be filtered to remove:

- Outliers: Unusually high or low readings not physically possible.
- Data Gaps: Periods where sensors lost connection or calibration.
- Startup and Shutdown Periods: These can skew average values and trends.

Modern SCADA software includes filters and statistical smoothing tools to ensure cleaner, more reliable trend analysis.

SCADA provides general operational data, while the Condition Monitoring System (CMS) offers higher-frequency data focused on vibration, acoustics, and oil quality.

By integrating both datasets, engineers can link electrical or thermal anomalies detected in SCADA to mechanical faults identified by the CMS. For example, a gearbox vibration increase recorded by CMS can be cross-validated with a temperature spike in SCADA data to confirm bearing wear.

Once analysis is complete, findings are documented in maintenance reports or entered into the Computerized Maintenance Management System (CMMS). Reports typically include:

- The fault description and SCADA event codes.
- A timeline of key data trends.

- The suspected root cause.
- Recommended corrective actions (e.g., inspection, part replacement, software update).

This information is then used to plan maintenance, prioritize turbine interventions, and refine predictive models.

Reading SCADA data for fault analysis involves interpreting alarms, analysing time-series data, cross-referencing parameters, and correlating trends with equipment behaviour. Effective SCADA interpretation allows technicians to move from reactive troubleshooting to proactive maintenance, minimizing downtime and maximizing turbine availability.

With advancements in AI and analytics, modern SCADA systems can even predict potential failures before alarms trigger — transforming the way wind farms are monitored and maintained.

Chapter 14

Preventive and Corrective Maintenance

After a wind turbine has been commissioned and entered regular operation, its long-term reliability, performance, and safety depend on one central factor — effective maintenance management. Wind turbines operate in harsh environments, continuously exposed to mechanical stress, extreme weather, and electrical load cycles. Without structured maintenance, even minor wear or imbalance can escalate into major failures that cause costly downtime and safety risks. Chapter 14, *Preventive and Corrective Maintenance*, examines how wind energy operators sustain turbine performance through a combination of proactive and reactive maintenance strategies.

This chapter begins by defining the types of maintenance used in modern wind energy operations — preventive, predictive, condition-based, and corrective. Each approach represents a different philosophy of asset care: preventive maintenance focuses on regular servicing and inspection schedules to avoid failures; predictive maintenance uses real-time data and advanced analytics to forecast potential issues before they occur; condition-based monitoring continuously tracks system parameters like vibration and temperature; and corrective maintenance involves restoring functionality after an unexpected breakdown. Together, these methods form an integrated system of turbine care that maximizes uptime, extends component lifespan, and ensures safety and regulatory compliance.

The second section explores specialized maintenance procedures that apply these strategies to critical turbine components. Readers will learn how technicians conduct blade inspections, identify erosion or structural damage, and carry out precision repair techniques using composite materials, resin systems, and advanced non-destructive testing (NDT) tools. Topics such as lightning damage repair, leading-edge protection (LEP) replacement, and full section replacement highlight the complexity and craftsmanship involved in maintaining aerodynamic efficiency and structural integrity. These procedures not only preserve turbine output but also maintain compliance with international standards like IEC 61400-23 and IEC 61400-24.

Subsequent sections detail gearbox, generator, tower, and foundation maintenance, emphasizing the importance of lubrication management, vibration analysis, corrosion prevention, and torque verification. These areas represent the mechanical and structural backbone of every wind turbine, where routine care prevents catastrophic failures. The chapter also explains how condition monitoring systems (CMS) and SCADA integration enable predictive maintenance by detecting early signs of wear through vibration, temperature, and oil debris sensors — transforming maintenance from reactive response to proactive precision.

Finally, the chapter discusses the corrective maintenance process, which addresses damage, fatigue, and corrosion when they occur. Procedures such as crack repair, grout replacement, and re-leveling demonstrate how technicians restore turbine stability and safety following structural or environmental degradation. Each task is presented with step-by-step explanations, safety considerations, and documentation requirements to ensure compliance and traceability under industry regulations.

Ultimately, *Preventive and Corrective Maintenance* provides wind turbine technicians, engineers, and maintenance planners with a comprehensive understanding of how to protect, repair, and optimize one of the world's most vital renewable energy technologies. It emphasizes that effective maintenance is not merely a technical duty — it is a critical component of sustainable energy production, operational safety, and environmental stewardship.

Types of Maintenance

Wind turbines operate continuously in challenging environmental conditions—exposed to wind, rain, lightning, and temperature fluctuations. To ensure maximum reliability, efficiency, and safety, operators employ several types of maintenance strategies. These strategies differ in timing, technology, and scope, but all share the goal of minimizing downtime and extending the turbine's lifespan. The four most common approaches are Preventive Maintenance, Predictive Maintenance, Condition-Based Monitoring, and Corrective Maintenance.

Preventive Maintenance

Preventive maintenance is the most widely used strategy in the wind energy industry. It involves performing regularly scheduled inspections, servicing, and replacements to prevent failures before they occur. The goal is to keep equipment in good working order and minimize the likelihood of unexpected breakdowns.

Typical preventive maintenance tasks include:

- Inspecting and cleaning rotor blades to remove dirt, ice, and debris that can reduce aerodynamic efficiency.
- Checking and tightening bolts and fasteners in the nacelle, tower, and hub.

- Lubricating bearings, yaw drives, and pitch systems to prevent friction and wear.
- Replacing filters and oil in the gearbox and hydraulic systems at regular intervals.
- Testing safety systems such as brakes and fire suppression units.

Example: A maintenance crew might perform a biannual service on a turbine, where technicians inspect the gearbox, check blade integrity, and replace gearbox oil. Keeping blades clean and smooth ensures efficient airflow, potentially improving turbine output by 2–4% annually.

Preventive maintenance is cost-effective in the long run, but it requires careful scheduling and planning to avoid unnecessary downtime during high-wind seasons when turbines generate the most power.

Predictive Maintenance

Predictive maintenance leverages advanced technologies and real-time data to forecast when a component is likely to fail. Instead of relying solely on fixed schedules, this method uses sensor-based monitoring systems and data analytics to assess component health dynamically. This allows operators to perform maintenance only when necessary, reducing both downtime and maintenance costs.

Common predictive monitoring tools include:

- Vibration sensors on the gearbox and generator to detect imbalance or bearing wear.
- Temperature sensors to identify overheating in bearings or electrical systems.
- Oil analysis sensors to detect metal particles or contamination.
- SCADA data analytics to track performance trends over time.

Example: A turbine's vibration sensor detects abnormal oscillations in the main bearing, suggesting early-stage fatigue. The operator schedules a bearing replacement during a low-wind period, preventing a catastrophic failure and saving thousands in emergency repair costs.

Predictive maintenance requires investment in Condition Monitoring Systems (CMS) and data analytics software, but it offers excellent long-term returns by reducing unplanned outages and extending equipment lifespan.

Condition-Based Monitoring (CBM)

Condition-Based Monitoring is closely related to predictive maintenance but focuses on continuous performance tracking rather than purely predictive analytics. It uses a network of sensors, SCADA systems, and remote diagnostic tools to provide real-time insights into turbine health and performance.

CBM relies on threshold limits—when parameters like vibration, temperature, or oil pressure deviate from their normal ranges, the system automatically alerts operators to take corrective action. The advantage of CBM is that it allows operators to make data-driven maintenance decisions without constant manual inspection.

Example: A wind farm's SCADA system continuously monitors power output and wind speed. If one turbine's power generation falls below the expected curve while wind conditions remain stable, the system flags it for inspection. The maintenance team discovers that the blade pitch sensor is malfunctioning—correcting it before it impacts energy production further.

Condition-Based Monitoring is ideal for large wind farms or offshore installations, where manual inspection is costly or difficult. It also enhances reliability by catching subtle performance drops that might otherwise go unnoticed.

Corrective Maintenance

Corrective maintenance, sometimes called reactive maintenance, occurs after a failure or malfunction has already happened. This strategy is generally considered a last resort because it leads to unplanned downtime, costly emergency repairs, and potential safety risks. However, it is sometimes unavoidable, especially when dealing with sudden or unpredictable failures.

Corrective maintenance tasks might include:

- Replacing a failed generator or gearbox after a breakdown.
- Repairing or replacing damaged blades following a lightning strike or impact.
- Fixing electrical faults in control systems or power converters.
- Replacing hydraulic pumps or seals that have failed unexpectedly.

Example: A turbine unexpectedly stops operating due to a generator bearing seizure. The maintenance team must deploy a heavy-lift crane to remove the nacelle cover and replace the generator, causing several days of downtime and significant repair costs.

While corrective maintenance is the most expensive and disruptive type, it remains necessary for **unforeseen failures** that preventive and predictive methods cannot always avoid. Operators strive to minimize its frequency through rigorous preventive and predictive programs.

Each maintenance strategy plays a crucial role in ensuring wind turbine reliability:

Table 4: Summary of maintenance types.

Maintenance Type	Approach	Timing	Tools/Techniques	Example
Preventive	Scheduled servicing and inspections	Regular intervals	Manual inspection, lubrication, part replacement	Cleaning blades, replacing oil filters
Predictive	Uses real-time data to anticipate failures	As needed (based on data)	Sensors, vibration and oil analysis, SCADA data	Detecting early bearing wear
Condition-Based Monitoring	Continuous performance tracking	Continuous	Remote sensors, automated alerts	Identifying pitch control anomalies
Corrective	Reactive repair after failure	After breakdown	Repair tools, cranes, component replacement	Replacing a failed gearbox

Blade Inspection and Repair Techniques

Blade Inspection and Repair Techniques are fundamental to both preventive and corrective maintenance in wind turbine operations. They ensure aerodynamic efficiency, structural integrity, and long-term reliability of the turbine. Because blades are constantly exposed to harsh environmental conditions—such as ultraviolet (UV) radiation, rain erosion, salt spray, icing, and lightning strikes—they are among the most vulnerable components of a turbine. Without regular inspection and timely repairs, minor surface defects can progress into major structural damage, resulting in costly downtime and significant energy losses.

The main goal of blade inspection is to identify early signs of deterioration and maintain aerodynamic performance. Even small imperfections, such as erosion along the leading edge or hairline cracks in the blade surface, can reduce energy output by several percentage points. Inspections focus on detecting various types of damage, including cracks and delamination in composite materials, erosion or wear along the leading and trailing edges, lightning strike damage near tips or receptors, contamination from dirt or oil, and structural deformation caused by fatigue or impact.

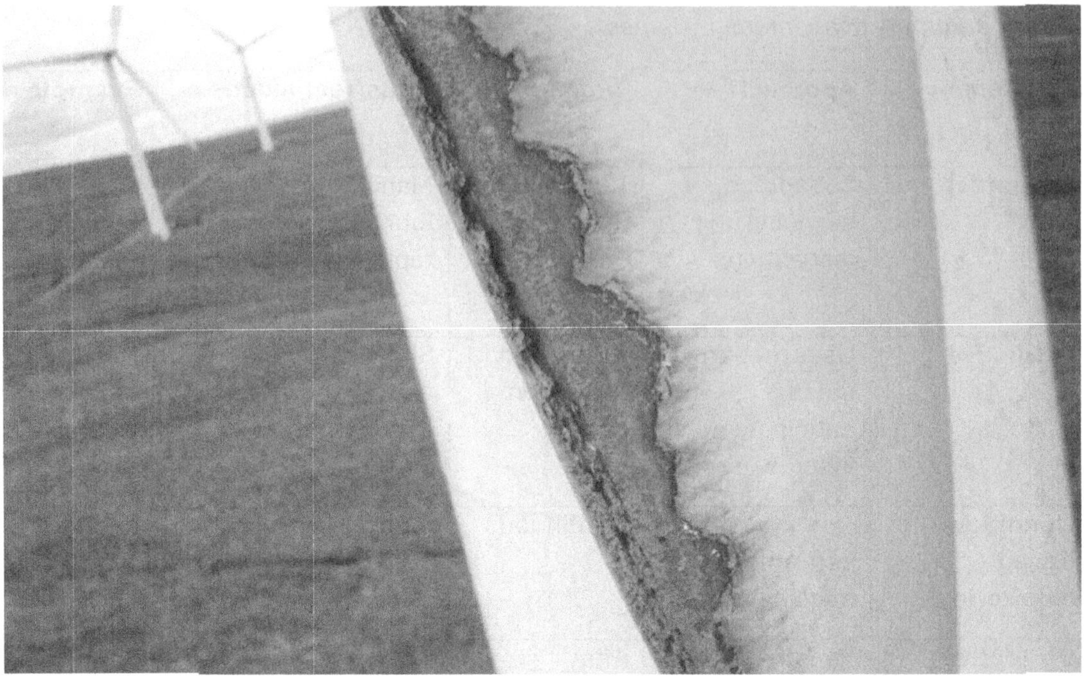

Figure 121: Blade leading edge damage. Belzona Polymerics Ltd., CC0, via Wikimedia Commons.

Routine inspections, usually performed once or twice a year, form a key part of preventive maintenance programs. Unscheduled inspections may also occur following operational alarms or irregularities detected by the turbine's SCADA (Supervisory Control and Data Acquisition) system or vibration monitoring sensors. By addressing early-stage defects, operators can avoid major blade failures and extend the lifespan of the asset.

A range of inspection methods is used to assess blade condition, combining traditional manual techniques with modern non-destructive testing (NDT) and digital imaging technologies.

Visual Inspection (Ground-Based or Rope Access): This is the most common and straightforward method. Technicians perform visual checks of the blade surface for signs of cracks, discoloration, delamination, or impact damage. Ground-based inspections may use telescopic lenses or high-resolution cameras to identify visible defects, while rope-access inspections allow technicians to perform close-up assessments for more detailed examination of problem areas.

Drone (UAV) Inspections: Unmanned aerial vehicles (UAVs) equipped with high-definition and thermal cameras have become the industry standard for wind turbine blade inspections. Drones can quickly and safely capture detailed imagery of all three blades from multiple angles without requiring technicians to climb the turbine. Artificial intelligence (AI) and image recognition software are often used to analyse these images automatically, identifying and classifying defects while reducing inspection time and cost.

Ultrasonic Testing (UT): This method is used to detect internal defects such as delamination, voids, or separation within composite layers. Ultrasonic probes transmit high-frequency sound waves through the blade structure, and any irregularities or changes in signal reflection reveal hidden flaws not visible on the surface.

Thermographic (Infrared) Inspection: Infrared thermography identifies variations in surface temperature caused by underlying defects. Uneven heat distribution can indicate issues such as moisture ingress, cracks, or bonding failures between composite layers. This technique is especially useful for detecting hidden damage in thick or layered materials.

Tap Testing: A simple but effective technique, tap testing involves lightly striking the blade surface with a hammer or tapping tool and listening to the resulting sound. A clear, ringing tone indicates solid material, while a dull or hollow sound suggests delamination or voids beneath the surface.

Laser Scanning and Photogrammetry: These advanced 3D inspection techniques use laser beams or multiple photographs to create a detailed digital model of the blade's geometry. By comparing current measurements to baseline data, engineers can detect deformation, erosion, or warping that may impact aerodynamic performance and energy capture.

Repair Techniques

Once blade damage is detected, technicians apply repair methods appropriate to the defect's type, size, and severity. The goal is to restore the blade's structural integrity, aerodynamic efficiency, and weather resistance while minimising downtime and ensuring long-term reliability.

Surface Repairs (Minor Damage): Minor surface defects such as small cracks, chips, or areas of leading-edge erosion can typically be repaired on-site. Technicians begin by sanding and cleaning the damaged area to remove contaminants and loose material. An epoxy or polyurethane-based filler is then applied to rebuild the surface, followed by sanding to restore the blade's aerodynamic shape. A final UV-resistant gelcoat or topcoat is applied to protect the repair from sunlight and moisture, ensuring durability and smooth airflow across the blade surface.

Composite Laminate Repair (Moderate Damage): For more serious issues such as deeper cracks or delamination within the blade structure, a composite laminate repair is required. In this process, the damaged material is carefully removed or ground out to expose intact layers. Technicians then rebuild the section using glass or carbon fibre mats impregnated with resin. Once the new layers are applied, the repair is cured (often using heat lamps or vacuum-bagging techniques), sanded to match the blade contour, and sealed to prevent moisture ingress. This restores the blade's load-bearing capacity and ensures continued aerodynamic efficiency.

Lightning Damage Repair: Lightning strikes can cause localized burns, melting, or delamination near blade tips or lightning receptors. Repairs typically involve removing the damaged section, inspecting and replacing the lightning receptor cable, and reapplying conductive materials to re-establish electrical continuity. The repaired area is then sealed and coated to protect against future strikes and environmental exposure.

Leading-Edge Protection (LEP) Replacement: The leading edge of the blade experiences constant erosion from rain, dust, and airborne particles. When the protective coating or erosion tape wears away, it must be replaced promptly to prevent structural damage. Modern maintenance programs increasingly use advanced polyurethane coatings or pre-formed LEP tapes that provide extended service life and easier installation compared to traditional coatings.

Full Section Replacement (Severe Damage): When a blade suffers extensive structural damage—such as core delamination, severe lightning impact, or fractures in load-bearing areas—a full section replacement may be required. This process involves cutting out and replacing the damaged portion or, in extreme cases, replacing the entire blade. It requires specialized cranes, heavy-lift equipment, and coordination with the original equipment manufacturer (OEM). Due to its complexity, this type of repair results in significant downtime and higher maintenance costs.

Surface Repairs

Surface Repairs (Minor Damage) on wind turbine blades are a key part of preventive and corrective maintenance, designed to restore aerodynamic performance and prevent the spread of small defects that could evolve into major structural issues. These repairs are generally conducted on-site, often using rope access or platform systems, and follow a carefully controlled sequence to ensure the repaired section matches the original blade surface both structurally and aerodynamically.

Before beginning any surface repair, technicians conduct a visual inspection to determine the type, size, and depth of the defect. Typical minor defects include:

- Small surface cracks or chips from debris impact.
- Localized leading-edge erosion from rain, sand, or hail.
- Minor gelcoat or paint loss due to UV exposure.

The area is then marked and photographed for documentation and to verify that the damage does not extend into structural layers. If deeper delamination or fibre exposure is detected, the repair is reclassified as moderate or major, requiring composite rebuilding.

Proper surface preparation is critical for achieving strong adhesion between the blade and the repair materials.

- **Sanding:** The damaged area is sanded down using fine-grit abrasive tools (usually 80–120 grit). This removes oxidation, contaminants, and loose or flaking composite material.
- **Cleaning:** The area is then wiped thoroughly with a solvent-based cleaner or acetone to eliminate dust, oil, and grease. Cleanliness is essential to ensure proper bonding between the old and new materials.
- **Tapering Edges:** For small chips or cracks, the surrounding surface is tapered slightly to create a smooth transition zone for the filler and to enhance structural blending.

During this stage, technicians wear protective gloves and respirators to prevent exposure to fiberglass dust and chemical fumes.

Once the surface is prepared, the defect is rebuilt using a repair compound, typically an epoxy or polyurethane-based filler.

- **Mixing:** The filler resin and hardener are mixed according to manufacturer specifications to ensure the correct chemical reaction and curing time.
- **Application:** The compound is applied in layers using a spatula or plastic spreader to fill cracks or voids. For leading-edge erosion, the filler may be applied along a broader contour to restore the original aerodynamic profile.
- **Curing:** The material is left to cure under controlled temperature and humidity. Portable heaters or infrared lamps may be used to accelerate curing, especially in cold or humid environments.

Technicians check for complete hardening before proceeding to the next step.

After the filler has cured, the surface is sanded again to achieve a smooth, aerodynamic contour.

- Coarse sanding (80–120 grit) removes excess filler.
- Fine sanding (240–400 grit) smooths the surface and prepares it for coating. Using templates or profile gauges, technicians verify that the blade's aerodynamic shape has been restored within manufacturer tolerances, ensuring efficient airflow and minimal drag.

To ensure long-term durability, the repaired area must be sealed and protected from environmental degradation.

- A UV-resistant gelcoat or polyurethane topcoat is applied evenly across the repaired section.
- This coating prevents UV radiation, moisture, and chemical exposure from weakening the underlying composite.

- For leading-edge repairs, some operators apply additional protective tapes or polyurethane coatings for erosion resistance.

Proper curing of the coating is verified before the turbine is returned to service.

Every surface repair concludes with a final inspection to confirm:

- The surface contour is restored.
- There are no air bubbles, voids, or visible imperfections.
- The coating has adhered uniformly and fully cured.

Technicians record the repair details, including materials used, curing times, environmental conditions, and inspection results, into the turbine's maintenance log. This documentation supports long-term performance monitoring and compliance with OEM and IEC 61400-23 blade repair standards.

Composite Laminate Repair

Composite Laminate Repair (Moderate Damage) is a specialised maintenance process used to restore the structural integrity of wind turbine blades that have sustained deeper damage, such as cracks, delamination, or fibre exposure. Unlike surface repairs that address only cosmetic or minor erosion issues, composite laminate repair deals with the blade's load-bearing layers—the glass or carbon fibre composites that give the blade its strength and rigidity. This repair must be conducted with precision to ensure the restored area regains its mechanical properties, aerodynamic shape, and long-term durability.

Before any repair begins, technicians conduct a comprehensive inspection to assess the extent of the damage. This typically involves:

- Visual examination to identify cracks, delamination, fibre breakage, or exposed laminate layers.
- Tap testing or ultrasonic scanning to detect hidden voids and delamination boundaries within the composite.
- Marking and documentation of the damaged zone, ensuring the full affected area is captured.

Once the repair area is defined, technicians establish a safe working environment, securing the turbine and ensuring the blade is stable—often using a suspended access platform or rope access system. The blade surface is cleaned and degreased to remove dirt, oil, or contaminants before grinding begins.

The first active step in a composite laminate repair is grinding out the damaged composite layers to expose clean, undamaged material.

- Using rotary sanders or grinders, technicians gradually remove cracked or delaminated laminate, feathering the edges outward to create a tapered bevel.
- The taper ratio is typically 1:20 or 1:30, depending on manufacturer recommendations, to ensure smooth load transfer between old and new material.
- Care is taken not to over-grind into undamaged fibres, as this could weaken the structure.

Once grinding is complete, the cavity is vacuumed and cleaned with acetone or a similar solvent to remove dust and debris. The clean and dry surface ensures proper adhesion of new composite layers.

The replacement material must match the original blade design specifications, which commonly use glass fibre-reinforced polymer (GFRP) or carbon fibre-reinforced polymer (CFRP).

- **Reinforcement Material:** Woven or unidirectional fibre mats are selected based on the blade's original laminate schedule.
- **Resin System:** A compatible epoxy or polyester resin is prepared according to manufacturer instructions, ensuring the correct mix ratio of resin to hardener.
- **Lay-Up Sequence:** The number and orientation of fibre layers are pre-determined to replicate the original laminate's structural properties (e.g., alternating 0°, ±45°, and 90° orientations for strength and stiffness).

All materials are prepared on-site, with pre-cut fibre mats and measured resin quantities to ensure efficiency during application.

The repair cavity is rebuilt layer by layer using the wet lay-up method:

1. **First Layer:** A thin resin layer is brushed onto the prepared surface to promote adhesion.
2. **Fibre Placement:** The first fibre mat is laid into the cavity, oriented to match the original laminate direction.
3. **Resin Impregnation:** Resin is applied evenly to fully saturate the fibres, eliminating air pockets.
4. **Subsequent Layers:** Additional layers are built up sequentially, each fully wetted with resin and compacted to remove trapped air using rollers or squeegees.

Technicians continue this process until the repaired laminate reaches the original thickness, ensuring all fibres are well bonded and compacted. For large or complex repairs, vacuum-bagging techniques may be used to improve compaction and remove air during curing.

The curing process allows the resin to harden and bond the new composite layers permanently.

- **Ambient Curing:** For small or moderate repairs, curing can occur naturally at ambient temperature, typically taking several hours.
- **Accelerated Curing:** In colder or high-humidity conditions, heat lamps, hot-air blowers, or infrared heaters are used to maintain an optimal curing temperature (usually 40–80°C).
- **Vacuum-Bagging Curing:** For advanced repairs, a vacuum bag is applied over the repair area, and negative pressure ensures tight fibre compaction and even resin distribution, enhancing the repair's mechanical performance.

Technicians monitor the curing time and temperature carefully to ensure the resin fully polymerizes without overheating or forming voids.

Once cured, the repaired area is ready for finishing.

- The surface is sanded smooth to match the blade's aerodynamic profile using fine-grit sandpaper.
- Technicians use contour gauges or aerodynamic templates to verify that the original blade curvature and thickness are restored.
- A final inspection checks for voids, cracks, or imperfections in the repair.

After sanding, the surface is cleaned again in preparation for sealing.

To prevent moisture ingress and UV degradation, the repaired area must be sealed and coated:

- A gelcoat or polyurethane topcoat is applied to provide environmental protection and a smooth aerodynamic surface.
- For leading-edge areas, additional erosion protection tapes or polyurethane coatings may be added to resist future wear from rain and debris.
- The coating is cured fully before the turbine is returned to operation.

Finally, the repair is subjected to a thorough inspection and quality check. This includes:

- Visual and tap testing to confirm structural soundness.
- Ultrasonic or thermographic testing (if required) to detect hidden delamination.
- Documentation of materials, methods, curing data, and inspection results in the turbine's maintenance record.

This ensures traceability and compliance with OEM specifications and industry standards such as IEC 61400-23 (blade structural testing) and IEC 61400-24 (lightning and composite protection).

Lightning Damage Repair

Lightning Damage Repair on wind turbine blades is a detailed and safety-critical procedure designed to restore the blade's structural integrity, aerodynamic performance, and lightning protection continuity after a strike. Because lightning strikes often occur near the blade tip or receptor, repairs must address both mechanical damage (such as burns, delamination, and melting) and electrical conductivity (to ensure future strikes are safely discharged through the blade's lightning protection system).

The process begins with a thorough inspection to determine the extent of both visible and internal damage.

- **Visual inspection:** Technicians look for signs of scorching, melting, cracking, or delamination near the blade tip or lightning receptor.
- **Tap testing and thermography:** These methods are used to detect subsurface delamination or voids not visible to the eye.
- **Electrical continuity testing:** Using a low-resistance ohmmeter, technicians verify that the lightning receptor and down conductor still provide a continuous path to ground.

If continuity is lost or readings exceed acceptable limits, the lightning protection system (LPS) must be repaired or replaced as part of the process.

Before beginning repairs, the turbine is locked out and tagged out (LOTO) to prevent accidental startup. The affected blade is positioned horizontally or safely accessed using rope access, suspended platforms, or cranes.

All repair materials and tools—including grinding equipment, resins, conductive paint, and protective coatings—are prepared on-site. The repair area is cleaned with solvents to remove contaminants such as carbonized resin, soot, or moisture.

All burnt, melted, or delaminated composite material must be removed until clean, undamaged laminate is exposed.

- Technicians use grinders or rotary tools to carefully remove the damaged zone, tapering the edges at a ratio of about 1:20 for proper bonding of new material.
- Any charred fibre, resin residue, or conductive debris is completely removed, as these can interfere with structural adhesion and electrical conductivity.

The area is then vacuumed, cleaned, and dried to ensure a smooth surface for subsequent repair.

If the lightning receptor (the metal strike point) or down conductor cable has been damaged, these must be repaired or replaced to re-establish the LPS pathway.

- **Receptor inspection:** The tip receptor is removed and visually examined for melting or pitting. If damaged, it is replaced with a new, manufacturer-approved receptor.
- **Cable testing:** The copper down conductor cable running inside the blade is checked for continuity and insulation integrity. If burnt or severed, it is replaced or reconnected using approved crimp or bonding methods.
- **Connector replacement:** Damaged terminal connectors or bonding points at the blade root are replaced, ensuring tight mechanical and electrical connections.

After ensuring mechanical integrity, the next step is to restore electrical conductivity within the repaired area:

- A conductive adhesive, copper mesh, or foil is applied to the repaired zone to reconnect the receptor to the down conductor.
- Conductive coatings or paints (silver or graphite-based) may be used over non-structural surfaces to ensure proper charge dissipation.
- The repaired circuit is tested again with a continuity meter, confirming that resistance values meet manufacturer specifications (typically below 10 ohms).

Once the LPS is reinstated, the structural composite repair proceeds similarly to a laminate rebuild:

- Layers of glass or carbon fibre mats impregnated with epoxy resin are applied over the ground-out area to rebuild the blade's original shape and strength.
- Each layer is compacted and smoothed to eliminate air bubbles and ensure proper adhesion.
- The repair is then cured using ambient conditions or controlled heating (e.g., heat blankets or infrared lamps) until full polymerization is achieved.

This process restores the blade's mechanical strength and aerodynamic profile.

After curing, the repaired area is sanded smooth to match the original blade contour.

- A UV-resistant gelcoat or polyurethane topcoat is applied to protect the repair from sunlight, moisture, and erosion.
- For the leading edge or tip, an additional erosion protection coating or tape may be applied to extend the lifespan of the repair.
- The area is then polished to ensure aerodynamic smoothness, preventing turbulence or drag.

Before the turbine is returned to service, technicians conduct final tests to confirm that both the mechanical and electrical systems are functioning correctly:

- Continuity and grounding tests verify that the LPS provides a low-resistance path from the receptor to the tower grounding system.
- Visual and ultrasonic inspection confirm that there are no voids, delamination, or surface imperfections.
- The repair area is logged and documented, including materials used, repair date, technicians involved, and test results, ensuring traceability for future maintenance.

Detailed repair records are uploaded into the Computerized Maintenance Management System (CMMS) or the turbine manufacturer's maintenance log. These include:

- Photos before and after repair.
- Test results confirming electrical continuity.
- Environmental conditions during repair and curing.
- Certification of materials used (resins, conductors, coatings).

This documentation supports warranty claims, compliance with IEC 61400-24 (Lightning Protection of Wind Turbines), and future inspection reference.

Leading-Edge Protection (LEP) Replacement

Leading-Edge Protection (LEP) Replacement is a critical maintenance task in wind turbine blade upkeep, designed to protect the blade's most vulnerable section—the leading edge—from erosion caused by rain, dust, sand, and other airborne particles. Over time, this constant abrasion degrades the protective layer, leading to surface roughness, efficiency loss, and eventually structural damage if left untreated. Replacing worn LEP systems—whether coatings or tapes—restores aerodynamic performance and prevents deeper laminate erosion. Below is a detailed explanation of how this procedure is performed.

The leading edge is the part of the blade that first meets the wind and is therefore exposed to the most aggressive environmental wear.

Erosion on this section can:

- Reduce aerodynamic efficiency and power output by up to 5–10%.
- Allow moisture ingress, leading to delamination or composite damage.
- Cause increased vibration and noise from uneven airflow.

LEP replacement forms part of both preventive and corrective maintenance, typically conducted every few years or when erosion depth exceeds manufacturer thresholds (e.g., 0.5 mm). By applying new polyurethane coatings or erosion protection tapes, technicians extend the service life of the blade and maintain its aerodynamic profile.

Before any work begins, the blade is thoroughly inspected to assess the extent of erosion and substrate condition.

- **Visual Inspection:** Technicians check for peeling, cracks, discolouration, or roughness along the leading edge.
- **Tactile Check:** Fingers are run along the surface to feel for irregularities and loss of smoothness.
- **Measurement Tools:** Digital microscopes or 3D scanners may be used to measure erosion depth.
- **Moisture and Adhesion Testing:** If the previous LEP coating has failed, moisture testing ensures that the substrate is completely dry before new material is applied.

Once inspection confirms that the underlying composite structure is intact, the replacement process begins. If deeper damage is present, composite repairs must be performed first.

Proper surface preparation is the most critical stage of LEP replacement and directly affects adhesion and durability.

- **Area Isolation:** The repair area is sectioned off using tape and plastic sheeting to protect unaffected areas.
- **Removal of Old Coating or Tape:** The worn LEP material is carefully removed using mechanical scraping or gentle heat (if tape-based).
- **Surface Sanding:** Technicians sand the leading edge lightly (typically using 180–320 grit sandpaper) to remove any remaining adhesive, oxidation, or loose fibres.
- **Cleaning:** The surface is cleaned with approved solvents such as isopropyl alcohol to eliminate dust, grease, and contaminants.
- **Drying:** The area must be completely dry before proceeding—any moisture can compromise adhesion and cause premature failure.

A surface cleanliness inspection is performed under proper lighting to confirm readiness for new LEP application.

There are two primary types of LEP replacement systems: polyurethane-based coatings and pre-formed erosion protection tapes. Each has a distinct application procedure.

a. Polyurethane Coating Application

Polyurethane coatings are liquid-applied systems that cure to form a tough, flexible barrier.

Steps:

1. **Primer Application:** A primer compatible with the polyurethane system is applied to promote adhesion between the composite surface and the coating.
2. **Mixing:** The polyurethane components (base and hardener) are mixed according to manufacturer ratios, ensuring homogeneity.
3. **Application:** The coating is applied using a brush, roller, or spray gun along the leading edge, typically in two to three thin layers for uniform coverage.
4. **Curing:** The coating is allowed to cure under controlled conditions (temperature 20–25°C, humidity <70%). Forced curing using heat lamps may be employed in cold or humid environments.
5. **Thickness Verification:** After curing, coating thickness (usually 250–500 microns) is measured using a wet film gauge or ultrasonic thickness tester to ensure compliance with manufacturer specifications.

Advantages:

- Seamless coverage (no joint lines).
- Strong chemical and UV resistance.
- Ideal for both small repairs and full-length blade protection.

b. Erosion Protection Tape Application

Pre-formed polyurethane tapes have become increasingly popular because they offer faster installation, consistent quality, and simplified field application.

Steps:

1. **Tape Cutting and Shaping:** Pre-cut LEP tape sections are trimmed to match the blade curvature and length using templates or manufacturer-supplied stencils.
2. **Adhesive Activation:** Some tapes come with pressure-sensitive adhesives, while others require a primer or adhesive activator applied to the blade surface.
3. **Alignment and Application:** The tape is carefully aligned along the leading edge, starting from the blade root toward the tip. Application rollers are used to eliminate air bubbles and ensure even pressure.
4. **Edge Sealing:** The tape's edges are sealed with a clear polyurethane sealant to prevent moisture or air ingress beneath the tape.
5. **Curing:** Depending on the adhesive system, curing may occur at ambient temperature or be accelerated with gentle heating.

Advantages:

- Quick, clean installation.
- Easy to replace during future maintenance cycles.
- No mixing or spraying equipment required.

After the LEP has been applied, technicians perform final finishing and quality checks:

- **Surface Inspection:** The coating or tape is inspected for uniform adhesion, smoothness, and absence of air bubbles or wrinkles.
- **Aerodynamic Contour Verification:** The repaired area is checked to ensure that the blade's original aerodynamic shape has been preserved.
- **Curing Confirmation:** If applicable, coating hardness or cure completeness is verified using durometer tests or solvent rub tests.
- **Sealant Application:** A UV-protective or anti-erosion topcoat may be applied over coatings or tape edges for added longevity.

Once the replacement is complete, all work is documented and logged in the turbine's maintenance management system. This includes:

- Type and batch number of coating or tape used.
- Environmental conditions during application (temperature, humidity).
- Curing time and inspection results.
- Before-and-after photos for maintenance records.

If required by the operator or manufacturer, a post-repair performance test may be conducted to monitor power output or vibration patterns to verify aerodynamic performance restoration.

Full Section Replacement

Full Section Replacement (Severe Damage) is the most complex and resource-intensive form of wind turbine blade repair, reserved for cases where the blade's structural integrity has been severely compromised. This may occur due to core delamination, major lightning strikes, impact from foreign objects, or fractures in the load-bearing spar caps or shear web. Because the blade is one of the most critical components of the turbine's aerodynamic and structural system, any major repair or replacement must be carried out with precision, under strict engineering and safety controls, and in close collaboration with the original equipment manufacturer (OEM).

Below is a detailed explanation of how this process is typically performed in the field or at a specialized repair facility.

Before any physical work begins, engineers and technicians conduct a comprehensive inspection to confirm that a full section replacement is necessary.

Assessment methods include:

- Visual and tactile inspection to identify visible fractures, cracks, or missing material.
- Ultrasonic or thermographic non-destructive testing (NDT) to locate internal delamination, crushed core material, or spar damage.
- Structural load analysis using OEM data to determine whether the blade can safely sustain designed operational loads after repair.
- Lightning damage evaluation if the strike path has compromised the lightning receptor or down-conductor network.

If the analysis shows that the structural core or main load paths have been compromised beyond repairable limits, engineers recommend sectional or full blade replacement. OEM approval is typically required to maintain turbine certification and warranty compliance.

Because a blade section replacement involves working with large, heavy components at height, strict safety and logistical planning are essential.

Key preparatory steps include:

- **Shutdown and Lockout/Tagout (LOTO):** The turbine is completely de-energized and locked out to prevent accidental startup.
- **Site setup:** Access platforms, scaffolding, or suspended work cages are installed if the work is done on the tower.
- **Cranes and lifting equipment:** Heavy-duty cranes (often 500- to 1000-ton class for utility-scale blades) are mobilized.
- **Weather monitoring:** Wind speeds must be below safe lifting limits (typically <9–10 m/s).
- **Safety perimeter:** Ground crews establish exclusion zones to prevent personnel entry during lifting operations.

All personnel involved—technicians, riggers, and crane operators—must be trained and certified under GWO (Global Wind Organisation) or equivalent work-at-height and lifting safety standards.

If the damage affects an area that cannot be accessed safely on the tower, or if the blade must be replaced entirely, it is removed from the hub.

Steps:

1. **Blade orientation:** The affected blade is positioned horizontally using the turbine's yaw and pitch control systems.
2. **Connection release:** Electrical cables (lightning conductor and pitch control wiring) are disconnected and labelled.
3. **Bolt removal:** The root bolts securing the blade to the hub flange are carefully loosened in sequence to prevent stress deformation.
4. **Lifting and transport:** The crane lifts the blade using padded lifting slings at designated balance points. Tag lines are used to control swing and rotation.
5. **Lowering to ground:** The blade is placed on custom cradles or supports to prevent bending or surface abrasion.

If only a section of the blade is being replaced (rather than the entire blade), the damaged portion is stabilized in position before being cut out.

For sectional replacements, technicians must remove the damaged segment precisely without affecting adjacent structural areas.

Typical procedure:

- **Marking the cut lines:** Engineers specify exact cut boundaries based on structural drawings and laminate lay-up details.
- **Cutting:** Diamond-tipped saws or water-jet cutting tools are used to make clean, controlled cuts through the composite layers.
- **Removal of damaged section:** The section is lifted away carefully using rigging points to avoid secondary cracking.
- **Surface preparation:** The remaining edges are sanded and bevelled to provide bonding surfaces for the new section.

The cut edges are inspected using NDT methods to ensure no hidden delamination or moisture remains before new bonding begins.

The replacement segment—either a pre-manufactured OEM section or a custom-fabricated repair piece—is fitted in place using high-strength structural adhesives and mechanical bonding techniques.

Process details:

1. **Alignment and positioning:** The new section is aligned using laser or mechanical jigs to ensure perfect aerodynamic and structural continuity.
2. **Bonding:** Two-part epoxy structural adhesives are applied to the joining surfaces. The section is pressed into position using clamps or temporary fasteners.

3. **Lamination:** Technicians apply additional glass or carbon fibre reinforcement layers over the joints to restore full structural strength.

4. **Curing:** The adhesive and laminate layers are cured under controlled temperature and humidity. In field repairs, portable heat blankets or vacuum-bagging systems are used to accelerate curing.

5. **Surface finishing:** Once cured, the joint area is sanded smooth and contoured to match the blade's aerodynamic profile.

For blades that sustained lightning damage, technicians must inspect and re-establish the Lightning Protection System (LPS).

Tasks include:

- Installing new lightning receptors or tip caps if they were destroyed.
- Replacing or splicing the down-conductor cable to maintain continuous electrical path from the receptor to the blade root.
- Testing electrical continuity and resistance (typically <10 Ω) between receptors, down conductor, and root grounding points.
- Applying conductive coatings or bonding materials to ensure proper energy dissipation during future strikes.

This ensures compliance with IEC 61400-24 lightning protection standards for wind turbine blades.

Once the repair or replacement section is complete and inspected, the blade is ready to be reattached to the turbine hub.

Procedure:

1. **Lifting:** The blade is lifted using cranes and attached rigging, ensuring proper balance.

2. **Flange alignment:** The root flange is aligned precisely with the hub interface.

3. **Bolt torquing:** Blade root bolts are tightened to OEM-specified torque values using calibrated hydraulic torque wrenches.

4. **Electrical reconnection:** Lightning protection cables and pitch control connectors are reattached.

5. **Inspection:** Technicians verify alignment, bolt tension, and electrical continuity before releasing the crane.

After reinstallation, comprehensive testing and inspection are performed to ensure safety and functionality.

- Visual and ultrasonic inspection of bonding joints and laminate transitions.

- Torque verification of all mechanical fasteners.
- Electrical continuity and insulation testing for LPS and internal wiring.
- **Balancing check:** The rotor is slowly rotated to confirm weight balance and vibration limits.
- **Functional testing:** The turbine is restarted in service mode to verify pitch and yaw response, power output, and SCADA fault-free operation.

If all results meet manufacturer and certification criteria, the turbine is cleared for normal operation.

All work is documented thoroughly, including:

- OEM engineering approvals and repair method statements.
- Materials used (adhesives, composites, coatings).
- Environmental conditions during curing.
- Test and inspection records.
- Photographic evidence before, during, and after repair.

This documentation is uploaded to the Computerized Maintenance Management System (CMMS) for traceability and regulatory compliance.

Safety and Environmental Considerations

Blade inspection and repair operations—especially those conducted at height—must follow strict safety procedures. All technicians performing work on towers or blades must comply with *Work at Height (WAH)* and *Lockout/Tagout (LOTO)* procedures to prevent accidental turbine movement or electrical hazards. Rope-access technicians are required to hold certifications from recognized bodies such as the *Industrial Rope Access Trade Association (IRATA)* or *Global Wind Organisation (GWO)*.

Additionally, environmental protection is an essential part of repair work. Resins, solvents, and fillers used in blade restoration must comply with environmental regulations. Waste materials, including used composite debris and contaminated rags, must be properly collected, contained, and disposed of in accordance with hazardous waste guidelines to prevent environmental contamination.

Gearbox and Generator Maintenance

Gearbox and Generator Maintenance are two of the most critical aspects of both preventive and corrective maintenance in wind turbine operation. These components form the mechanical and electrical heart of the turbine—converting kinetic wind energy into electrical power. Because of their high mechanical and thermal loads, they require meticulous inspection, lubrication, monitoring, and repair to prevent catastrophic failures and ensure continuous, efficient performance.

The gearbox converts the slow rotational speed of the turbine rotor into the high-speed rotation required by the generator. The generator, in turn, transforms mechanical energy into electrical energy for grid delivery. Any malfunction in these systems can cause extended downtime, expensive repairs, and substantial revenue loss due to halted power production. Routine gearbox and generator maintenance allows operators to:

- Detect wear and degradation early through vibration and oil analysis.
- Prevent overheating and mechanical stress failures.
- Maintain optimal lubrication and alignment.
- Extend component lifespan and improve turbine availability.
- Reduce unplanned maintenance interventions.

Gearbox Maintenance

Preventive gearbox maintenance focuses on early detection of wear and ensuring proper lubrication and alignment. Key activities include:

- **Oil Sampling and Analysis:** Gearbox oil is sampled at regular intervals (typically every 6–12 months) and tested for viscosity, water content, oxidation, and metal particles. Increased metal content (such as iron or copper) indicates bearing or gear wear. Oil analysis helps detect early-stage faults before mechanical symptoms appear.
- **Oil Replacement and Filtration:** Oil is drained, filtered, and replaced as per the manufacturer's schedule. High-quality synthetic oils or lubricants with anti-foaming and anti-oxidation properties are used to minimize friction and heat. Some turbines use offline filtration systems to continuously remove contaminants and extend oil life.
- **Temperature and Vibration Monitoring:** Sensors continuously measure temperature, vibration, and noise levels. Any deviation from baseline readings can indicate misalignment, lubrication loss, or bearing damage. These data are logged in the SCADA or Condition Monitoring System (CMS) for trend analysis.
- **Inspection of Seals and Gaskets:** Gearbox seals are checked for oil leaks that can cause lubrication loss or contamination ingress. Worn seals are replaced promptly to prevent gear and bearing damage.

- **Visual and Endoscopic Inspection:** During scheduled maintenance, technicians inspect internal gear and bearing surfaces using borescopes to identify pitting, scoring, or fatigue cracks.

Generator Maintenance

The generator converts mechanical torque from the gearbox into electrical power. Preventive maintenance ensures electrical efficiency and mechanical reliability.

- **Bearing Inspection and Lubrication:** Generator bearings are greased or oil-lubricated at regular intervals to prevent overheating and premature wear. Grease quality and volume are carefully controlled to avoid insulation breakdown due to contamination.
- **Electrical Testing:** Resistance, insulation, and continuity tests are performed on windings to detect degradation. Instruments like megohmmeters and partial discharge testers assess the health of insulation systems.
- **Alignment Checks:** Misalignment between the generator and gearbox shafts can lead to excessive vibration and bearing stress. Regular alignment verification using laser alignment tools ensures smooth torque transmission.
- **Cooling System Maintenance:** Generator air or liquid cooling systems are inspected for blockages, leaks, and fan functionality. Clean cooling pathways are vital for temperature control and insulation longevity.
- **Brushes and Slip Ring Inspection (for DFIG systems):** In Doubly-Fed Induction Generators (DFIGs), slip rings and brushes are checked for wear, arcing, or carbon build-up. Proper maintenance reduces electrical noise and improves power quality.

Corrective Maintenance Tasks

Corrective actions are taken when faults or failures are detected through condition monitoring or inspections.

<u>Gearbox Corrective Maintenance</u>

- **Bearing or Gear Replacement:** If vibration analysis or oil sampling confirms advanced wear, technicians replace damaged bearings or gears. This may involve partial or full gearbox disassembly.
- **Realignment and Balancing:** After component replacement, shafts and couplings are realigned, and the system is balanced to restore smooth operation.
- **Seal Replacement and Contamination Cleanup:** Oil leaks and contamination are corrected by replacing seals and cleaning internal surfaces. Damaged oil filters or breathers are also replaced.

- **Retrofits and Upgrades:** In older turbines, upgraded bearings, seals, or gear materials may be installed to improve reliability and extend service life.

Generator Corrective Maintenance

- **Rewinding or Insulation Repair:** If electrical testing reveals degraded insulation or winding faults, the generator may require partial rewinding or varnish re-impregnation to restore insulation integrity.
- **Bearing Replacement:** Failed bearings are replaced, and the lubrication system is flushed and refilled. Precision balancing and vibration checks follow.
- **Component Overhaul or Replacement:** In cases of severe rotor or stator damage, the generator may be overhauled or replaced. This involves significant downtime and may require crane operations, especially offshore.

Condition Monitoring Integration

Modern turbines integrate Condition Monitoring Systems (CMS) with SCADA to enable predictive maintenance.

- **Vibration Sensors:** Detect imbalances, misalignment, and bearing wear.
- **Temperature Sensors:** Identify overheating due to friction or electrical faults.
- **Oil Debris Sensors:** Detect metallic particles indicating wear.
- **Electrical Signature Analysis:** Tracks generator load and power quality for early fault detection.

This data-driven approach allows operators to schedule maintenance proactively, minimizing unplanned shutdowns.

Tower and Foundation Maintenance

The tower and foundation form the structural backbone of a wind turbine, supporting the nacelle, rotor, and blades while transmitting mechanical loads safely to the ground. Because these components experience continuous stress from wind loading, vibration, and environmental exposure, regular maintenance is essential to ensure structural integrity, safety, and long-term reliability. Maintenance activities focus on corrosion control, bolt tensioning, alignment verification, and the detection of cracks or settlement in the foundation.

The tower and foundation are designed to withstand complex forces — including bending, torsion, compression, and fatigue — caused by wind gusts and rotor dynamics. Over time, these stresses can lead to fatigue cracking, corrosion, grout degradation, or bolt loosening, all of which compromise structural safety. Regular inspections and maintenance prevent small defects from evolving into catastrophic structural failures.

Preventive maintenance ensures:

- Structural stability and proper load transfer from tower to foundation.
- Corrosion resistance, particularly in marine or coastal environments.
- Early detection of cracks, deformation, or settlement.
- Safe access and operational continuity of the turbine.

Corrective maintenance addresses any defects or damage identified during inspections or after extreme weather events such as storms, earthquakes, or lightning strikes.

Preventive Maintenance Activities

Preventive maintenance for the tower and foundation of a wind turbine involves a structured schedule of inspections, coating applications, and hardware checks to maintain structural integrity and prevent environmental degradation. These activities are designed to identify and address potential issues early, ensuring the longevity and stability of the turbine's supporting structures.

Routine visual and structural inspections are essential to detect early signs of deterioration or damage. Technicians thoroughly examine both the interior and exterior of the tower for corrosion, paint peeling, dents, cracks, or any deformation that could compromise its strength. Welded joints and flange connections are inspected for signs of fatigue using non-destructive testing (NDT) methods such as ultrasonic or magnetic particle inspection. Special attention is given to areas around access doors, joints, and rain-prone regions where coating damage is common. In the case of lattice or tubular steel towers, inspectors also check for loose rivets, missing bolts, or rust accumulation that may affect stability.

The tower's structural integrity relies heavily on properly torqued flange bolts connecting each section. During preventive maintenance, these bolts are checked using calibrated hydraulic tensioners or torque wrenches to verify they meet the manufacturer's specifications. Any loose bolts are re-torqued to prevent misalignment or fatigue caused by vibration. Anti-corrosion lubricants or coatings are also reapplied to protect fasteners from moisture and oxidation, particularly in humid or coastal environments.

Corrosion control is a critical part of tower maintenance. Steel towers are regularly inspected for coating damage and, where necessary, epoxy or polyurethane protective coatings are repaired or reapplied. Offshore wind turbines receive additional attention due to salt exposure — their cathodic protection systems (either sacrificial anodes or impressed current systems) are tested and maintained to ensure effective performance. Rust inhibitors or protective wax coatings are also applied to vulnerable components such as door frames, ladders, and bolted connections.

The foundation supports the entire turbine structure, making its inspection a top priority. Maintenance teams check the grout interface between the tower and the foundation for cracks, delamination, or voids caused by dynamic loads or freeze-thaw effects. Signs of settlement, soil erosion, or water pooling around the base are also monitored closely. Concrete foundations undergo integrity testing using ultrasonic pulse velocity methods or crack gauges. Anchor bolts are checked for corrosion, correct tension, and alignment to ensure stable load transfer from tower to foundation.

Preventive maintenance also includes inspecting the systems housed inside the tower. Lighting, communication, and safety systems are tested to confirm functionality. Ladders, internal platforms, fall arrest systems, and cable trays are examined for corrosion, wear, and secure attachment. Finally, drainage systems at the tower base are inspected to ensure proper water removal and prevent flooding or corrosion near electrical components.

Perform Bolt Tension Checks on Wind Turbine Towers

Bolt tension checks are a critical part of preventive maintenance for wind turbine towers. These checks ensure that all tower section flange bolts maintain the proper preload tension to prevent structural misalignment, vibration fatigue, or catastrophic connection failures. The process involves inspection, measurement, adjustment, and protection of fasteners according to manufacturer specifications.

1. Preparation and Safety Measures

Before beginning the tension check, technicians must isolate the turbine from operation and follow all lockout/tagout (LOTO) safety procedures to eliminate electrical and mechanical hazards. The tower interior should be well-lit and properly ventilated. Technicians should use appropriate personal protective equipment (PPE) such as gloves, hard hats, safety harnesses, and protective eyewear. All torque tools must be calibrated and certified before use to ensure accuracy.

2. Identifying and Accessing Flange Bolts

Wind turbine towers are typically constructed in sections connected by flanges. These flanges are secured by a circular array of high-strength bolts and nuts. The technician must access each flange level using the internal ladders or platforms and mark the bolts to track progress during the inspection. It's essential to check that no bolts, nuts, or washers are missing or damaged before proceeding with torque verification.

3. Checking Bolt Torque or Tension

Bolt tension can be measured directly using a hydraulic tensioner or indirectly using a torque wrench.

- **Hydraulic Tensioners:** These apply controlled hydraulic pressure to stretch the bolt to a precise elongation, ensuring uniform preload without frictional error. This method is preferred for large-diameter bolts on modern turbines.
- **Torque Wrenches:** For smaller bolts or quick checks, a calibrated torque wrench is used to verify that the torque matches the manufacturer's specified value. The wrench should be applied smoothly, and readings must be recorded systematically for quality assurance.

4. Re-Torquing Loose Bolts

If a bolt is found to be under-tensioned, it must be re-torqued or re-tensioned to the correct value as specified by the original equipment manufacturer (OEM). Torque patterns are usually applied in a star or cross-sequence to evenly distribute stress and prevent flange distortion. For hydraulic systems, all bolts in a flange are tensioned uniformly to maintain consistent preload. After re-torquing, the technician should confirm that the bolt head and nut are seated properly without signs of deformation.

5. Reapplication of Anti-Corrosion Protection

After bolt checks and adjustments, technicians must inspect the fasteners for corrosion. Any exposed metal surfaces are cleaned and treated with anti-corrosion coatings, protective waxes, or lubricants. Offshore wind turbines often require specialized marine-grade greases to protect against saltwater exposure. Bolts near tower doors, external platforms, or cable penetrations should receive extra attention due to increased exposure to moisture and weather.

Perform Corrosion Prevention and Coating Maintenance on Wind Turbine Towers

Corrosion prevention and coating maintenance are essential for preserving the structural integrity and longevity of wind turbine towers. Given their exposure to harsh weather, moisture, and salt-laden air (especially offshore), these structures rely heavily on high-quality coatings and protective systems to resist corrosion. The following steps outline how to perform effective corrosion prevention and coating maintenance.

1. Preparation and Safety

Before beginning any corrosion control work, ensure the turbine is safely shut down, locked out, and tagged out (LOTO). Access platforms or rope access systems must be securely installed, and all technicians should wear appropriate personal protective equipment (PPE), including gloves, respirators, goggles, and protective suits. For offshore turbines, additional fall protection and marine safety equipment are required.

The surface area to be treated should be clearly marked and isolated to prevent contamination of nearby components. Environmental conditions — particularly temperature, humidity, and wind speed — should be within the limits specified by the coating manufacturer for optimal adhesion and curing.

2. Surface Inspection and Assessment

A detailed visual inspection is the first step in corrosion control. Technicians check the entire tower exterior and interior for signs of:

- Paint peeling, blistering, or chalking
- Rust streaks, pitting, or scaling
- Coating delamination or discoloration
- Corrosion near bolts, welds, door frames, and access platforms

Corroded or damaged areas should be mapped and classified according to severity. For critical zones, non-destructive testing (NDT) techniques such as ultrasonic thickness measurement or magnetic particle inspection may be used to evaluate material loss beneath the coating.

3. Surface Preparation

Surface preparation is crucial for ensuring coating adhesion and long-term protection. The preparation process depends on the degree of corrosion:

- **Minor surface rust:** Remove by wire brushing, sanding, or mechanical grinding.
- **Moderate to heavy rust:** Use abrasive blasting (sandblasting or grit blasting) to achieve the required surface cleanliness, typically ISO 8501-1 Sa 2½ or equivalent.
- **Localized corrosion pits:** Fill with approved metal fillers or epoxy repair compounds before coating.

After cleaning, the surface must be wiped with a solvent to remove grease, salt deposits, or dust. Moisture and contaminants will compromise coating performance, so surfaces must remain dry and clean until coating application.

4. Reapplication of Protective Coatings

Reapply or repair anti-corrosion coatings using epoxy, polyurethane, or zinc-rich primer systems recommended by the turbine manufacturer. These coatings act as both a physical barrier and, in some cases, a galvanic protection layer.

- **Epoxy coatings** provide excellent adhesion and chemical resistance, ideal for tower interiors and flange areas.
- **Polyurethane topcoats** offer superior UV and weather resistance for external tower surfaces.

Coating is applied using brushes, rollers, or airless spray systems, depending on the repair area size. Each layer must cure fully before the next is applied. For optimal durability, apply coatings in multiple layers — typically a primer, intermediate coat, and topcoat — following the manufacturer's specified dry film thickness (DFT).

5. Offshore Turbine Corrosion Control

For offshore installations, cathodic protection systems provide an additional defence against corrosion. Two main systems are used:

- **Sacrificial Anodes:** Made of zinc or aluminium, these anodes corrode preferentially, protecting the steel structure. Inspect them regularly for material loss and replace them when 50–70% depleted.
- **Impressed Current Cathodic Protection (ICCP):** Uses a controlled electrical current to prevent corrosion. Technicians must inspect control units, measure current outputs, and verify that reference electrode readings remain within the designed potential range.

Regular seawater washdowns can also help remove salt deposits that accelerate corrosion near the tower base and splash zones.

6. Protective Treatments for Exposed Components

Even with coatings and cathodic systems in place, exposed metal parts require extra attention. Technicians apply rust inhibitors, protective waxes, or corrosion-blocking greases to areas prone to moisture and mechanical wear, including:

- Bolt heads and flanges
- Access door frames and hinges
- Cable entries and ladder rungs
- Platform joints and fastener recesses

These treatments create a hydrophobic barrier, preventing oxidation and corrosion creep in vulnerable spots.

7. Final Inspection and Documentation

After all coating and protection work is complete, perform a final inspection to verify uniform coating coverage and thickness using a dry film thickness gauge (DFT meter). Record all maintenance activities, materials used, and environmental conditions during application. This documentation helps track corrosion trends and supports warranty or compliance reporting.

Corrective Maintenance Activities

Corrective maintenance is carried out when structural or mechanical defects are detected during regular inspections, or following unexpected events such as accidents, lightning strikes, severe storms, or seismic activity. The purpose of this maintenance is to restore the structural integrity and operational reliability of the wind turbine tower and foundation.

When corrosion is identified on tower surfaces or structural components, the affected areas are cleaned using sandblasting, wire brushing, or grinding to remove rust and loose material. Once the surface is properly prepared, a primer and corrosion-resistant paint or coating system — typically epoxy or polyurethane — is applied to prevent further degradation. In cases where corrosion has severely compromised the metal structure, the affected steel sections may need to be repaired through welding or replaced entirely. These measures restore protective barriers and maintain the tower's structural strength.

Cracks in welded joints or steel tower sections are serious defects that require immediate attention. Repairs are conducted using certified weld repair procedures that follow the original equipment manufacturer's (OEM) specifications and are verified by a structural engineer. For concrete foundations, cracks are treated with epoxy injection or polyurethane grout to bond the separated surfaces, restore load-bearing capacity, and prevent water ingress that could lead to freeze-thaw damage or rebar corrosion.

The interface between the tower and foundation is often grouted to ensure even load transfer. Over time, this material can deteriorate due to vibration, dynamic loading, or environmental effects. Damaged or degraded grout is carefully removed and replaced with high-strength, non-shrink grout to restore full contact and stability. Laser alignment tools are then used to verify tower verticality and ensure uniform load distribution after re-grouting.

Loose or damaged bolts in tower flange connections can lead to misalignment, vibration, and fatigue cracking. During corrective maintenance, defective bolts are replaced and torqued to the manufacturer's precise specifications using calibrated torque wrenches. If flange deformation has occurred due to uneven tightening, partial disassembly may be necessary. In such cases, all bolts are re-torqued in a cross-pattern sequence to evenly distribute stress and restore the connection's integrity.

When severe settlement or soil instability is detected, corrective measures such as compaction grouting, underpinning, or micro-piling are employed to strengthen the foundation base. These ground improvement techniques stabilize the soil and prevent further subsidence. Additionally, drainage improvements — such as installing French drains or grading the surrounding soil — are implemented to prevent water accumulation near the foundation, which can cause erosion or frost heave.

<u>Perform Corrosion and Coating Repairs</u>

Corrosion and coating repairs are critical parts of corrective maintenance for wind turbine towers. These procedures restore structural integrity, prevent further material loss, and extend the service life of the tower — especially for steel structures exposed to harsh weather or marine environments. The process involves inspection, surface preparation, coating restoration, and, in severe cases, welding or component replacement.

1. Preparation and Safety

Before beginning corrosion repair work, technicians must ensure all safety protocols are in place. The wind turbine should be fully shut down, isolated, and locked out/tagged out (LOTO) to prevent accidental startup. Work areas should be safely accessed using certified ladders, scaffolding, or rope-access systems, depending on the tower height and location of the damage.

Personal protective equipment (PPE) is essential — this includes protective clothing, gloves, eye protection, respirators, and fall arrest systems. For offshore turbines, technicians must also wear harnesses, flotation devices, and follow marine safety procedures.

2. Inspection and Assessment of Corrosion

A detailed inspection is the first step to determine the extent and type of corrosion. Corrosion is commonly found around:

- Tower flanges and bolts
- Access doors and joints
- Weld seams and fasteners

- Base sections exposed to standing water or salt spray

Technicians should assess whether the corrosion is surface-level (affecting only the coating) or structural (affecting the metal thickness). Ultrasonic thickness testing (UT) and visual inspection are often used to determine how much material has been lost. The results will dictate whether coating repair is sufficient or whether welding and replacement are required.

3. Surface Preparation – Removing Corrosion

The first and most important step in repair is removing all rust, scale, and loose paint:

- **Minor to moderate corrosion:** Remove using wire brushing, sanding, or power tools (such as needle scalers or grinders). This prepares a clean, smooth surface for coating adhesion.

- **Severe corrosion:** Use abrasive blasting (sandblasting or grit blasting) to achieve a near-white metal finish (typically ISO 8501-1 Sa 2½). This method removes all rust and old coatings efficiently.

After cleaning, the surface should be wiped with an approved solvent or degreaser to eliminate any remaining contaminants such as oil, dirt, or salt deposits. The surface must be dry before applying any primer or coating, as trapped moisture can cause blistering or premature failure.

4. Priming and Repainting

Once the corroded area is cleaned, it must be primed immediately to prevent flash rusting (which can occur within minutes in humid environments).

- Apply a zinc-rich epoxy primer or another corrosion-inhibiting primer compatible with the existing coating system. This primer forms the first barrier against oxidation and provides adhesion for subsequent layers.

- Follow with one or more intermediate coats (typically epoxy-based) to build up the protective layer.

- Finish with a topcoat of polyurethane or fluoropolymer paint, chosen for its resistance to UV radiation, moisture, and environmental degradation.

Each layer must be applied to the manufacturer's specified dry film thickness (DFT) and allowed to cure under proper environmental conditions. For offshore or coastal applications, high-performance marine-grade coating systems are recommended.

5. Welding Repair or Replacement (Severe Corrosion)

If inspection reveals that corrosion has compromised the structural integrity of the steel, coating repair alone is insufficient. In these cases, the damaged section may need to be welded or replaced:

- **Weld repair:** The affected area is ground to remove rust and contaminants before welding. A qualified welder then restores the metal section using compatible filler material. The repaired area is subsequently smoothed, cleaned, and recoated.
- **Component replacement:** Severely corroded bolts, plates, or brackets that cannot be safely welded are removed and replaced with OEM-approved components made from corrosion-resistant steel (such as hot-dip galvanized or stainless steel).

After welding or replacement, the new metal must undergo surface preparation and coating similar to other repaired areas to restore full corrosion protection.

6. Final Inspection and Quality Verification

After repairs are completed, a final inspection ensures that the coatings are uniform, the surface finish meets specifications, and no unprotected metal is exposed. Thickness measurements using a dry film thickness (DFT) gauge confirm proper coating build-up.

In offshore environments, technicians may also measure coating adhesion strength and surface potential (for cathodic systems) to ensure continued protection. All findings and materials used should be documented in maintenance records for traceability and future inspection planning.

<u>Perform Crack and Fatigue Repair</u>

Crack and fatigue repairs are critical corrective maintenance procedures that address material failures caused by repeated loading, cyclic stresses, or environmental degradation. Both steel tower sections and concrete foundations can develop cracks over time, and each requires a specific repair approach to restore structural strength, prevent water ingress, and ensure long-term stability.

1. Crack and Fatigue Repair in Steel Towers

Cracks in steel wind turbine towers usually occur at welded joints, flange connections, or high-stress areas, such as door frames and bolted flanges. These cracks can result from vibration-induced fatigue, uneven torque on bolts, or dynamic loading from high winds.

Step 1 – Inspection and Assessment

Technicians start by inspecting the affected area visually and using non-destructive testing (NDT) methods such as:

- Ultrasonic testing (UT) to detect internal flaws not visible on the surface.
- Magnetic particle inspection (MPI) or dye penetrant testing (DPT) to identify surface or near-surface cracks.

Once the extent of the crack is known, a structural engineer evaluates the damage and determines whether repair by welding is permissible or if component replacement is required.

Step 2 – Preparation for Welding

Before welding, the damaged section must be prepared by:

- Grinding or gouging out the crack to remove all weakened material.
- Cleaning the area to eliminate rust, oil, or coatings using an approved solvent or abrasive tools.
- Ensuring all adjacent surfaces are dry and free of contaminants.

The tower's load should be reduced if necessary (for example, by halting turbine operation) to minimize stress on the cracked section during repair.

Step 3 – Weld Repair Procedure

Welding is performed in accordance with the Original Equipment Manufacturer (OEM) specifications and must be approved by a qualified structural engineer. The procedure includes:

- Preheating the steel surface to the correct temperature to prevent cold cracking.
- Performing multi-pass welds using suitable filler material that matches the tower's steel grade.
- Allowing the weld to cool gradually to avoid residual stress buildup.

After welding, the repaired area is inspected again using NDT methods to confirm that the crack has been fully eliminated and that the weld quality meets engineering standards.

Step 4 – Surface Protection

Once verified, the welded area is ground smooth and recoated with a corrosion-resistant paint system (typically a zinc-rich primer followed by epoxy and polyurethane layers). This step restores both the mechanical strength and corrosion resistance of the structure.

2. Crack Repair in Concrete Foundations

Cracks in concrete turbine foundations can form due to dynamic loading, freeze-thaw cycles, settlement, or improper curing during construction. If left unaddressed, cracks can allow water ingress, which accelerates reinforcement corrosion and weakens the structure.

Step 1 – Crack Evaluation

The first step is to determine the cause and severity of the crack.

- Hairline surface cracks may be cosmetic and only require sealing.
- Structural cracks that extend deep into the concrete require injection repair to restore integrity.

Technicians use crack gauges and ultrasonic pulse velocity testing to assess the crack's depth and monitor whether it is active (still expanding) or stable.

Step 2 – Surface Preparation

Before repair, the crack must be cleaned and dried:

- Remove dust, loose debris, and any efflorescence using compressed air or a vacuum.
- For wide cracks, surface notches may be created to improve adhesion during sealing.

Step 3 – Crack Injection

There are two primary materials used for foundation crack repair: epoxy resin and polyurethane grout, selected based on crack behaviour and environmental conditions.

- Epoxy Injection: Epoxy is used for structural cracks that require strength restoration.
    - Injection ports are installed along the length of the crack using epoxy paste.
    - The crack is then sealed on the surface to contain the resin.
    - Using a low-pressure pump, epoxy is injected into the ports until it fills the entire void.
    - Once cured, epoxy forms a rigid bond that restores the foundation's original load-bearing capacity.
- Polyurethane Injection: Polyurethane grouting is used for non-structural or water-bearing cracks, as it expands upon contact with moisture to fill gaps and stop leaks.
    - The material remains flexible after curing, accommodating minor future movements in the foundation.

Step 4 – Post-Repair Inspection

After the injection process, the ports are removed, and the surface is ground smooth. The repaired area is inspected to ensure that no cracks remain open or leaking.

For exposed foundations, a protective waterproof coating or sealant may be applied to prevent future moisture ingress.

<u>Perform Grout Replacement or Re-Leveling</u>

Grout replacement and re-leveling are essential corrective maintenance procedures performed when the grout at the tower-to-foundation interface deteriorates, cracks, or loses adhesion. The grout layer provides critical load transfer between the wind turbine tower flange and the concrete foundation. Over time, cyclic loading, thermal expansion, or poor initial installation can lead to voids, settlement, or cracking — compromising structural alignment and stability. The repair process restores full contact and ensures even load distribution, preventing long-term fatigue or bolt stress.

1. Damage Assessment and Preparation

Before any repair work begins, a detailed inspection of the grout bed and tower base is performed.

Technicians look for:

- Cracking, spalling, or delamination of the grout.
- Signs of voids or uneven settlement around the base plate.
- Evidence of water ingress or corrosion near anchor bolts.

Non-destructive testing (NDT) techniques such as hammer sounding, ultrasonic pulse velocity, or ground-penetrating radar may be used to detect hidden voids under the base plate.

Once the extent of damage is confirmed, the turbine is safely shut down, and the tower section may be partially unloaded to reduce stress on the grout layer.

2. Removal of Damaged Grout

The defective grout is carefully removed using mechanical chipping tools, rotary hammers, or hydro-demolition to avoid damaging the surrounding concrete foundation or tower base plate.

All loose material, dust, and contaminants must be removed thoroughly to ensure a clean bonding surface.

The substrate is then cleaned with compressed air or high-pressure water jets and allowed to dry.

If the removal exposes corroded anchor bolts or reinforcement, these components are cleaned, treated with anti-corrosion compounds, or replaced if necessary.

3. Surface Preparation and Formwork Installation

After cleaning, the foundation surface is roughened or lightly scarified to enhance bonding with the new grout.

Formwork or grout containment shutters are then installed around the perimeter of the base plate. These forms are sealed to prevent grout leakage and designed to maintain the required pour thickness — typically between 25 mm and 100 mm, depending on the turbine design.

Bonding agents or primers recommended by the grout manufacturer may be applied to the prepared surfaces to improve adhesion.

4. Mixing and Placement of Grout

High-strength, non-shrink cementitious or epoxy grout is used to ensure stability and prevent void formation during curing.

Key characteristics of suitable grout include:

- Compressive strength typically above 60 MPa.
- Non-shrink properties to maintain full contact after curing.
- Flowable consistency for even distribution beneath the tower base plate.

The grout is mixed according to the manufacturer's specifications, ensuring uniform consistency without entrapped air. It is then placed or pumped under the base plate using one-side entry and opposite-side venting to ensure complete filling.

Continuous monitoring during pouring ensures there are no air pockets or gaps under the tower flange.

5. Leveling and Alignment Using Laser Tools

After grout placement, laser alignment tools or theodolites are used to check and correct the tower's vertical alignment.

Precise re-leveling is critical to ensure even load transfer across all anchor bolts and prevent eccentric stresses that can lead to tower flange cracking or bolt fatigue.

Adjustments are made while the grout remains in its plastic (workable) state, ensuring that the base plate sits evenly before curing begins.

6. Curing and Final Inspection

Once the alignment is verified, the grout is left to cure under controlled conditions — usually protected with moist coverings or curing compounds to prevent rapid dehydration.

Curing periods typically last between 24 and 72 hours, depending on ambient temperature and the grout material used.

After curing:

- Formwork is removed, and the grout surface is inspected for cracks or shrinkage.
- Anchor bolts are re-torqued to the manufacturer's specifications.
- A protective sealant or epoxy coating may be applied around the base to prevent water ingress and freeze-thaw damage.

Final laser alignment verification confirms that tower verticality and load distribution have been restored within OEM tolerances.

Part 6

Advanced Topics

Chapter 15

Offshore Wind Turbine Operations

As the global wind energy industry expands beyond the limitations of land-based installations, offshore wind turbines have emerged as one of the most dynamic and technically advanced frontiers in renewable energy. Offshore sites offer stronger and more consistent wind resources, enabling the deployment of larger turbines and the generation of vast amounts of clean electricity. However, these benefits come with substantial engineering, logistical, and safety challenges. Operating far from shore in the harsh and unpredictable marine environment demands specialised infrastructure, vessels, technologies, and expertise. Chapter 15, *Offshore Wind Turbine Operations*, explores these unique operational realities and the systems that keep offshore wind farms safe, efficient, and reliable.

The chapter begins by examining marine safety and crew transfer operations, the cornerstone of offshore wind work. Unlike onshore turbines that can be accessed by road, offshore turbines require careful coordination of vessel logistics, weather forecasting, and personnel safety procedures. Crew Transfer Vessels (CTVs) and Service Operation Vessels (SOVs) are the primary lifelines for offshore maintenance teams, transporting technicians, equipment, and spare parts across open seas. The text explains how these vessels are designed for stability, speed, and safety, and how motion-compensated gangways, dynamic positioning systems, and personal safety gear protect workers during high-risk transfers. It also discusses the role of helicopter operations in emergency response and rapid intervention, emphasizing the need for precision planning, meteorological awareness, and air–sea coordination.

As offshore wind farms grow in size and distance from the coast, advanced transfer systems and technologies have become vital to daily operations. The chapter explores innovations such as motion-compensated walk-to-work (W2W) gangways, automated lifting systems, and fall-arrest mechanisms that allow technicians to safely access turbines even in challenging sea states. These technologies, once confined to the offshore oil and gas sector, have been adapted to the renewable energy industry to improve operational efficiency and reduce weather-related downtime.

The discussion then turns to saltwater corrosion prevention, a critical engineering and maintenance challenge in offshore environments. Constant exposure to salt spray, humidity, and ultraviolet radiation accelerates corrosion, threatening both structural and electrical integrity. Readers will learn about multi-layer coating systems, cathodic protection, material selection, and inspection regimes that protect turbines from degradation. By combining preventive coatings, sacrificial anodes, and controlled electrical protection systems, operators ensure the long-term safety and stability of turbine foundations and towers in the corrosive marine climate.

Finally, the chapter addresses logistics and emergency response offshore, an area where precision planning and robust safety management systems are essential. Offshore wind operations depend on the seamless coordination of vessels, crews, and supply chains, supported by 24/7 Marine Coordination Centres and advanced digital monitoring systems. The section outlines how operators prepare for and respond to incidents such as man-overboard events, fires, equipment failures, or extreme weather, ensuring that every offshore technician has access to rapid medical support, evacuation procedures, and continuous communication with onshore command teams.

Throughout this chapter, readers gain insight into how marine operations, safety systems, and maintenance logistics combine to form the backbone of offshore wind energy. From daily crew transfers to large-scale emergency coordination, each aspect reflects the industry's ongoing commitment to operational excellence, worker safety, and environmental stewardship. Offshore wind turbine operations represent the convergence of maritime expertise, renewable energy engineering, and advanced technology — a combination that is driving the next generation of sustainable power production on a truly global scale.

Marine Safety and Transfer Vessels

Marine safety and the effective use of transfer vessels are fundamental aspects of offshore wind turbine installation, operation, and maintenance. The harsh marine environment poses significant challenges — including high winds, waves, currents, and limited accessibility — which require specialized safety procedures, vessel designs, and transfer systems to protect personnel and equipment. These measures ensure that technicians can safely reach turbines, perform maintenance, and return to shore without incident, even in difficult sea conditions.

Offshore wind farms are typically located many kilometres from shore, often in areas with unpredictable weather and rough seas. This makes marine safety a top priority. Every transfer, whether for installation or maintenance, involves risks such as:

- Slips, trips, and falls during boarding or disembarkation.
- Crush hazards between vessels and turbine transition pieces.
- Capsizing or man-overboard incidents due to wave motion.

- Fatigue and hypothermia from exposure to cold, wet conditions.

To mitigate these risks, offshore wind operators follow strict safety management systems based on international maritime and occupational health standards, including IMO (International Maritime Organization) codes, SOLAS (Safety of Life at Sea), and GWO (Global Wind Organisation) training standards.

Personnel working offshore must complete mandatory safety training covering topics such as:

- Personal Survival Techniques (sea survival, man-overboard recovery)
- Working at Heights
- First Aid and Fire Awareness
- Transfer and Evacuation Procedures

Crew Transfer Vessels (CTVs) are the primary mode of transportation for offshore wind technicians, tools, and small cargo between the shore and wind turbines situated up to around 30 nautical miles offshore. These vessels play a vital role in daily operations, ensuring the safe and efficient movement of personnel and equipment in challenging marine environments.

CTVs are typically between 20 and 30 metres in length and are constructed from lightweight aluminium or composite materials to optimise speed and fuel efficiency. They are powered by high-speed engines capable of achieving 20–30 knots, enabling rapid transit between the port and wind farm. The vessels are equipped with fendered bows, which allow them to perform a "push-on" docking manoeuvre against the turbine's transition piece without damaging the hull or the turbine. To maintain stability during personnel transfers, many CTVs are fitted with dynamic positioning systems (DPS) that keep the vessel steady in variable sea and wind conditions. Internally, they are designed for crew comfort and safety, featuring shock-absorbing seating, tool storage compartments, and emergency equipment.

The transfer process involves the CTV carefully manoeuvring against the turbine's boat landing platform using its engine thrust. Once positioned securely, technicians step across a narrow gap from the vessel to the turbine's ladder platform. This step must be carefully timed with the wave motion to maintain balance and prevent accidents. As a result, this operation requires a combination of professional training, experience, and suitable weather conditions. Typically, safe transfer limits are set for wave heights below 1.5 metres to reduce the risk of slips or vessel movement during boarding.

To ensure the highest levels of safety, all personnel and crew on CTVs must wear Personal Flotation Devices (PFDs) at all times and, in colder climates, immersion suits to protect against hypothermia in the event of a man-overboard situation. Additional safety gear includes safety harnesses for secure movement on deck, two-way radios and Personal Locator Beacons (PLBs) for communication and tracking, and Man Overboard (MOB) recovery systems for quick retrieval in emergencies.

As offshore wind farms continue to expand in size and are constructed farther from the coast — often 60 to 100 kilometres offshore — Service Operation Vessels (SOVs) have become a critical asset for supporting long-term maintenance and operational activities. Unlike Crew Transfer Vessels (CTVs), which return to shore daily, SOVs are designed to stay offshore for extended periods, enabling continuous support for wind farm operations even in challenging sea conditions.

Figure 122: Service Operation Vessels (SOV), Windea Leibniz. Thomas Dahlstrøm Nielsen, CC BY-SA 4.0, via Wikimedia Commons.

SOVs are large, purpose-built vessels typically measuring between 70 and 100 metres in length. They are designed to remain offshore for two to four weeks at a time, allowing maintenance teams to work efficiently without frequent transits to shore. Onboard, they can accommodate up to 60 technicians and crew members, providing comfortable living quarters, workshops, spare parts storage areas, and offices for project coordination. These vessels are equipped with advanced motion-compensated gangways, also known as Walk-to-Work (W2W) systems, which allow technicians to safely transfer between the vessel and turbines even in wave heights of up to three metres. In addition, most SOVs feature Dynamic Positioning (DP2) systems for precise station-keeping, onboard cranes for lifting heavy components, and helidecks to facilitate crew changes and emergency evacuations.

The use of SOVs offers several key advantages. They significantly reduce downtime by allowing technicians to live and work offshore, meaning maintenance activities can continue during periods of poor weather when daily transfers by smaller vessels would be unsafe. The motion-compensated gangway system greatly enhances safety by providing stable and reliable turbine access, even in rougher sea states. Furthermore, SOVs enable round-the-clock operations, supporting night-time maintenance and 24-hour work schedules, which increases the overall efficiency of offshore wind farm management.

Helicopter transfers play a specialised but important role in offshore wind operations, primarily used for urgent maintenance tasks, crew changes, or medical evacuations when sea conditions make vessel access unsafe or impossible. While not as common as vessel-based transfers, helicopters provide a rapid and flexible means of reaching turbines or offshore substations, particularly during severe weather or when time-sensitive repairs are needed to restore lost generation capacity.

Access points for helicopter operations are carefully designed for safety and precision. Landing pads or hoist platforms are typically installed on turbine nacelles or offshore substations, allowing technicians or medics to be safely lowered via winch systems. Every helicopter operation requires strict coordination with air traffic control (ATC), vessel traffic management systems, and the wind farm control centre to ensure full situational awareness and avoid conflicts with other air or sea movements.

However, helicopter transfers are highly dependent on weather conditions, including visibility, wind speed, and turbulence limits. Flights can only proceed when meteorological and operational safety criteria are met, and the cost of each transfer is considerably higher than that of a vessel operation. For this reason, helicopter access is typically reserved for high-priority interventions — such as repairing a turbine that is offline during a period of peak wind generation or responding to a medical emergency.

In summary, while helicopters are not used for routine maintenance, they offer a vital rapid-response capability for offshore wind farms. Their ability to reach remote installations quickly and independently of sea state makes them an essential component of the industry's emergency and high-value maintenance strategies.

Transfer Systems and Technologies

The offshore wind industry employs a range of advanced transfer systems and technologies designed to enhance safety, efficiency, and reliability during personnel and equipment transfers. These innovations have become essential as turbines increase in size and are installed farther from shore, often in harsher marine environments.

One of the most significant advancements is the motion-compensated gangway — a hydraulically or electronically stabilised bridge system, such as the Ampelmann or Uptime platform. These gangways automatically adjust in real time to compensate for vessel movement, enabling safe walk-to-work (W2W) transfers between Service Operation Vessels (SOVs) and turbine platforms, even in sea states with wave heights of up to 3 metres. This technology allows technicians to access turbines more frequently, reducing weather-related downtime.

Boat landing platforms remain the standard access point for Crew Transfer Vessel (CTV) operations. These steel structures, equipped with vertical ladder systems, allow vessels to perform controlled "push-on" manoeuvres, enabling technicians to step safely from the bow to the ladder during calm sea conditions.

For transferring heavy or bulky items, davit cranes and lifting platforms are used to hoist tools, spare parts, and small components from vessels onto the turbine transition piece. This eliminates the need for manual lifting and reduces the risk of injury or equipment damage.

Figure 123: Ampelmann platform system.

Finally, fall arrest systems integrated into turbine ladders and internal platforms provide an essential layer of protection for technicians during vertical climbs and maintenance activities. These systems ensure that in the event of a slip or misstep, workers remain securely attached and protected from falls.

Saltwater Corrosion Prevention

Saltwater corrosion prevention is one of the most critical aspects of offshore wind turbine maintenance and design. The harsh marine environment—characterised by constant exposure to salt-laden air, high humidity, and seawater spray—creates extremely corrosive conditions for metallic structures and electrical components. Without robust protection, corrosion can compromise the structural integrity of towers, foundations, and fasteners, while also degrading electrical connections and reducing turbine reliability. Effective corrosion control strategies combine material selection, protective coatings, cathodic protection, and regular inspection regimes.

Saltwater contains sodium chloride (NaCl), which dissociates into ions that accelerate electrochemical reactions on metal surfaces. When oxygen and moisture are present, these ions promote oxidation, forming rust or other corrosion byproducts. Offshore turbines are particularly vulnerable at splash zones—the area where seawater repeatedly wets and dries the structure—and at crevices or joints where moisture can accumulate.

One of the primary defences against saltwater corrosion is the use of high-performance coating systems. Offshore turbine towers and foundations are typically protected with multiple layers of epoxy primers and polyurethane or fluoropolymer topcoats, which provide both chemical resistance and UV protection. The coatings act as a physical barrier, isolating steel surfaces from oxygen and moisture. In the splash zone, polyamide-cured epoxy coatings or glass-flake reinforced coatings are often applied for enhanced abrasion resistance. Regular coating inspections and touch-ups during preventive maintenance are essential to maintain effectiveness.

In addition to coatings, cathodic protection (CP) is widely used for submerged and splash-zone components. Two main types of CP systems are employed:

- Sacrificial Anode Systems – Made from more reactive metals such as zinc or aluminium alloys, these anodes corrode preferentially, protecting the steel structure through galvanic action.
- Impressed Current Cathodic Protection (ICCP) – In this system, a controlled electrical current is applied using inert anodes and a DC power supply, ensuring uniform protection across large structures like monopiles or jacket foundations.

Both systems require regular monitoring of potential differences and current flow to ensure optimal protection levels.

Material selection plays a vital role in corrosion prevention. Marine-grade stainless steels, duplex steels, and hot-dip galvanized components are often chosen for areas exposed to seawater or high humidity. Design features also help prevent corrosion—such as avoiding sharp crevices where water can collect, using closed welds instead of bolted joints where possible, and applying sealants around penetrations and flanges.

Inside the turbine tower and nacelle, corrosion can be reduced through humidity control systems, such as dehumidifiers and desiccant-based dryers, which maintain relative humidity below 60%. Routine cleaning, salt removal, and the reapplication of protective waxes or inhibitors are standard preventive maintenance activities.

Regular inspection of coatings, fasteners, and anodes ensures early detection of corrosion before it becomes structural. Ultrasonic thickness testing, visual inspection, and potential measurement surveys are used to assess protection performance. In advanced systems, corrosion sensors integrated into the structure provide real-time data to maintenance teams.

Saltwater corrosion prevention in offshore wind turbines relies on a multi-layered defence strategy—combining protective coatings, cathodic systems, corrosion-resistant materials, and vigilant maintenance. By implementing these measures, operators can extend component life, minimise costly repairs, and ensure the long-term safety and efficiency of offshore wind energy installations.

Logistics and Emergency Response Offshore

Effective logistics and emergency response management are essential for maintaining safe, reliable, and efficient offshore wind farm operations. Offshore environments are inherently challenging—remote locations, unpredictable weather, and complex access requirements mean that both day-to-day logistics and emergency planning must be meticulously coordinated. A well-organised offshore logistics system ensures that personnel, equipment, and materials are transported safely and efficiently, while a robust emergency response framework protects workers and assets in the event of accidents, severe weather, or technical failures.

Offshore logistics covers the planning, coordination, and execution of all transport and supply activities necessary for wind farm construction, operation, and maintenance. These systems ensure that technicians, spare parts, and critical tools reach turbines on time and in good condition.

Crew and Equipment Transport: The backbone of offshore logistics is the Crew Transfer Vessel (CTV) and Service Operation Vessel (SOV) network. CTVs handle daily transfers for nearshore farms (up to ~30 nautical miles), while SOVs remain offshore for extended maintenance missions, typically 2–4 weeks. SOVs are equipped with accommodation, workshops, and motion-compensated gangways for safe access to turbines in rough seas. For urgent situations, helicopters are used for rapid personnel movement or emergency medical evacuation (medevac).

Supply Chain Coordination: Offshore operations depend on a tightly managed supply chain, including the storage, packaging, and dispatch of spare parts, consumables, and tools. Key logistical bases—often called Offshore Coordination Centres (OCCs) or Port Operations Hubs—serve as the staging point for all maintenance operations. Here, materials are checked, loaded onto vessels, and tracked via digital logistics platforms to ensure traceability and compliance with safety and environmental regulations.

Weather and Sea-State Planning: Weather conditions strongly influence offshore logistics. Marine operations are limited by wave height, wind speed, and visibility, known collectively as the *weather window*. Operators use advanced meteorological forecasting and marine traffic management software to plan vessel movements, ensuring safe transfer conditions and efficient scheduling.

Because offshore environments are remote and high-risk, emergency response (ER) systems must be proactive, well-rehearsed, and compliant with international maritime and occupational safety regulations (e.g., IMO, ISO, GWO, and local WHS laws).

Emergency Response Planning (ERP): Each offshore wind farm has a site-specific Emergency Response Plan detailing roles, communication protocols, and evacuation procedures. The ERP covers scenarios such as:

- Medical emergencies (injuries, illnesses, or accidents during maintenance)
- Man Overboard (MOB) incidents
- Fires or explosions in the turbine or vessel
- Structural damage or collisions
- Severe weather events requiring evacuation
- Search and Rescue (SAR) coordination with maritime authorities

These plans are regularly updated and tested through emergency drills involving vessel crews, turbine technicians, and onshore coordination centres.

Communication and Command Systems: During an emergency, coordination is managed through a 24/7 Marine Coordination Centre (MCC) or Operations Control Centre (OCC). These centres use real-time communication systems—VHF/UHF radio, AIS tracking, satellite communication, and digital emergency alert software—to maintain situational awareness. Each technician and vessel carries Personal Locator Beacons (PLBs) and Automatic Identification Systems (AIS) for rapid location tracking.

Evacuation and Medical Support: In serious cases, evacuation may be performed via:

- Helicopter hoisting from nacelle or offshore substation platforms.
- Rescue boats deployed from SOVs or CTVs.
- Offshore medical rooms on SOVs or substations, where trained medics provide first aid or advanced care until evacuation is possible.

Many SOVs are equipped with onboard medical facilities, stretchers, and telemedicine links to onshore doctors.

All offshore personnel must complete Global Wind Organisation (GWO) safety training, which includes:

- Sea Survival and Emergency Response
- First Aid and Fire Awareness

- Working at Heights
- Manual Handling and Transfer Procedures

These qualifications ensure that all workers can respond appropriately in emergencies and operate safely during routine offshore transfers.

Regular emergency drills (e.g., Man Overboard recovery, turbine evacuation, fire simulation) reinforce preparedness. Compliance is monitored by safety officers and verified through audits by authorities such as maritime safety agencies and energy regulators.

Modern offshore operations rely increasingly on digital safety management systems and real-time monitoring technologies. Examples include:

- Digital Twin models of wind farms for simulation and risk assessment.
- IoT-based tracking systems for personnel and equipment.
- AI-driven weather forecasting to anticipate emergency conditions.
- Drones and remotely operated vehicles (ROVs) for rapid inspection following incidents or storms.

These technologies enhance situational awareness and decision-making, ensuring faster, safer responses when incidents occur.

Chapter 16

Emerging Technologies

Hybrid Systems (Solar + Wind)

Hybrid renewable energy systems, particularly solar-wind integrations, have emerged as pivotal solutions for achieving reliable and sustainable energy production. These systems capitalize on the inherent complementarity between solar and wind energy, which often peak at different times of the day, thus facilitating enhanced energy stability. Solar power typically reaches its maximum output during peak daylight hours, while wind energy generally generates more power during the night or in overcast conditions. By combining these two sources, hybrid systems reduce the intermittency of energy production, leading to a more stable energy supply and reducing reliance on energy storage systems [261-263].

The concept of hybrid systems revolves around maximizing energy availability and minimizing gaps in power production. For instance, in various climatic conditions—such as coastal or desert areas—solar generation can be maximized during sunny days, complemented by increased wind generation at night [263, 264]. Hybrid systems promise improved reliability and enhance overall economic viability by optimizing the use of shared infrastructure like transformers and inverters, potentially reducing project costs [265, 266].

Operational advantages of hybrid solar-wind systems include improved reliability through enhanced power continuity. By leveraging their complementary strengths, these systems can provide smoother power outputs, effectively reducing issues such as blackouts and voltage fluctuations on the grid [266, 267]. Furthermore, the efficiency of energy yield is elevated, as solar panels and wind turbines can compensate for each other's downtime, thereby maximizing total energy output over time [261, 266]. The reduced reliance on large-scale storage solutions, such as batteries, minimizes both operational costs and environmental impacts—issues that have historically plagued renewable energy systems [268].

The system components integral to solar–wind hybrids include wind turbines, solar photovoltaic panels, hybrid inverters or power controllers, and optionally, energy storage solutions [269, 270]. Each component plays a critical role in ensuring efficient energy conversion into usable forms and smoothing the potential variability in power generation. The hybrid inverter, for instance, efficiently synchronizes the energy outputs from both solar and wind sources into a consistent AC power suitable for local or grid use [266, 271].

Design considerations for these systems are paramount and involve meticulous analysis of local resource conditions, including daily solar irradiance and seasonal wind speed profiles [272, 273]. Advanced energy management systems are employed to manage and optimize real-time power flows, ensuring a balance between generation and consumption. This optimization is facilitated through algorithms that account for variable resource availability, enhancing the reliability of the power supply [262, 274].

Despite their numerous advantages, hybrid systems also face challenges, notably in power management and initial setup costs. Coordinating outputs from both solar and wind generation requires sophisticated forecasting and control systems, while initial investments can be higher compared to single-source systems [275, 276]. However, innovations such as AI-driven controls and modular system designs are proving essential in overcoming these obstacles, thereby making hybrid systems more attractive and feasible for broader adoption [277].

Future trends indicate that hybrid solar-wind systems will play a critical role in the transition towards a carbon-neutral energy landscape. With declining costs in renewable technologies and improvements in energy management solutions, there is a strong potential for these systems to become mainstream in both utility-scale projects and decentralized applications. As energy demands heighten globally, the integration of solar and wind resources through hybrid systems will be fundamental in achieving sustained electricity generation and reducing dependency on fossil fuels [278].

Floating Wind Turbines

A floating wind turbine is an advanced form of offshore wind technology that enables power generation in deep waters where conventional fixed-bottom foundations are impractical or uneconomical. Unlike traditional offshore turbines anchored directly to the seabed, floating turbines are installed on floating platforms that are anchored to the ocean floor using mooring lines and anchors. This design allows them to operate in water depths beyond 60 metres, where stronger and more consistent winds are available, significantly expanding the potential for offshore wind energy development.

Floating wind turbines make it possible to harness wind energy far from shore, where winds are typically faster and steadier, resulting in higher energy yields and greater efficiency. Because they are situated further offshore, they also minimize visual impact from the coast and better accommodate marine traffic and fisheries. For countries with limited shallow continental shelves—such as Japan, Portugal, Spain, France, and the U.S. West Coast—floating wind farms offer an ideal renewable energy solution.

Most floating turbines are pre-assembled onshore in port facilities and then towed to their installation site, reducing the need for expensive offshore construction. This logistical advantage makes them easier to install and maintain, although their initial cost remains higher due to the complexity of floating foundations and anchoring systems. As of October 2024, approximately 245 MW of floating wind power is operational worldwide, with a development pipeline exceeding 266 GW, highlighting their growing global importance.

Notable examples include the Hywind Tampen project in Norway—the world's largest floating wind farm—which began operating in August 2023. It comprises 11 turbines located 140 km offshore and supplies roughly 35 % of the electricity needed by nearby oil and gas platforms. Another milestone project, the Green Volt Wind Farm off Scotland (approved in 2024), features 35 turbines rated at 16 MW each, with a total capacity of 560 MW. This project is expected to decarbonize offshore oil operations and supply additional power to the national grid.

Figure 124: Blue H Technologies - World's First Floating Wind Turbine. Green Storm 7, CC BY-SA 4.0, via Wikimedia Commons.

How floating wind turbines work:

- **Floating Structure:** The wind turbine is mounted on a floating platform, which can be one of several types:
    - Spar buoy (a long vertical cylinder providing stability through ballast weight).
    - Semi-submersible platform (supported by multiple buoyant columns and pontoons).
    - Barge (a flat-bottomed structure suitable for calmer waters).
- **Mooring System:** The floating platform is stabilized using anchored mooring lines—chains, steel cables, or synthetic ropes—that connect to seabed anchors. These maintain the turbine's position while allowing slight movement to absorb wave and wind forces. Power generated is transmitted to shore via a subsea cable connected to the grid.
- **Electricity Generation:** The operation mirrors that of traditional wind turbines: wind rotates the blades, driving a generator in the nacelle that produces electricity. The power is then converted and transmitted through underwater cables to onshore substations.

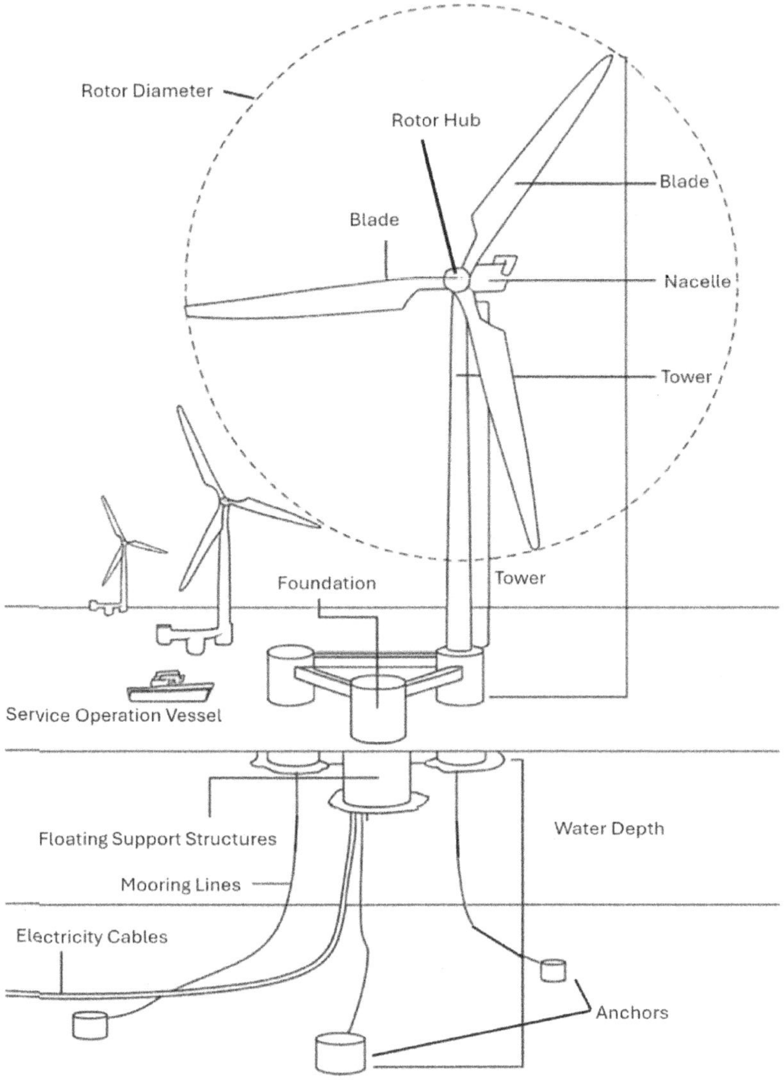

Figure 125: Floating wind turbine.

The mooring system is critical for maintaining stability and position. There are three main configurations, each suited to different water depths and seabed conditions:

- **Catenary System:** Uses heavy chain mooring lines that hang in a curved shape, providing restoring force through their weight. The anchor is loaded almost horizontally, and the line typically extends six to eight times the water depth. This system is the most common and is used in projects like Hywind Tampen, which also pioneered shared anchoring, where multiple turbines share common anchors to reduce cost and seabed impact.

- **Taut-Leg System:** Employs synthetic ropes or steel lines under tension, angled between 30° and 45° to the seabed. It offers better station-keeping with less seabed footprint and is effective in deeper waters.
- **Tension-Leg Platform (TLP):** Utilizes vertical, pre-tensioned tethers that provide high stiffness and minimize vertical movement, ensuring strong resistance to pitch and roll motions even under severe conditions.

According to IEC 61400-3-2, floating wind turbines must undergo site-specific load analyses that account for environmental conditions such as wind speed, wave height, and ocean currents to ensure safe and efficient operation.

Floating wind turbines provide several significant advantages that make them a key innovation in the offshore renewable energy sector. One of the most important benefits is their ability to access deeper waters. Floating technology enables wind farms to be installed far beyond the depth limitations of fixed-bottom foundations, unlocking vast offshore areas with enormous wind energy potential. This expansion dramatically increases the total capacity available for global wind power generation, especially for nations with deep coastal shelves.

Another advantage is access to stronger and more consistent winds. Offshore wind speeds increase with distance from land and are less affected by turbulence. As a result, floating turbines operating in deeper waters can achieve higher capacity factors and generate more electricity over time. Their ability to harness steadier winds translates directly into improved reliability and efficiency compared to nearshore or onshore installations.

A further strength lies in simplified assembly and deployment. Floating wind turbines are largely constructed and pre-assembled in port facilities, where conditions are controlled and safer for workers. Once completed, they are towed to their operating sites using conventional vessels, reducing the need for expensive offshore construction vessels and minimizing time spent at sea. This approach helps lower installation costs and risks associated with adverse weather.

In addition, floating turbines have reduced visual and spatial impact. Their greater distance from shore makes them less visible from coastal areas, easing public concerns about visual intrusion. Their offshore placement also allows for better integration with existing maritime activities, such as fishing and shipping, without causing major disruptions.

Despite these advantages, floating wind turbines face several engineering and economic challenges that currently limit large-scale adoption. The most significant barrier is high cost. Floating platforms require complex mooring systems, durable materials, and specialized electrical infrastructure to withstand harsh marine environments. Because supply chains and fabrication facilities are still developing, production costs remain higher than those for traditional fixed-bottom turbines.

Another challenge is stability and structural load management. Floating platforms must maintain balance while subjected to simultaneous wind, wave, and current forces. Designing systems that resist excessive motion or tilting without compromising efficiency requires advanced hydrodynamic modelling, robust materials, and precise control systems.

The environmental impact of floating wind technology is also an area of active study. Potential issues include underwater noise during installation, electromagnetic fields from power cables affecting marine species, and possible disruptions to migration or navigation routes. Ongoing environmental assessments and technological improvements aim to mitigate these effects.

Floating offshore wind energy represents the next frontier in renewable power generation. As research and development continue, economies of scale, mass production, and standardized designs are expected to bring costs down substantially. The use of shared anchoring systems, modular floating platforms, and expanded industrial fabrication facilities will further enhance cost efficiency and speed of deployment.

In the coming decades, floating wind farms are projected to play a crucial role in achieving global decarbonization targets by expanding access to offshore wind resources previously out of reach. With continued innovation and supportive policy frameworks, floating wind technology is poised to become a cornerstone of the world's sustainable energy future.

Advanced Monitoring (AI, Drones, Predictive Analytics)

Modern wind turbine maintenance and performance management increasingly leverage advanced monitoring technologies, which include artificial intelligence (AI), unmanned aerial vehicles (UAVs or drones), and predictive analytics. The integration of these technologies is transforming traditional maintenance practices into intelligent, data-driven operations that enhance the reliability and efficiency of wind turbine systems.

AI is pivotal for the surveillance and analysis of wind turbine performance, processing vast amounts of operational data obtained from sensors, Supervisory Control and Data Acquisition (SCADA) systems, and Condition Monitoring Systems (CMS). Machine learning algorithms facilitate the detection of minute changes in parameters such as vibration, temperature, and power output, which may indicate underlying mechanical or electrical issues. For instance, studies illustrate how machine learning models like artificial neural networks can predict wind turbine performance and thereby support efficient anomaly detection [279, 280]. Specific fault indicators such as bearing wear or gearbox failures can be anticipated through AI analysis, allowing for proactive intervention [281, 282].

Furthermore, the application of predictive maintenance methodologies has been emphasized, where AI continues to learn from incoming data to refine its predictive abilities. This transition from reactive to predictive maintenance is underscored by research indicating that AI-enabled systems can substantially reduce maintenance costs and downtime, thereby optimizing operational efficiency [280, 283].

The advent of drones has significantly revolutionized wind turbine inspections, particularly for evaluating blade conditions. Traditional inspection methods, which often require manual access or cranes, have been replaced by UAVs equipped with high-resolution cameras and thermal imaging capabilities, enabling comprehensive and rapid assessments of turbine blades from various angles [146, 284]. A drone can complete a full turbine inspection in a fraction of the time compared to conventional methods, enhancing efficiency and safety by minimizing the need for technicians to work at hazardous heights [285].

Figure 126: Drone photograph taken at the Lomont wind farm site, showing the rear view of a wind turbine. Koestel, CC BY-SA 4.0, via Wikimedia Commons.

Additionally, the integration of AI with drone technologies allows for automated image processing to identify defects, ascertain their severities, and recommend maintenance actions [75, 286]. Such advancements underscore the critical role of drones in routine maintenance, especially in challenging environments like offshore wind farms, which present accessibility hurdles [287].

Predictive analytics amalgamates real-time sensor data, historical maintenance records, and environmental variables to forecast equipment failures and optimize turbine performance. This data-driven modelling reveals significant correlations between operational parameters and the degradation of components. For instance, vibration data can forecast bearing failures, and power curve analyses can expose inefficiencies stemming from blade fouling [288, 289]. By employing predictive analytics, wind farm operators can strategically plan maintenance during periods of low energy demand, effectively mitigating unexpected downtimes [290, 291].

Modern SCADA systems now incorporate real-time dashboards that visualize predictive models, keeping maintenance teams alert to potential faults before they manifest into significant issues [283, 292]. The efficacy of these predictive tools illustrates their capacity to prolong equipment lifespan and enhance continuous performance.

Emerging technologies such as digital twins—virtual representations of wind turbines powered by AI and predictive analytics—are becoming increasingly commonplace in maintenance strategies. Digital twins provide a platform for simulating turbine performance under varying conditions and evaluating maintenance actions in a risk-free environment [293, 294]. By integrating live operational data, these models assist in precise lifecycle management, optimizing decision-making processes regarding maintenance and performance monitoring [186, 295].

Adopting advanced monitoring technologies in wind turbine maintenance yields numerous advantages:

- **Reduced Maintenance Costs**: Early detection of faults can decrease the likelihood of extensive repairs [75].
- **Improved Safety**: Automation of inspection processes reduces the need for human intervention in high-risk scenarios [146, 284].
- **Higher Availability**: Predictive insights minimize unplanned downtime, ensuring turbines operate efficiently [287, 295].
- **Extended Asset Life**: Proactive maintenance interventions can slow down component degradation [186, 289].
- **Optimized Energy Output**: Continued performance tuning maximizes power generation, aligning with renewable energy goals [146, 279].

As the reliance on renewable energy expands, the incorporation of advanced monitoring technologies is poised to become standard practice across wind farms. Future innovations in AI, machine learning algorithms, edge computing, and drone automation are anticipated to further streamline inspection and diagnostics. The synergy with cloud-based analytics platforms and IoT-enabled sensors will eventually empower wind turbines to self-diagnose, facilitating autonomous communication with maintenance systems to coordinate repairs autonomously [75, 293].

These comprehensive advancements establish a solid foundation for the evolution of wind turbine maintenance, where operational integrity and efficiency are heightened through intelligent, technology-driven approaches.

Part 7

Professional Practice

Chapter 17

Documentation and Reporting

The wind energy industry stands at the forefront of technological innovation, driven by the need to enhance efficiency, reliability, and sustainability in a rapidly evolving global energy landscape. As renewable energy adoption accelerates worldwide, emerging technologies are reshaping how wind power is generated, integrated, and maintained. Chapter 16, *Emerging Technologies*, explores the next generation of advancements transforming the sector — from hybrid renewable systems and floating wind turbines to intelligent monitoring powered by artificial intelligence (AI), drones, and predictive analytics. Together, these innovations mark a decisive shift toward smarter, more adaptable, and interconnected energy solutions.

The chapter begins with Hybrid Systems (Solar + Wind), which represent one of the most promising pathways for achieving continuous and stable renewable power generation. By integrating solar and wind energy, hybrid systems leverage the natural complementarity between sunlight and wind patterns — where solar power peaks during the day and wind power often rises at night. This combination reduces intermittency, improves energy security, and maximizes the use of shared infrastructure such as transformers, inverters, and grid connections. Readers will explore how hybrid system design requires precise environmental assessment, advanced control algorithms, and energy management systems capable of balancing real-time fluctuations in generation and demand. These hybrid configurations are not only improving grid stability but are also driving cost reductions and supporting the global transition toward decarbonized energy networks.

The discussion then turns to Floating Wind Turbines, one of the most transformative developments in offshore renewable energy. Unlike traditional fixed-bottom turbines limited to shallow waters, floating turbines are designed to operate in deep-sea environments, unlocking vast wind resources previously beyond reach. Mounted on buoyant platforms anchored to the seabed, these systems enable deployment in water depths exceeding 60 metres, where winds are stronger and more consistent. The section examines the different platform types — including spar buoys, semi-submersibles, and tension-leg platforms — and explains how mooring systems and dynamic positioning technologies maintain stability under challenging ocean conditions. Floating wind technology is redefining offshore wind potential, particularly for nations with deep continental shelves, and plays a pivotal role in global efforts to expand renewable energy capacity in deeper waters.

The final section, Advanced Monitoring (AI, Drones, Predictive Analytics), explores how digital transformation is revolutionizing wind turbine operation and maintenance. Modern wind farms now rely heavily on intelligent systems that continuously analyse data from sensors, SCADA networks, and condition monitoring systems. AI and machine learning algorithms can detect subtle anomalies in vibration, temperature, and power output, predicting failures before they occur and allowing for proactive intervention. Drones equipped with high-resolution and thermal imaging cameras have replaced many traditional inspection methods, providing faster, safer, and more comprehensive assessments of turbine blades and structures. Predictive analytics further enhances operational efficiency by correlating real-time and historical data to anticipate maintenance needs and optimise energy production. Collectively, these technologies enable a transition from reactive to predictive maintenance — reducing downtime, lowering operational costs, and extending turbine lifespan.

Throughout this chapter, readers will gain insight into how technological convergence — where digital intelligence, renewable integration, and advanced materials intersect — is shaping the future of the wind energy industry. The innovations explored here not only improve performance and reliability but also strengthen the economic and environmental viability of wind power on a global scale. *Emerging Technologies* underscores a fundamental truth: the evolution of renewable energy depends not just on harnessing the wind, but on continuously advancing the tools, systems, and intelligence that make it possible.

Maintenance Logs and Digital Work Orders

Maintenance logs and digital work orders form the backbone of an effective wind turbine operation and maintenance (O&M) system. They serve as the central recordkeeping and workflow management tools that ensure reliability, regulatory compliance, and traceability in both onshore and offshore wind operations.

Maintenance logs document all maintenance activities performed on a wind turbine, from routine inspections and lubrication to major component replacements. They provide a complete operational history, enabling technicians and asset managers to track component wear, identify recurring issues, and plan future interventions more efficiently.

Digital work orders, on the other hand, transform this process into a real-time, paperless system. By using tablets or mobile devices, technicians can receive, complete, and close out maintenance tasks directly from the field. This digital workflow improves coordination between maintenance crews, control centres, and asset management systems.

Together, maintenance logs and digital work orders:

- Improve asset reliability through accurate tracking of preventive and corrective actions.

- Ensure compliance with manufacturer and regulatory maintenance schedules.
- Support warranty claims and audits by maintaining a verifiable maintenance history.
- Enhance data-driven decision-making by integrating with condition monitoring and SCADA systems.

A comprehensive maintenance log typically includes:

- **Date and Time:** When maintenance was performed.
- **Technician Details:** Names, certifications, and shift information.
- **Work Order Reference:** Unique identifier linking to the digital or paper-based work order.
- **Turbine Details:** Location, turbine ID, and model.
- **Description of Work:** Preventive tasks (inspection, lubrication, testing) or corrective actions (repairs, replacements).
- **Parts and Materials Used:** Components replaced or consumed during the activity.
- **Findings and Observations:** Notes on anomalies, wear patterns, or safety hazards.
- **Signatures or Digital Verification:** Confirmation of task completion by authorized personnel.

Digital systems often incorporate photo uploads, sensor readings, and voice notes to enhance documentation accuracy.

Digital Work Orders and Workflow Integration

Modern wind farms depend heavily on Computerized Maintenance Management Systems (CMMS) or Enterprise Asset Management (EAM) software to organize, schedule, and track all maintenance activities in a centralized digital environment. These systems—such as SAP, IBM Maximo, or WindESCo—form the digital backbone of turbine maintenance operations, replacing traditional paper-based records with real-time, automated workflows. Their primary role is to streamline communication between monitoring systems, maintenance teams, and asset managers, ensuring that every detected issue results in a documented and properly executed maintenance task.

The process begins with fault detection, where abnormalities are identified by the wind farm's Supervisory Control and Data Acquisition (SCADA) system or by condition monitoring sensors installed in the turbines. These sensors continuously collect data on vibration, temperature, oil condition, and electrical performance, detecting early signs of wear or failure. When a deviation is identified—such as an unusual vibration in the gearbox or overheating in the generator—the system automatically flags the issue and transmits it to the CMMS or EAM for further action.

Next, the system initiates automatic work order generation. The CMMS creates a digital work order that contains detailed information about the problem, including the affected turbine, the suspected fault, recommended actions, priority level, and safety requirements. This automation ensures that no detected fault is overlooked and that every issue is logged and traceable within the maintenance system. Preventive tasks can also be generated on a scheduled basis, such as inspections, lubrication, or filter changes.

Once the digital work order is created, it moves to the assignment stage, where it is allocated to a specific field technician or maintenance crew. Assignment is based on the team's availability, skill set, and location within the wind farm. For offshore or large-scale projects, the CMMS may even optimize crew logistics—such as assigning technicians who are already on-site or scheduling maintenance windows that align with vessel or helicopter transfers.

The execution phase occurs when technicians access the digital work order via mobile tablets or handheld devices. These devices provide them with real-time instructions, turbine schematics, safety checklists, and spare parts data. As technicians perform the work, they update the work order by recording observations, entering part numbers, capturing photos, and noting time spent on each task. This live data entry eliminates delays, reduces paperwork, and ensures that supervisors have an immediate overview of maintenance progress.

Once the task is completed, the work order moves to the verification and closure stage. A supervisor or asset manager reviews the work details, verifying that all maintenance actions were performed correctly and that relevant safety procedures were followed. If the repair involves warranty-covered components or critical infrastructure, additional approval steps may be required. Once verified, the work order is officially closed in the CMMS, updating the turbine's digital maintenance history and performance record.

This end-to-end integration—from fault detection to work order closure—creates a highly efficient and traceable maintenance process. By linking SCADA systems, predictive analytics, and technician workflows, modern CMMS and EAM solutions ensure that even minor faults detected remotely are quickly translated into actionable maintenance tasks. The result is reduced downtime, faster response times, optimized resource allocation, and a stronger foundation for data-driven decision-making across the wind farm's entire operational lifecycle.

Sample Digital Systems

SAP (Enterprise Asset Management / Intelligent Asset Management)

SAP is a leading enterprise software suite widely used across industries to manage business operations and assets. One of its key modules, Enterprise Asset Management (EAM), is designed to oversee the entire lifecycle of physical assets such as wind turbines, power equipment, and infrastructure. SAP EAM supports asset planning, maintenance, performance monitoring, and eventual decommissioning, ensuring organizations maintain high operational efficiency and compliance across all assets [296, 297].

SAP EAM enables maintenance planning, scheduling, and execution for both preventive and corrective maintenance tasks. It also integrates closely with procurement and inventory systems, allowing seamless tracking and ordering of spare parts and maintenance materials. Another critical function is asset health and performance monitoring, achieved through integration with IoT sensors and real-time data systems. This allows maintenance teams to predict potential failures and optimize service schedules.

Additional features include risk assessment and reliability-centred maintenance (RCM) strategies, which help identify critical components and prioritize maintenance based on failure risk and operational impact. SAP also provides mobile and field integration, enabling technicians to access digital work orders, update progress, attach photos, and record maintenance data in real time using tablets or mobile devices.

SAP is known for its robustness, scalability, and integration across business lines such as finance, supply chain, and operations. It supports complex workflows and regulatory compliance requirements, making it suitable for large-scale renewable energy and infrastructure projects. However, it also presents challenges, including implementation complexity, customization costs, and the need for significant user training. Integration with operational systems such as SCADA or condition monitoring platforms can also be technically demanding but is crucial for maximizing efficiency in wind farm maintenance.

IBM Maximo (Maximo Application Suite)

IBM Maximo is one of the most established and widely used Enterprise Asset Management (EAM) and Computerized Maintenance Management System (CMMS) solutions globally. It is particularly suited to asset-intensive industries such as energy, utilities, transportation, and manufacturing. The latest version, Maximo Application Suite (MAS), brings together asset management, performance monitoring, predictive maintenance, and AI/IoT integration under a unified platform [298, 299].

Maximo provides comprehensive work order management, covering the creation, assignment, progress tracking, and closure of maintenance activities. It includes a detailed asset registry and hierarchy management system to track equipment, components, and their maintenance histories. Through integration with IoT and SCADA systems, Maximo enables real-time condition monitoring, anomaly detection, and predictive maintenance.

Other key features include mobile technician access (with offline functionality), analytics dashboards, and KPI tracking for maintenance performance. The platform supports both cloud and on-premises deployment, using containerized architecture (via Red Hat OpenShift) for scalability and flexibility.

Maximo's main strengths lie in its maturity, flexibility, and advanced analytics capabilities. When combined with Maximo Predict and Maximo Monitor modules, it can provide powerful predictive maintenance insights based on sensor and performance data. However, similar to SAP, the system's complexity requires careful configuration and ongoing data management.

Implementation challenges often include data migration, integration with SCADA systems, and user adoption among field teams.

WindESCo

WindESCo is a specialized analytics and optimization company dedicated to the wind energy sector. Unlike general EAM systems, WindESCo focuses on improving wind turbine performance, reliability, and energy output through AI-driven analytics. Its platform uses SCADA data, sensor inputs, and condition monitoring information to detect underperformance, diagnose issues, and recommend corrective actions.

The WindESCo Pulse platform analyses turbine data to identify actionable performance and reliability issues, offering prioritized recommendations and quantifying potential energy gains. Its eCMS (Electrical Condition Monitoring System) focuses on detecting faults in key electrical subsystems such as generators and converters using high-frequency data sampling. Another innovation, WindESCo Swarm™, introduces cooperative control between turbines, optimizing wake steering and boosting total wind farm output through coordinated turbine adjustments.

WindESCo fills a crucial gap in the wind industry by providing deep subsystem visibility that many OEM control systems lack. It converts raw turbine data into actionable maintenance and performance insights, enabling operators to move from reactive troubleshooting to proactive, performance-based O&M strategies. The result is improved turbine availability, reduced operational costs, and measurable gains in energy production.

Comparison and Integration in Wind Operations

SAP and IBM Maximo serve as broad enterprise asset management platforms, managing work orders, spare parts, maintenance schedules, and regulatory compliance across various industries. In contrast, WindESCo provides specialized, data-driven intelligence for wind turbines, focusing on performance optimization and predictive diagnostics.

In practice, many wind farm operators integrate these systems:

- SAP or Maximo acts as the core EAM or CMMS, handling scheduling, logistics, and maintenance documentation.
- WindESCo interfaces with SCADA data to provide advanced diagnostics, which then trigger automated work orders in SAP or Maximo.

This hybrid approach allows operators to combine enterprise-level asset management with real-time, turbine-specific performance intelligence, creating a powerful ecosystem for predictive maintenance and operational optimization in modern wind farms.

Compliance with Regulatory Audits

Compliance with regulatory audits is a fundamental aspect of wind turbine operation, ensuring that every turbine is designed, constructed, maintained, and operated in line with applicable international, national, and local legislation. These audits serve as a structured verification process that confirms the turbine's adherence to safety, environmental, and grid performance requirements. While the principles are universal, the specific audit procedures and standards vary depending on jurisdictional frameworks, certification bodies, and the type of installation — onshore or offshore.

The primary purpose of regulatory audits in the wind energy sector is multifaceted. First and foremost, audits provide safety assurance by verifying that turbines meet design and operational safety standards, thereby protecting maintenance personnel, the surrounding environment, and nearby communities. They also ensure environmental compliance, confirming that turbines operate within acceptable thresholds for noise, wildlife interaction, and visual impact. Another essential goal is grid integration verification, which checks that turbines and associated substations meet regional grid codes for frequency response, voltage control, and fault ride-through capability. Additionally, audits validate performance and reliability, confirming that turbines generate electricity efficiently and in accordance with contractual guarantees such as power purchase agreements (PPAs). Finally, they support legal and financial accountability, helping operators demonstrate compliance for insurance coverage, warranties, and investor confidence.

Regulatory audits are usually conducted by independent third-party assessors, certification bodies, or government agencies. The frequency of these audits can vary depending on regional legislation and project size but typically occurs annually or biennially. In certain cases, audits may also be triggered by significant events such as equipment upgrades, repowering, or incidents affecting turbine integrity. Through these recurring audits, wind farm operators maintain transparency, uphold regulatory standards, and ensure that their assets continue to operate safely, efficiently, and sustainably.

International compliance standards establish the foundation for safe, reliable, and environmentally responsible wind turbine design, construction, and operation. These standards ensure consistency across global markets, allowing turbines to be certified and operated under recognised technical, safety, and quality frameworks.

One of the most important sets of international regulations is the IEC 61400 series, developed by the *International Electrotechnical Commission (IEC)*. These standards define the global benchmarks for wind turbine performance, design, and safety. Key components include IEC 61400-1, which specifies design requirements for mechanical integrity and load calculations; IEC 61400-22, which outlines conformity testing and certification for turbines and major components; IEC 61400-24, which establishes lightning protection standards; and IEC 61400-25, which governs communication and data exchange between turbine control systems. Compliance with these IEC standards is often mandatory for certification and international project approvals.

In addition to the IEC framework, broader international standards such as ISO 9001, ISO 14001, and ISO 45001 are applied across the wind industry. These cover quality management, environmental management, and occupational health and safety, respectively. Together, they ensure that wind turbine manufacturers and operators maintain robust management systems that are auditable, transparent, and aligned with continuous improvement principles.

Independent certification bodies such as UL Solutions and DNV also play a key role in validating compliance. These organisations conduct audits and certification processes to confirm turbine safety, reliability, and environmental performance. Their assessments are widely recognised across regions, particularly in Europe, North America, and emerging renewable energy markets.

Overall, international compliance standards create a unified framework that supports technical excellence, operational safety, and environmental stewardship in wind energy projects worldwide.

Regional Compliance Frameworks

Europe

Europe has one of the most comprehensive regulatory frameworks for wind energy:

- The European Union Renewable Energy Directive (RED II) governs renewable energy targets and sustainability reporting.
- Member states implement national permitting and audit requirements, often managed by environmental agencies and grid operators.
- Regular Health, Safety, and Environment (HSE) audits are mandated for both onshore and offshore operations.
- Independent certification (from bodies like TÜV, DNV, or SGS) is required to validate compliance with IEC and EU standards.

United States

In the U.S., compliance is guided by:

- The Federal Energy Regulatory Commission (FERC) and North American Electric Reliability Corporation (NERC), which enforce grid reliability and interconnection standards.
- Occupational Safety and Health Administration (OSHA) oversees worker safety compliance.
- State environmental agencies regulate noise, wildlife protection, and land use permits.

- Certification is often carried out under UL and AWEA (American Clean Power Association) frameworks.

Australia

In Australia:

- Compliance falls under Clean Energy Regulator (CER) and Clean Energy Council (CEC) guidelines, ensuring projects meet renewable energy targets and technical standards.
- Turbines must meet AS/NZS standards for electrical and mechanical safety.
- Environmental audits are conducted in line with state-based EPA regulations and biodiversity protection acts.
- Offshore projects are audited under the Offshore Electricity Infrastructure Act 2021.

Asia-Pacific

Countries such as China, Japan, and India have rapidly developing regulatory systems:

- China enforces GB/T national standards, aligned with IEC guidelines, through agencies like CNCA (Certification and Accreditation Administration).
- Japan follows JET (Japan Electrical Safety & Environment Technology Laboratories) for turbine certification and seismic compliance.
- India's Central Electricity Authority (CEA) and National Institute of Wind Energy (NIWE) regulate turbine testing, certification, and performance validation.

Middle East & Africa

In emerging wind markets such as Saudi Arabia, South Africa, and Morocco, compliance frameworks are modelled on European IEC and DNV standards. Local energy regulators (e.g., NERSA in South Africa) audit projects for environmental and grid compliance.

Key Audit Procedures

During a regulatory audit, wind farm operators undergo a comprehensive examination of both documentation and physical systems to verify compliance with technical, environmental, and safety standards. Auditors, who may represent government agencies, certification bodies, or independent assessors, evaluate the turbine's performance and management systems to ensure that all operations align with international and national regulations. The process is designed not only to identify compliance gaps but also to improve operational reliability and sustainability across the wind farm.

A key component of the audit involves the review of technical documentation. Inspectors examine design drawings, engineering specifications, and certification reports to verify that each turbine meets approved design and construction standards, such as those outlined in IEC 61400 or national equivalents. Maintenance records, inspection logs, and SCADA (Supervisory Control and Data Acquisition) data are also scrutinized to confirm that regular maintenance, fault rectification, and performance monitoring are being conducted as required. SCADA data, in particular, helps auditors verify operational uptime, performance consistency, and incident reporting accuracy over time.

The next phase involves operational inspections, where auditors physically assess turbine components such as nacelles, blades, transformers, and substations. These inspections check for mechanical integrity, corrosion, wear, and any visible damage that may compromise performance or safety. Inspectors may climb turbines or use drones to assess blade condition, checking for cracks, erosion, or lightning damage. Electrical systems, including transformers and switchgear, are examined for proper insulation, grounding, and overload protection.

Attention is also given to safety systems, as they are essential for protecting personnel and equipment. Inspectors verify that emergency stop functions are fully operational and accessible, that fire suppression systems (such as extinguishers, alarms, or automatic systems) are correctly maintained, and that lightning protection networks are functional and properly bonded. Additionally, fall-arrest systems and personal protective equipment (PPE) protocols are checked to ensure technicians can safely access towers and nacelles in accordance with occupational safety standards.

Environmental compliance is another critical focus area. Auditors review environmental data such as noise level measurements, bird and bat collision monitoring reports, and documentation related to spill prevention and hazardous material handling. This ensures that the wind farm adheres to environmental permits and local biodiversity protection laws. For offshore installations, audits also examine oil spill prevention systems, waste disposal procedures, and the potential impact on marine ecosystems.

Finally, auditors assess grid performance to ensure that turbines are operating within grid code requirements. This includes checking compliance with frequency response, voltage stability, fault ride-through capabilities, and reactive power control. These parameters confirm that the wind farm contributes safely and reliably to the wider electrical network without causing disturbances or power quality issues.

At the conclusion of the audit, inspectors compile a detailed compliance report summarizing findings, including any non-conformances or areas for improvement. If deficiencies are identified, a corrective action plan is issued, specifying measures that must be taken and setting deadlines for completion. Follow-up inspections or documentation reviews may be required to verify that corrective actions have been properly implemented. This structured process ensures that wind farms maintain continuous improvement, uphold regulatory standards, and operate safely and efficiently throughout their service life.

Importance of Ongoing Compliance

Maintaining continuous compliance in wind turbine operations is not simply a regulatory obligation — it is a cornerstone of operational, financial, and reputational sustainability. Wind farms operate within a tightly regulated framework that governs safety, environmental protection, and grid performance. Failing to maintain compliance can have far-reaching consequences, from operational shutdowns to financial and legal liabilities. Regular adherence to regulatory requirements ensures that wind farms can operate safely, maintain stakeholder confidence, and contribute reliably to renewable energy generation.

One of the most immediate reasons for maintaining compliance is operational licensing. Wind turbines and farms must hold valid permits issued by national or regional authorities, often subject to periodic review. If a site fails to meet technical, environmental, or safety obligations, regulators may suspend or revoke its operational license. This can result in turbines being taken offline until deficiencies are corrected, leading to lost production and revenue. In extreme cases, repeated non-compliance can result in permanent closure or penalties. Therefore, keeping up with audit schedules, implementing corrective actions, and documenting all maintenance and safety activities are essential to sustaining legal authorization to operate.

Compliance also plays a critical role in insurance and warranty validity. Turbine manufacturers and insurers require operators to maintain up-to-date inspection, maintenance, and audit records to ensure that the equipment is being operated within specified parameters. Failure to meet these obligations can void warranties or insurance coverage, leaving operators financially exposed in the event of component failure, environmental incidents, or accidents. For example, if a gearbox failure occurs after missed scheduled maintenance, the insurer or manufacturer may refuse to cover repair or replacement costs. Regular audits, therefore, act as proof that the wind farm adheres to its operational and safety commitments.

From a financial perspective, investors and lenders depend on compliance as part of their due diligence and risk management processes. Regulatory audits and certification reports serve as assurance that a wind farm is performing within acceptable limits and that its revenue generation is sustainable. Non-compliance or unresolved audit findings can jeopardize financing agreements, delay funding, or trigger breach-of-contract clauses with investors. In contrast, demonstrating a consistent record of compliance enhances investor confidence, facilitates access to capital for future expansion, and strengthens the project's long-term financial stability.

Finally, maintaining compliance supports public trust and Environmental, Social, and Governance (ESG) reporting. As renewable energy projects are increasingly scrutinized for their sustainability credentials, transparency in regulatory adherence reinforces a company's reputation for responsible operation. Publicly disclosing compliance with environmental and safety standards helps demonstrate commitment to sustainability goals, worker safety, and community engagement. In the context of ESG reporting, this transparency is not just beneficial — it is often required by corporate governance frameworks and sustainability rating agencies.

Communication with Stakeholders

Effective communication with stakeholders is a core responsibility of wind turbine technicians and is essential for ensuring safe, efficient, and transparent operations within a wind farm. Because technicians work at the intersection of technical maintenance, safety compliance, and operational performance, their ability to convey information accurately and professionally has a direct impact on decision-making, maintenance planning, and overall project success. Stakeholder communication encompasses a wide range of interactions, including coordination with supervisors, engineers, contractors, control centre operators, regulatory authorities, and even local communities.

Within the wind farm team, technicians must maintain clear communication with their supervisors, maintenance planners, and fellow team members. Before starting work, they participate in toolbox talks, handover briefings, and safety meetings to discuss current weather conditions, operational status, and any hazards or outstanding maintenance tasks. During maintenance or troubleshooting, technicians provide live updates via radio or digital communication platforms to ensure that others are aware of their status — particularly important when lockout/tagout (LOTO) procedures are active. Clear, consistent internal communication helps prevent misunderstandings, ensures safety compliance, and supports seamless workflow transitions between shifts or teams.

Wind turbine technicians frequently interact with remote operations or control centres (ROCs/NCCs) that monitor turbine performance in real time. These centres oversee turbine operation across multiple sites and rely on technicians in the field to verify, diagnose, or resolve issues reported through SCADA systems. Communication between the field and control centre must be precise and timely. Technicians report turbine status, confirm system resets, or provide details after performing fault rectifications. They also coordinate remote restarts or shutdowns, ensuring that safety protocols are followed at both ends. Miscommunication in this context can result in operational errors or safety risks, so technicians use standardized terminology and structured communication formats such as SBAR (Situation, Background, Assessment, Recommendation).

After completing maintenance, inspection, or repair work, technicians are responsible for producing detailed written reports. These include findings, measurements, torque values, test results, and photographs where applicable. Reports are entered into the Computerized Maintenance Management System (CMMS) or digital work order platform, ensuring traceability and compliance with quality standards. Communication with engineers often involves describing symptoms, discussing possible root causes, and recommending corrective actions. Accurate technical communication allows engineers to analyse recurring issues, update maintenance procedures, and plan component replacements. Managers also rely on technician feedback to schedule maintenance, manage budgets, and coordinate supply chain logistics.

Technicians often collaborate with external service providers such as crane operators, blade repair teams, high-voltage specialists, and condition monitoring experts. In these situations, clear coordination is vital to maintain safe work practices and efficient task execution. Technicians communicate site access requirements, safety protocols, and operational constraints. They may also act as points of contact between the site management team and third-party contractors, ensuring that work is completed according to technical specifications and within agreed timeframes.

During audits or inspections by regulatory authorities, technicians may be asked to provide operational records, maintenance documentation, or firsthand explanations of site procedures. Their ability to clearly communicate maintenance history, safety practices, and operational data helps demonstrate compliance with local and international standards such as ISO, IEC, or occupational health and safety regulations.

In some cases, technicians may engage indirectly with local communities, particularly during site visits, outreach events, or when addressing public concerns about noise, visual impact, or environmental protection. Maintaining professionalism and transparency in such interactions helps foster public trust and supports the social licence to operate — an important factor in renewable energy development.

Technicians use a combination of communication tools, including handheld radios, mobile phones, digital tablets, and specialized software applications. SCADA interfaces, CMMS dashboards, and work order tracking systems serve as central communication hubs between technicians, engineers, and management. Increasingly, augmented reality (AR) and remote support platforms are being used to enhance real-time collaboration between field technicians and remote experts.

Strong communication skills are as critical as technical proficiency. Technicians must be able to translate complex technical findings into clear, concise, and accurate messages suited to different audiences — from engineers requiring technical precision to managers or regulators who need concise summaries. They must also be capable of active listening, situational awareness, and non-verbal communication, particularly during safety-critical operations.

Communicating with property owners and stakeholders who oppose the installation of wind farms on or near their land requires empathy, professionalism, and a well-balanced understanding of both the technical and social dimensions of renewable energy development. Wind energy professionals — including technicians, site representatives, and project managers — may encounter individuals or communities who express frustration, distrust, or opposition to proposed projects. The key is to engage respectfully, listen actively, and respond with factual, transparent information rather than defensiveness or technical jargon. The goal is to build trust, reduce tension, and ensure that stakeholders feel acknowledged and respected, even if they ultimately remain opposed to the project.

When meeting with concerned or resistant stakeholders, the first and most important step is to listen without interruption. Allow them to express their views fully, whether their concerns relate to noise, visual impact, property values, environmental effects, or community consultation. People are more likely to engage constructively when they feel heard and understood. Maintain open body language, steady eye contact, and a calm, courteous tone. Avoid interrupting or responding with technical arguments too early. Phrases such as:

- "I understand why you'd be concerned about that."
- "That's a valid point — let's talk through it together."

help demonstrate empathy and prevent discussions from becoming confrontational.

Emotions can run high when people feel that their property, lifestyle, or local environment is being affected. Stakeholders may express anger or distrust toward the project or its representatives. It is essential for wind energy professionals to remain calm, respectful, and neutral, even under pressure. Avoid raising your voice, contradicting forcefully, or becoming defensive. Maintaining a composed, professional demeanour helps de-escalate tension and reflects the organisation's commitment to respectful community engagement. If a discussion becomes overly heated, politely suggest rescheduling with a community liaison officer or project manager present to ensure concerns are properly addressed.

Each stakeholder's objections are shaped by personal, economic, and environmental factors. Common concerns include:

- **Land and property impacts:** Worries about property values, visual intrusion, or restricted land use.
- **Noise and shadow flicker:** Concerns about turbine noise levels or flickering shadows affecting homes and businesses.
- **Environmental and wildlife impacts:** Fear that turbines may harm birds, bats, or natural landscapes.
- **Community consultation and trust:** Scepticism about whether the project developers are transparent or responsive to community input.

Understanding the root of these concerns allows for more effective, tailored communication. Acknowledge their perspective and explain — clearly and factually — what measures are in place to minimise impacts, such as setback distances, environmental monitoring, or visual screening options.

Avoid using overly technical or scientific terms that can confuse or alienate people. Instead, explain complex concepts in plain, relatable language. For example, instead of saying,

"The turbine emits 108 dB(A) at hub height, which attenuates exponentially with distance,"

you could say,

"By the time the sound reaches your property line, it's about the same level as the background noise from a household refrigerator."

Using relatable comparisons helps stakeholders better understand the information and reduces the perception that technical language is being used to obscure the facts.

Transparency is essential to building trust. When property owners question noise levels, environmental effects, or visual impact, provide them with real data and credible, third-party reports — such as environmental impact assessments, acoustic studies, or government approvals. Offer to show where they can independently verify information or invite them to visit an operational wind farm to see the technology in action. Openness and factual communication demonstrate accountability and help dispel misinformation or exaggerated claims circulating in the community.

Some questions — particularly those involving planning approvals, compensation, or land access agreements — may be beyond the authority of technicians or field staff. In these cases, refer the stakeholder respectfully to the appropriate contact, such as a community engagement manager, project liaison officer, or legal representative. You might say:

"I'd like to make sure your concern is addressed by the right person — would you like me to connect you with our community liaison officer who handles these matters?"

This ensures accurate communication, prevents misinformation, and reassures stakeholders that their issue will receive proper attention.

Once the conversation becomes more balanced, outline — without exaggeration — the local and regional benefits of the project. These may include job creation, road and infrastructure improvements, land lease payments, or community investment funds that support local schools or environmental initiatives. However, it is important to remain factual rather than persuasive. The purpose is not to "sell" the project but to provide a clear understanding of its potential contributions to the community.

Trust is built through consistency and reliability. If you promise to provide more information, raise an issue internally, or arrange a follow-up meeting, ensure it is done promptly. Keep communication channels open — through email, phone, or scheduled visits — to show ongoing accountability. Even small actions, such as providing updates or sharing documents, demonstrate respect and commitment to resolving concerns.

Wind farms typically operate for 20 years or more, making nearby residents and property owners long-term neighbours. Establishing constructive, respectful relationships early in the project lifecycle helps prevent conflicts later. Site representatives and technicians should view each interaction as an opportunity to strengthen cooperation, trust, and mutual understanding. Simple gestures — greeting residents courteously, being visible in the community, and responding promptly to inquiries — contribute to a culture of respect and transparency.

Chapter 18

Career Development

The wind energy industry is not only a symbol of global progress toward clean power but also one of the most dynamic and rapidly expanding employment sectors in the modern world. As nations commit to ambitious net-zero targets and invest in large-scale renewable infrastructure, the demand for skilled professionals across all aspects of the wind energy lifecycle continues to surge. Chapter 18, *Career Development*, examines how individuals can build, advance, and sustain rewarding careers within this growing industry — from entry-level technical roles to advanced engineering, supervisory, and leadership positions. It highlights how continuing education, professional development, and ethical practice form the cornerstone of long-term success in this innovative and sustainability-driven field.

This chapter begins by exploring the multidisciplinary nature of wind energy careers, where collaboration between engineers, technicians, environmental scientists, planners, financiers, and policy specialists ensures that projects move seamlessly from conception to operation. Each stage of a wind farm — from site selection and environmental assessment to construction, operation, and maintenance — requires diverse expertise and cooperation across technical, environmental, social, and economic domains. Readers will discover how global wind projects create a wide spectrum of career opportunities, not only in engineering and technical trades but also in logistics, stakeholder engagement, health and safety, and environmental compliance.

The chapter then focuses on Continuing Education and Upskilling, a critical aspect of maintaining competence in a fast-evolving technological environment. Wind energy professionals must adapt continuously as turbines become larger, smarter, and more integrated with digital monitoring systems and hybrid energy networks. Lifelong learning ensures that technicians and engineers remain proficient in areas such as high-voltage systems, SCADA data analysis, predictive maintenance, and offshore operations. International standards established by the Global Wind Organisation (GWO) and national vocational frameworks underpin global mobility and professional recognition, enabling workers to apply their skills across borders. This section also discusses how employers, training institutions, and governments collaborate to create structured pathways for professional growth, including advanced certifications, mentorship programs, and industry scholarships.

Next, the chapter addresses Leadership and Supervisory Roles, which are vital for maintaining safety, efficiency, and team cohesion in complex operational environments. Effective leaders in the wind industry combine deep technical knowledge with communication, decision-making, and organisational skills to manage diverse teams and ensure compliance with safety and quality standards. The text explores how leadership roles develop through hands-on experience, formal training, and mentorship — showing how many supervisors and managers begin their careers as technicians and progress through consistent performance and leadership development. These leaders play a crucial role in fostering a culture of safety, accountability, and continuous improvement across all wind farm operations.

The final section explores Sustainability and Ethical Considerations, emphasising how modern wind energy careers are inseparable from environmental and social responsibility. Technicians, engineers, and managers alike contribute not only to producing clean power but also to ensuring that renewable energy development aligns with ethical labour practices, environmental protection, and community engagement. As the renewable energy transition accelerates, employers increasingly value professionals who demonstrate integrity, environmental awareness, and a commitment to fair and transparent work practices. Ethical conduct — from maintaining accurate maintenance records to respecting community stakeholders — underpins the industry's reputation and long-term social licence to operate.

Throughout *Chapter 18: Career Development*, readers will gain a comprehensive understanding of the human dimension of the wind energy industry. Beyond turbines, data systems, and technology, it is people — their skills, values, and dedication — that drive sustainable progress. The chapter underscores that a successful career in wind energy depends not only on technical competence but also on a lifelong commitment to learning, leadership, and ethical practice. In doing so, it presents the wind industry not just as a field of employment, but as a global movement — empowering individuals to contribute to a cleaner, safer, and more equitable energy future.

Continuing Education and Upskilling

The development, construction, and operation of wind farms around the world represent one of the most complex and multidisciplinary undertakings in modern energy infrastructure. While engineers play a central role in the technical design and specification of turbines, electrical systems, and site layouts, they are only one part of a vast professional network. The global wind energy industry relies on planners, environmental scientists, stakeholder engagement specialists, financial experts, legal advisors, and skilled tradespeople — each contributing essential expertise to bring a wind farm from concept to operation.

Across every region — from Europe and North America to Asia, Africa, and Oceania — wind energy projects begin long before construction. Engineers work alongside urban and regional planners, environmental scientists, and geospatial analysts to identify sites with strong, consistent wind resources, minimal ecological impact, and feasible access to transmission infrastructure. Environmental consultants conduct biodiversity assessments, avian and bat impact studies, noise modelling, and visual impact analyses to ensure projects comply with environmental regulations and international best practices.

Simultaneously, stakeholder engagement officers and social impact specialists work with local communities, landowners, and traditional custodians to gain consent and address concerns about noise, visual intrusion, or land use. Lawyers and policy experts navigate land rights, planning approvals, and power purchase agreements (PPAs), while financial analysts and investment managers secure funding through green bonds, institutional investors, or international development banks. In emerging markets, this stage often involves collaboration with organisations like the World Bank, International Finance Corporation (IFC), or Asian Development Bank (ADB) to ensure compliance with global sustainability frameworks.

This multidisciplinary collaboration ensures that wind farms meet both technical feasibility and social licence to operate — two equally vital pillars of modern renewable energy projects.

Once all approvals are secured and financing is confirmed, a wind farm project moves into the construction phase, which typically lasts between 12 and 24 months depending on site size, terrain, and logistics. During this stage, the focus shifts from design and planning to heavy civil and mechanical activity.

Access roads are built to accommodate oversized vehicles carrying massive turbine components such as towers, blades, and nacelles. Foundations — often involving reinforced concrete slabs weighing hundreds of tonnes — are laid to anchor each turbine. The construction process mobilises hundreds of skilled tradespeople and operators, including:

- Civil engineers and construction workers for earthworks and concrete foundations.
- Mechanical and electrical trades for turbine assembly, cabling, and commissioning.
- Crane operators, truck drivers, and riggers to transport and position turbine components.
- Health, safety, and environmental officers (HSE) to oversee compliance during construction.

This phase also supports local and regional employment — a key economic benefit of renewable energy investment. In many countries, wind projects boost rural economies by creating temporary construction jobs and long-term technical roles that remain after completion.

Once a wind farm is commissioned, a smaller but highly skilled workforce takes over to manage ongoing operation and maintenance (O&M) activities. Globally, O&M roles are in high demand due to the rapid expansion of wind capacity and the ageing of existing fleets.

A typical operational team includes a mix of degree-qualified engineers and vocationally trained technicians, such as:

Degreed roles:

- SCADA Engineers – monitor data and performance remotely using Supervisory Control and Data Acquisition systems.
- Reliability Engineers – analyse failure data to improve turbine availability and efficiency.
- Blade Engineers – oversee aerodynamic performance, structural integrity, and repair programs.

Vocational or trade-qualified roles:

- Wind Turbine Technicians (Electrical or Mechanical) – perform inspections, maintenance, and repairs.
- Blade Technicians – specialise in composite materials, blade inspection, and structural repairs.
- High-Voltage (HV) Operators and Cable Jointers – maintain electrical substations and underground cabling.
- Linespersons – maintain overhead transmission connections.
- Site Supervisors or Managers – coordinate field teams and manage safety compliance.
- Control Room Operators – oversee turbine performance and coordinate maintenance responses.

Globally, there is a shortage of qualified technicians, making this one of the most in-demand technical professions in renewable energy.

Becoming a wind turbine technician or engineer requires both technical competence and a strong safety foundation. Candidates often come from electrical, mechanical, or industrial trades, or have experience in related industries such as mining, oil and gas, automotive, or aerospace. To transition into the wind sector, they complete specialist training under the Global Wind Organisation (GWO), which sets international safety and technical standards.

Typical mandatory certifications include:

- Construction industry white card or equivalent safety induction.
- High-risk work licences (e.g., dogging, rigging, crane operation).
- Working at heights and confined space entry training.
- First aid and emergency response training.
- GWO Basic Safety Training (BST) and GWO Technical Training (BTT) modules.

Some technicians also complete IRATA rope access certification to perform blade inspections and repairs on suspended lines — particularly for offshore and tall onshore turbines.

The global nature of the wind industry creates opportunities for international mobility. Blade technicians, for example, often follow seasonal repair cycles between hemispheres — working in Europe or North America during their summer, then moving to Australia, South Africa, or Latin America for the southern summer season. This makes the role not only technically demanding but also highly adventurous and globally connected.

Globally, the International Renewable Energy Agency (IRENA) estimates that more than 1.4 million people are employed in the wind sector, with strong growth projected as countries pursue net-zero energy targets. Technicians and engineers represent the fastest-growing job categories, especially in regions expanding their renewable infrastructure such as Asia-Pacific, North America, and sub-Saharan Africa.

Career progression can lead from entry-level technician to lead technician, site supervisor, asset manager, or technical specialist roles. For those with advanced analytical or management skills, pathways extend into SCADA system analysis, project engineering, HSE management, or renewable energy consultancy.

In every country where wind energy operates — from the offshore farms of Denmark and the UK to the vast onshore arrays of Texas, China, and Australia — successful projects depend on a diverse, global workforce. Engineers provide the technical backbone, but planners, financiers, environmental experts, and tradespeople all contribute essential expertise.

The construction of a wind farm typically spans 18 months, employing hundreds of skilled workers and mobilising sophisticated machinery. Once operational, highly trained technicians and engineers ensure reliability, safety, and efficiency over decades of service life. With increasing global investment in renewable energy and the continued retirement of fossil fuel infrastructure, demand for skilled professionals in every part of the wind energy lifecycle — from planning to maintenance — will continue to grow worldwide.

Wind energy, therefore, is not just about turbines and technology — it's about people: a global network of experts, trades, and technicians working together to power a sustainable future.

Continuing education and upskilling are essential components of a successful and sustainable career in the global wind energy industry. As technology evolves rapidly and turbines grow in size, complexity, and automation, technicians, engineers, and other professionals must continually enhance their knowledge and technical capabilities to remain competent, compliant, and employable. The industry's strong focus on safety, innovation, and efficiency means that lifelong learning is not just encouraged — it is an operational requirement for anyone working on or around wind turbines.

The wind energy sector is dynamic and heavily technology-driven. Modern turbines are now integrated with advanced control systems, SCADA platforms, predictive maintenance algorithms, and smart sensors that require a deeper understanding of electronics, data analysis, and digital communication networks than in previous decades. Because of this, even experienced technicians and engineers must engage in regular training to stay current with new systems, standards, and regulatory requirements.

Continuing education ensures that workers understand evolving safety protocols, environmental regulations, and technical advancements, such as hybrid renewable systems, battery storage integration, and offshore wind technology. It also helps professionals maintain compliance with national and international certifications — a vital requirement for operating in highly regulated environments where safety and quality are paramount.

The global wind industry has adopted internationally recognised frameworks to support lifelong learning and skills renewal. The Global Wind Organisation (GWO) sets the benchmark for safety and technical competence training across the industry. GWO courses must be refreshed at regular intervals, usually every two years, ensuring all personnel remain competent in working at heights, first aid, fire awareness, manual handling, and other critical safety procedures.

Beyond GWO, organisations such as DNV, IEC (International Electrotechnical Commission), and ISO contribute to technical training and professional certification programs that ensure workers align with global standards. In addition, national frameworks — such as TAFE and Registered Training Organisations (RTOs) in Australia, community colleges in the United States, and technical institutes across Europe and Asia — offer tailored vocational courses for wind energy careers.

These frameworks ensure that a technician trained in one country can apply their skills internationally, supporting global workforce mobility.

For wind turbine technicians, upskilling focuses on improving technical proficiency, troubleshooting ability, and safety performance. Common areas for continuing education include:

- **Advanced electrical and mechanical systems:** Training in programmable logic controllers (PLCs), fault diagnostics, and high-voltage switching.
- **Hydraulic and lubrication systems:** Understanding of fluid power, filtration, and maintenance best practices.
- **Blade inspection and repair:** Advanced courses in composite materials, aerodynamics, and resin application.
- **Rope access and rescue training:** Certifications such as IRATA or SPRAT for technicians performing elevated or offshore maintenance.
- **Digital competencies:** Using SCADA systems, data analytics tools, and condition monitoring systems to interpret turbine performance data.

Upskilling in these areas not only enhances technical capability but also allows technicians to take on **lead or supervisory roles**, increasing job security and earning potential.

For engineers and managers, continuing education focuses on innovation, systems integration, and leadership. As wind projects become larger and more interconnected, engineers are required to understand not only mechanical and electrical systems but also grid integration, cybersecurity, project management, and sustainability principles.

Continuing education opportunities for these professionals may include:

- Postgraduate degrees or microcredentials in renewable energy systems, data science, or power engineering.
- Project management certifications, such as PRINCE2, PMP, or Agile frameworks for managing large-scale renewable projects.
- Courses in asset management and reliability engineering, focusing on lifecycle cost optimisation and predictive maintenance.
- Leadership and communication training to manage multidisciplinary teams and community stakeholders effectively.

Many universities now offer online or blended-learning programs tailored for working professionals, allowing them to balance ongoing study with full-time employment in the sector.

Employers play a central role in supporting ongoing education and professional development. Leading wind energy companies maintain in-house training academies and partnerships with vocational institutions and universities to deliver accredited programs.

These partnerships ensure that employees receive practical, hands-on learning directly aligned with the company's operational requirements and the latest technological innovations. Some companies also offer mentorship programs, apprenticeships, and rotational placements that help workers broaden their skill sets across different turbine models, technologies, and geographies.

Government programs and industry associations also contribute by funding training grants, scholarships, and apprenticeship incentives to attract and retain a skilled workforce in the renewable energy sector.

In the global wind energy industry, continuing education is directly linked to career advancement. Technicians who gain additional certifications can progress to senior technician, lead engineer, or site supervisor positions. Similarly, engineers who pursue advanced studies or management training can transition into roles such as asset manager, project director, or renewable energy consultant.

Continuous upskilling also allows professionals to move horizontally across related fields such as offshore wind, solar, or energy storage, expanding career versatility in a rapidly growing clean energy economy.

Wind energy technologies are evolving faster than ever — with innovations in aerodynamics, digital twins, AI-driven predictive maintenance, and floating offshore platforms changing how turbines are designed and maintained. To remain effective, professionals must commit to lifelong learning as an integral part of their careers.

Online learning platforms, virtual reality (VR) simulators, and digital twins are increasingly being used for remote training and competency assessment, making continuing education more accessible and realistic. These tools allow workers to practice high-risk procedures in virtual environments before performing them in the field, reducing accidents and improving confidence.

Leadership and Supervisory Roles

Leadership and supervisory roles in the wind energy sector are essential to ensuring that wind farms operate safely, efficiently, and sustainably. These positions form the bridge between corporate management and on-site operations, blending technical expertise with strong interpersonal and organisational skills. Effective leaders in this industry not only direct daily activities but also motivate their teams, uphold safety and quality standards, and drive continuous improvement throughout all stages of a wind farm's life cycle — from construction and commissioning to maintenance and performance optimisation.

The wind energy sector operates within high-risk environments that involve heavy machinery, complex electrical systems, and exposure to challenging weather conditions. In such settings, leadership is critical to maintaining safety, productivity, and morale among field teams. Supervisors and site managers ensure that all operations align with company policies, regulatory frameworks, and international best practices.

Strong leadership supports safety compliance by enforcing safe work procedures, conducting regular briefings, and ensuring risk controls are applied in every task. It improves operational efficiency by coordinating personnel and resources to complete maintenance and construction activities on time and within budget. Effective leaders also maintain the technical integrity of turbines, substations, and grid systems, ensuring they meet performance targets and maintenance schedules. Finally, they promote team cohesion by fostering open communication, trust, and a shared commitment to excellence. In this way, leadership in the wind sector is about balancing technical decision-making with people management — leading by example in environments that demand precision, discipline, and teamwork.

Wind farms operate under a structured leadership hierarchy to manage technical, safety, and administrative responsibilities. Common supervisory roles include the Site Supervisor or Team Leader, who oversees daily maintenance, inspections, and technician assignments, ensuring work meets company and manufacturer standards. The Site or Operations Manager is responsible for the broader coordination of workforce management, logistics, budget control, and communication with corporate management.

The Health, Safety, and Environment (HSE) Coordinator leads site-wide safety programs, audits, and incident investigations, maintaining compliance with local and international regulations. The Commissioning Lead manages turbine testing and final acceptance, confirming that systems meet operational specifications before entering service. Meanwhile, the Control Room or Operations Centre Supervisor monitors turbine performance through SCADA systems, coordinates troubleshooting, and dispatches field crews when faults occur.

Each of these positions demands not only strong technical expertise but also organisational, analytical, and leadership skills to ensure efficient and safe operations.

Supervisory leaders in wind energy are accountable for both operational results and team wellbeing. Their responsibilities include planning and scheduling maintenance or construction activities, allocating resources, and managing logistics for large equipment such as cranes and transport vehicles. They monitor turbine efficiency and fault trends using SCADA data and analytics to optimise performance.

Leaders also act as mentors, guiding less experienced technicians, conducting toolbox meetings, and promoting skills development. Safety leadership is another central duty — supervisors enforce permit-to-work systems, conduct risk assessments, and lead emergency drills. They maintain communication with contractors, grid operators, landowners, and internal departments to ensure operational harmony. When unexpected issues arise, such as mechanical failures or weather delays, effective leaders respond with calm, informed decision-making.

Essential leadership competencies include strong communication, empathy, decision-making under pressure, technical understanding, and adaptability — qualities that are vital in a fast-paced and evolving industry.

Many leaders in the wind industry begin their careers as field technicians. Progressing into a supervisory role requires both technical mastery and the development of leadership skills. Common pathways include internal promotions based on consistent performance and initiative, participation in formal leadership or management training (such as Certificate IV in Leadership and Management in Australia), and mentorship programs where senior supervisors guide emerging leaders in communication and project coordination.

Technicians often prepare for leadership roles by taking on additional responsibilities, such as coordinating shift work, reviewing documentation, or overseeing team safety procedures. Over time, these experiences build the foundation for effective supervisory leadership.

One of the most critical aspects of leadership in wind energy is the ability to cultivate a strong safety culture. Given that turbine maintenance involves working at heights, handling high-voltage systems, and operating in remote or offshore environments, leaders must model and reinforce safe behaviours at all times.

Effective safety leaders hold daily pre-start meetings and toolbox talks to review hazards and controls, encourage open communication where workers can report risks without fear of blame, and demonstrate commitment by following safety procedures themselves. They also promote a “stop work authority” culture — empowering anyone on site to halt unsafe activities immediately. Such leadership fosters both safety and trust, laying the foundation for a resilient, high-performing team.

While leadership principles in the wind industry are universal — focusing on accountability, safety, and performance — cultural and regional variations shape how they are applied. In Europe, leadership often centres on collaboration, regulatory compliance, and environmental stewardship. In North America, supervisors may emphasise innovation, productivity, and data-driven management. In Asia, leadership tends to follow hierarchical structures but increasingly values communication and skill development. Meanwhile, in Australia and New Zealand, leadership prioritises teamwork, safety, and adaptability in remote and challenging environments.

As the wind energy industry becomes increasingly globalised, leaders must also manage multinational teams and navigate cultural differences. This requires flexibility, cross-cultural awareness, and excellent communication skills to build cohesive, high-performing international teams.

The future of leadership in wind energy will be shaped by digitalisation, sustainability, and global collaboration. Supervisors are increasingly expected to combine traditional leadership skills with digital proficiency — using tools such as predictive analytics, AI-based maintenance systems, and digital twins to make informed operational decisions.

In addition, the global shift toward decarbonisation and sustainability places new demands on leaders to integrate environmental, social, and governance (ESG) principles into operations. Future leaders will not only manage turbines but also serve as ambassadors for ethical, inclusive, and community-driven renewable energy development.

Leadership and supervisory roles are vital to maintaining safe, efficient, and sustainable wind energy operations. Leaders coordinate teams, enforce safety protocols, and ensure turbines perform to their full potential. Their responsibilities extend far beyond technical management — they inspire confidence, foster collaboration, and nurture a culture of excellence.

As the global wind industry continues to expand, the demand for skilled, adaptable leaders will grow in parallel. Those who combine technical expertise with empathy, communication, and strategic thinking will drive the next generation of clean energy operations — ensuring that wind farms remain reliable, safe, and productive contributors to a sustainable global future.

Sustainability and Ethical Considerations

Sustainability and ethical considerations are increasingly central to career development and opportunities in the wind energy sector, particularly for wind turbine technicians (windtechs). As the global transition to renewable energy accelerates, technicians are not only maintaining and operating turbines but also becoming key contributors to environmental stewardship, responsible energy production, and community engagement. A career in wind energy therefore demands both technical expertise and an understanding of the broader sustainability principles and ethical responsibilities that define the modern renewable workforce.

Wind Turbine Fundamentals for Technicians

Wind energy is a cornerstone of global sustainability efforts because it provides clean, renewable power with minimal greenhouse gas emissions. However, sustainability in the wind industry extends beyond generating low-carbon electricity — it encompasses the entire lifecycle of a wind farm, from site selection and manufacturing to operation, maintenance, and decommissioning. For windtechs, this means that their daily work contributes directly to reducing the environmental footprint of energy production. Every turbine they service helps displace fossil fuel generation, cut emissions, and advance national and international climate targets. As a result, wind technicians are not just tradespeople — they are sustainability professionals helping to drive the world's energy transition. Career development in this sector is therefore tied to a growing demand for technicians who understand sustainable practices such as efficient resource use, waste minimisation, and responsible maintenance planning that extends turbine life and reduces the need for new materials.

Ethics play a major role in maintaining trust, transparency, and accountability within the renewable energy industry. Windtechs and other professionals must operate according to high ethical standards, as their work directly affects the environment, local communities, and workplace safety. Ethical practice begins with environmental integrity — ensuring that maintenance activities do not cause pollution, habitat disruption, or unnecessary waste. This includes following strict protocols for handling lubricants, hydraulic fluids, and composite materials. Workplace safety and duty of care are equally vital. Ethical technicians always prioritise safety for themselves, their colleagues, and the public, refusing to cut corners even under time pressure. Transparency and honesty are also critical. Technicians must accurately report turbine conditions, maintenance outcomes, and incident data, as falsifying records or ignoring faults can lead to serious operational and ethical consequences. Respect for communities and landowners is another essential principle, as technicians represent both their employer and the renewable energy industry. Their conduct influences local attitudes toward wind power. Finally, fair labour practices — including equitable treatment, diversity, and inclusion — are key to a just and sustainable transition to renewable energy.

Sustainability knowledge is becoming a key differentiator in career progression for wind turbine technicians. Those who understand how their work affects energy efficiency, environmental protection, and resource management are increasingly valued by employers. Developing sustainability-related skills enhances employability across many areas. Technicians who understand the lifecycle of turbine components and support circular economy principles contribute to reducing waste and improving design efficiency. Using SCADA data and diagnostic tools to maximise turbine output while minimising wear reflects an understanding of energy efficiency and performance optimisation. Knowledge of environmental compliance, such as adherence to noise limits, wildlife protection, and waste management laws, further strengthens professional credibility. Applying sustainable maintenance practices — extending component lifetimes and reducing replacement costs — and possessing carbon literacy help technicians align their work with broader national and corporate carbon reduction goals.

A sustainable wind energy sector also depends on the wellbeing and ethical treatment of its workforce. Ethical employment practices ensure that technicians work in safe environments, receive fair wages, and have opportunities for ongoing professional growth. Many employers in the wind industry are committed to corporate social responsibility (CSR) and Environmental, Social, and Governance (ESG) frameworks that promote fairness, inclusion, and worker safety. Technicians can expect structured career development supported by globally recognised training and certification programs, such as GWO-accredited safety and technical courses. Equal opportunity initiatives encourage gender diversity and inclusion, while employee wellbeing programs address the physical and mental demands of remote, high-risk work. Global mobility also provides opportunities for technicians to gain valuable experience across international wind farms. This ethical and sustainable approach ensures that the benefits of renewable energy extend beyond environmental gains to include the people who make the energy transition possible.

The rapid growth of both onshore and offshore wind energy is creating a global surge in demand for skilled and ethically minded professionals. Governments, investors, and corporations are increasingly prioritising sustainability-focused projects that require workers who understand environmental compliance, circular design, and social responsibility. This global trend is expanding opportunities for technicians to specialise in emerging areas such as offshore operations, where environmental and safety standards are especially high. It also includes blade recycling and end-of-life management, which focus on sustainable materials and circular economy solutions. Hybrid renewable systems that combine wind, solar, and energy storage are another growing area, as is environmental monitoring and ESG reporting that connects technical performance to sustainability metrics. Professionals who combine strong technical ability with ethical and environmental awareness are becoming increasingly valuable in this evolving industry.

Community support plays a critical role in the long-term success of wind energy projects, and ethical behaviour is key to maintaining that support. Respectful communication and engagement with local stakeholders, including landowners, traditional custodians, and residents, help build trust and strengthen a project's social licence to operate. Technicians often act as the visible representatives of their companies, especially during site visits and maintenance work. By demonstrating professionalism, cultural awareness, and courtesy, they help foster positive relationships and dispel misconceptions about wind energy. Ethical engagement is therefore a daily practice that shapes public perception and promotes long-term community acceptance of renewable energy.

Sustainability and ethics are not static — they evolve alongside technological progress, new regulations, and shifting public expectations. This means that wind professionals must commit to lifelong learning to stay informed and effective. Continuous professional development in environmental management systems such as ISO 14001, sustainable operations, and corporate ethics can complement technical qualifications and open career pathways into leadership, environmental management, or compliance roles. By continually reflecting on the ethical dimensions of their work and updating their skills, technicians can ensure their careers remain both socially responsible and professionally fulfilling.

Sustainability and ethics form the foundation of a responsible and forward-looking career in the wind energy industry. Wind turbine technicians and other professionals play a direct role in helping to achieve global climate targets while maintaining environmental, social, and operational integrity. Upholding ethical standards — including workplace safety, transparency, and community respect — ensures that renewable energy delivers benefits without compromising human or ecological wellbeing. Embracing sustainability and ethical practice is both a moral duty and a strategic advantage. As the renewable energy transition accelerates worldwide, the most successful wind professionals will be those who combine technical skill with environmental awareness, ethical integrity, and a commitment to building a cleaner, fairer, and more sustainable future.

References

1. Shouman, E.R.M., *Global Prediction of Wind Energy Market Strategy for Electricity Generation.* 2020.
2. Zeng, Z., et al., *A Reversal in Global Terrestrial Stilling and Its Implications for Wind Energy Production.* Nature Climate Change, 2019. **9**(12): p. 979-985.
3. Perera, S.M.H.D., et al., *Wind Energy Harvesting and Conversion Systems: A Technical Review.* Energies, 2022. **15**(24): p. 9299.
4. Dahlke, S., J. Sterling, and C. Meehan, *Policy and Market Drivers for Advancing Clean Energy.* 2019.
5. Shen, C., et al., *Increases of Offshore Wind Potential in a Warming World.* Geophysical Research Letters, 2024. **51**(14).
6. Song, D., et al., *Review on the Application of Artificial Intelligence Methods in the Control and Design of Offshore Wind Power Systems.* Journal of Marine Science and Engineering, 2024. **12**(3): p. 424.
7. Tang, M., Z. Peng, and H. Wu, *Fault Detection for Pitch System of Wind Turbine-Driven Doubly Fed Based on IHHO-LightGBM.* Applied Sciences, 2021. **11**(17): p. 8030.
8. Tuinema, B.W., et al., *Reliability Analysis of Offshore Grids—An Overview of Recent Research.* Wiley Interdisciplinary Reviews Energy and Environment, 2018. **8**(1).
9. Koivisto, M., et al., *North Sea Region Energy System Towards 2050: Integrated Offshore Grid and Sector Coupling Drive Offshore Wind Power Installations.* Wind Energy Science, 2020. **5**(4): p. 1705-1712.
10. Onea, F., et al., *Evaluation of the Wind Energy Potential Along the Mediterranean Sea Coasts.* Energy Exploration & Exploitation, 2016. **34**(5): p. 766-792.
11. Koengkan, M., Y.E.M. Poveda, and J.A. Fuinhas, *Globalisation as a Motor of Renewable Energy Development in Latin America Countries.* Geojournal, 2019. **85**(6): p. 1591-1602.
12. Anangapal, H.B. and V. Kirubakaran, *Towards Sustainable Energy: Integrating ERA5 Data for Offshore Wind Farm Planning in India.* Engineering Research Express, 2024. **6**(3): p. 035570.
13. Wei, W., A. Malekjafarian, and M. Salauddin, *Scour Protection Measures for Offshore Wind Turbines: A Systematic Literature Review on Recent Developments.* Energies, 2024. **17**(5): p. 1068.
14. Tümse, S., et al., *Comparative Analysis of Global Onshore and Offshore Wind Energy Characteristics and Potentials.* Sustainability, 2024. **16**(15): p. 6614.
15. Haghi, A. and M. Rahimi, *Control and Stability Analysis of VSC-HVDC Based Transmission System Connected to Offshore Wind Farm.* Scientia Iranica, 2019. **0**(0): p. 0-0.

16. Barbosa, L.d.S.N.S., et al., *Hydro, Wind and Solar Power as a Base for a 100% Renewable Energy Supply for South and Central America*. Plos One, 2017. **12**(3): p. e0173820.
17. Agudelo, M.S., et al., *Post-Construction Bird and Bat Fatality Monitoring Studies at Wind Energy Projects in Latin America: A Summary and Review*. Heliyon, 2021. **7**(6): p. e07251.
18. Li, J., et al., *A Review on Development of Offshore Wind Energy Conversion System*. International Journal of Energy Research, 2020. **44**(12): p. 9283-9297.
19. Stede, J., M. Blauert, and N. May, *Way Off: The Effect of Minimum Distance Regulation on the Deployment and Cost of Wind Power*. SSRN Electronic Journal, 2021.
20. Weinand, J.M., et al., *The Impact of Public Acceptance on Cost Efficiency and Environmental Sustainability in Decentralized Energy Systems*. Patterns, 2021. **2**(7): p. 100301.
21. Telesetsky, A., *Multilateral Debt Relief for Clean Ocean Energy*. Sustainability, 2023. **15**(20): p. 14702.
22. Beiter, P., et al., *Expert Perspectives on the Wind Plant of the Future*. Wind Energy, 2022. **25**(8): p. 1363-1378.
23. Charhouni, N., A. Arbaoui, and M. Sallaou, *Analysis of Wake Impact on Wind Farm Performance Using Two Analytical Models*. 2014: p. 323-327.
24. Caduff, M., et al., *Wind Power Electricity: The Bigger the Turbine, the Greener the Electricity?* Environmental Science & Technology, 2012. **46**(9): p. 4725-4733.
25. Jonkman, J., et al., *Definition of a 5-Mw Reference Wind Turbine for Offshore System Development*. 2009.
26. González, J.S. and L.A. Roberto, *Technological Evolution of Onshore Wind Turbines—a Market-based Analysis*. Wind Energy, 2016. **19**(12): p. 2171-2187.
27. Muljadi, E. and C.P. Butterfield, *Pitch-Controlled Variable-Speed Wind Turbine Generation*. Ieee Transactions on Industry Applications, 2001. **37**(1): p. 240-246.
28. Fu-xing, L.I., L. Li, and Y. Peng, *Research on Digital Twin and Collaborative Cloud and Edge Computing Applied in Operations and Maintenance in Wind Turbines of Wind Power Farm*. 2021.
29. Haghshenas, A., et al., *Predictive Digital Twin for Offshore Wind Farms*. Energy Informatics, 2023. **6**(1).
30. Pacheco-Blazquez, R., et al., *A Digital Twin for Assessing the Remaining Useful Life of Offshore Wind Turbine Structures*. Journal of Marine Science and Engineering, 2024. **12**(4): p. 573.
31. Vickerman, J., *What is the difference between a megawatt and a megawatt hour?* 2024, Enverus: RatedPower.
32. Keyser, D.J. and S. Tegen, *The Wind Energy Workforce in the United States: Training, Hiring, and Future Needs*. 2019.
33. Brown, J.P., et al., *Ex Post Analysis of Economic Impacts From Wind Power Development in U.S. Counties*. Energy Economics, 2012. **34**(6): p. 1743-1754.
34. Jaraitė, J., A. Karimu, and A. Kažukauskas, *Policy-Induced Expansion of Solar and Wind Power Capacity: Economic Growth and Employment in EU Countries*. The Energy Journal, 2017. **38**(5): p. 197-222.

35. Mikulić, D., Ž. Lovrinčević, and D. Keček, *Economic Effects of Wind Power Plant Deployment on the Croatian Economy.* Energies, 2018. **11**(7): p. 1881.
36. Deksnys, R.P. and A. Stankevičius, *Investigation of Wind Power Generation in Lithuania.* Oil Shale, 2011. **28**(1S): p. 203.
37. Yeh, T.-H. and L. Wang, *Benefit Analysis of Wind Turbine Generators Using Different Economic-Cost Methods.* 2007: p. 1-6.
38. Armeanu, D., G. Vintilă, and Ș.C. Gherghina, *Does Renewable Energy Drive Sustainable Economic Growth? Multivariate Panel Data Evidence for EU-28 Countries.* Energies, 2017. **10**(3): p. 381.
39. Gunn, C., et al., *A Framework for Effective Science Communication and Outreach Strategies and Dissemination of Research Findings for Marine Energy Projects.* Journal of Marine Science and Engineering, 2022. **10**(2): p. 130.
40. Jan, A.A., et al., *Solar Photovoltaic Technology and Its Impact on Environmental, Social and Governance (ESG) Performance: A Review.* Kne Social Sciences, 2023.
41. Victoria, M., et al., *Solar Photovoltaics Is Ready to Power a Sustainable Future.* Joule, 2021. **5**(5): p. 1041-1056.
42. Tusar, M.I.H. and B.R. Sarker, *Maintenance Cost Minimization Models for Offshore Wind Farms: A Systematic and Critical Review.* International Journal of Energy Research, 2021. **46**(4): p. 3739-3765.
43. U.S. Bureau of Labor Statistics, *Wind Turbine Technicians*. 2025, Bureau of Labor Statistics, U.S. Department of Labor,.
44. GEV Wind Power, *Global Technician Demand For 2027 As Wind Industry Continues To Grow*. 2024, GEV Wind Power,.
45. Fluke, *Global wind energy growth driving demand for wind turbine technicians*. 2025.
46. Weather Guard, *The Complete Guide to Wind Turbine Technician Jobs, Salary and More in 2024*. 2023.
47. U.S. Depatement of Energy, *Career Map: Wind Technician*. 2025, U.S. Depatement of Energy,.
48. Yahalom, A., *How Can Wind Turn Into Electricity?* Frontiers for Young Minds, 2023. **11**.
49. Yohana, E., et al., *Experimental Study of Wind Booster Addition for Savonius Vertical Wind Turbine of Two Blades Variations Using Low Wind Speed.* E3s Web of Conferences, 2019. **125**: p. 14003.
50. Paranchis, M.I. and H. Pratama, *The Wind Turbine Design and Test.* Teknomekanik, 2022. **5**(1): p. 13-27.
51. Didane, D.H., et al., *Performance Investigation of a Small Savonius-Darrius Counter-rotating Vertical-axis Wind Turbine.* International Journal of Energy Research, 2019. **44**(12): p. 9309-9316.
52. Liu, C., et al., *A Wind Power Plant With Thermal Energy Storage for Improving the Utilization of Wind Energy.* Energies, 2017. **10**(12): p. 2126.
53. Ortiz, M., J.A.M. Ríos, and M. Acosta, *Wind Generation and Power System Interaction Analysis Using Probabilistic Techniques.* Renewable Energy and Power Quality Journal, 2024. **6**(1).

54. Ruzi, N.A. and N.H. Rosly, *Conceptual Design and Analysis of Small-Scale Wind Turbine for Household Usage in Malaysia.* Progress in Aerospace and Aviation Technology, 2023. **3**(2).
55. Jiang, G.J., et al., *Reliability Prediction Method for Offshore Wind Turbines Based on Time Series and Data Migration Models.* Quality and Reliability Engineering International, 2025. **41**(6): p. 2610-2623.
56. Abhinav, K.A. and N. Saha, *Stochastic Response of Jacket Supported Offshore Wind Turbines for Varying Soil Parameters.* Renewable Energy, 2017. **101**: p. 550-564.
57. Seixas, M., R. Melício, and V.M.F. Mendes, *Simulation by Discrete Mass Modeling of Offshore Wind Turbine System With DC Link.* International Journal of Marine Energy, 2016. **14**: p. 80-100.
58. Ren, Z., et al., *Development and Application of a Simulator for Offshore Wind Turbine Blades Installation.* Ocean Engineering, 2018. **166**: p. 380-395.
59. Verma, A.S., et al., *Impact Assessment of a Wind Turbine Blade Root During an Offshore Mating Process.* Engineering Structures, 2019. **180**: p. 205-222.
60. Hsu, C.-Y., et al., *Study on Dynamic Characteristics of Wind Farm With Different Wind Turbine Types.* 2015.
61. Zhang, Z., et al., *Distributed Model Predictive Control of All-dc Offshore Wind Farm for Short-term Frequency Support.* Iet Renewable Power Generation, 2022. **17**(2): p. 458-479.
62. Hsu, W.K. and C.-K. Yeh, *Offshore Wind Potential of West Central Taiwan: A Case Study.* Energies, 2021. **14**(12): p. 3702.
63. Kim, H.G., et al., *Development of a Time-Domain Simulation Tool for Offshore Wind Farms.* Journal of Power Electronics, 2015. **15**(4): p. 1047-1053.
64. Tsai, L., et al., *Life Cycle Assessment of Offshore Wind Farm Siting: Effects of Locational Factors, Lake Depth, and Distance From Shore.* Journal of Industrial Ecology, 2016. **20**(6): p. 1370-1383.
65. Afridi, S.K., et al., *Winds of Progress: An in-Depth Exploration of Offshore, Floating, and Onshore Wind Turbines as Cornerstones for Sustainable Energy Generation and Environmental Stewardship.* Ieee Access, 2024. **12**: p. 66147-66166.
66. Pugh, K., G. Rasool, and M.M. Stack, *Some Thoughts on Mapping Tribological Issues of Wind Turbine Blades Due to Effects of Onshore and Offshore Raindrop Erosion.* Journal of Bio- And Tribo-Corrosion, 2018. **4**(3).
67. Júnior, M.J.R., D.L. Medeiros, and E.d.S. Almeida, *Blade Manufacturing for Onshore and Offshore Wind Farms: The Energy and Environmental Performance for a Case Study in Brazil.* Gestão & Produção, 2023. **30**.
68. Schulz, C., et al., *CFD Study on the Impact of Yawed Inflow on Loads, Power and Near Wake of a Generic Wind Turbine.* Wind Energy, 2016. **20**(2): p. 253-268.
69. Fleming, P., et al., *Wind Plant System Engineering Through Optimization of Layout and Yaw Control.* Wind Energy, 2015. **19**(2): p. 329-344.
70. Chen, H., et al., *Early Fault Warning Method of Wind Turbine Main Transmission System Based on SCADA and CMS Data.* Machines, 2022. **10**(11): p. 1018.

71. Yu, X., et al., *Study on the Yaw-Based Wake Steering Control Considering Dynamic Flow Characteristics for Wind Farm Power Improvement.* Journal of Physics Conference Series, 2023. **2505**(1): p. 012010.
72. Gebraad, P., et al., *Wind Plant Power Optimization Through Yaw Control Using a Parametric Model for Wake Effects-a CFD Simulation Study.* Wind Energy, 2014. **19**(1): p. 95-114.
73. Dai, J., et al., *Further Study on the Effects of Wind Turbine Yaw Operation for Aiding Active Wake Management.* Applied Sciences, 2020. **10**(6): p. 1978.
74. Radhakrishnan, S., et al., *Intelligent Control System for Wind Turbine Farms Using IoT and Machine Learning.* E3s Web of Conferences, 2023. **387**: p. 04004.
75. Prabhavathi, K., et al., *Design and Implementation of a Machine Learning-Based Wind Turbine Control System.* E3s Web of Conferences, 2023. **387**: p. 01008.
76. Liu, H., et al., *Research on Key Technology of Wind Turbine Drive Train Fault Diagnosis System Based on Digital Twin.* Applied Sciences, 2024. **14**(14): p. 5991.
77. Peng, *Aerodynamic Characteristics and Structure of the Blade*. 2025.
78. Jenkins, B., et al., *Limiting Wave Conditions for the Safe Maintenance of Floating Wind Turbines.* Journal of Physics Conference Series, 2021. **2018**(1): p. 012023.
79. Roberts, R. and R. Flin, *Safety on the Horizon: Mapping Human Factors in Offshore Wind.* Proceedings of the Human Factors and Ergonomics Society Annual Meeting, 2025.
80. Lawani, K., B. Hare, and I. Cameron, *Integrating Early Refresher Practice in Height Safety and Rescue Training.* Safety Science, 2018. **110**: p. 411-417.
81. Wambura, J., *A Literature Review on the Influence of Training on Improving Employee Performance.* African Journal of Empirical Research, 2025. **6**(1): p. 554-559.
82. Neitzel, D.K., *Controlling Electrical Hazards Through Effective Risk Management.* 2011: p. 1-8.
83. Pande, S. and G. Mahajan, *Electrical Safety Audit and Power Quality Analysis of Electric Installation at Religious Place.* 2025: p. 80-90.
84. Reyes, M., et al., *Identifying Electrical Risks and Enhancing Safety Training for Non-Electrical Construction Workers.* **6**: p. 875-866.
85. Hadžiefendić, N., J. Trifunović, and K.N. M, *Improvement of the Standardized Procedure for Verifying the Quality of Low-Voltage Electrical Installations.* Tehnika, 2022. **77**(2): p. 215-222.
86. Taguada, D., et al., *Electrical Risk Management in the Work Environment: A Comprehensive Review.* 2025. **3**: p. 402.
87. Mukhametyanov, V.A., et al., *Interactive Training of Personnel on Electrical Safety Issues for an Oil Pumping Company.* Oil and Gas Business, 2024(6): p. 42-58.
88. Shakil, F., E. Hasan, and S.C. Banik, *Safety and Risk Management System at RPCL Power Plant in Bangladesh.* International Journal of Recent Trends in Engineering and Research, 2020. **Special Issue 7**: p. 263-272.
89. Sanchez, L.M., et al., *Safety in Photovoltaic Systems Applied to Electric Arc.* Dyna Ingenieria E Industria, 2024. **99**(4): p. 343-343.
90. You, F., et al., *Fire Risk Assessments and Fire Protection Measures for Wind Turbines: A Review.* Heliyon, 2023. **9**(9): p. e19664.

91. Uadiale, S., et al., *Overview of Problems and Solutions in Fire Protection Engineering of Wind Turbines.* Fire Safety Science, 2014. **11**: p. 983-995.
92. Kim, J.-H., et al., *Wind Turbine Fire Prevention System Using Fuzzy Rules and WEKA Data Mining Cluster Analysis.* Energies, 2023. **16**(13): p. 5176.
93. Ehlers, P., *A Success Story? Fifty Years of MARPOL.* The International Journal of Marine and Coastal Law, 2024. **40**(1): p. 57-112.
94. Hussain, N., et al., *Addressing Marine Pollution: An Analysis of MARPOL 73/78 Regulations and Global Implementation Efforts.* Journal of Social Sciences Review, 2023. **3**(1): p. 572-589.
95. Rusli, M.H.M., P. Дремлюга, and D. Puspitawati, *Pollution Control Mechanism for the Straits of Malacca and Singapore: The Proposed Designation as a "Special Area" Under MARPOL 73/78.* Asian Politics & Policy, 2020. **12**(4): p. 502-515.
96. Ferdi, F., et al., *The Difference Between Threshold of Oil Sludge Discharge to the Sea Based on MARPOL 1973/1978 and the Provision of Oil Sludge Threshold in Indonesia's Seawater Quality Regulation That Causes Pollution in Bintan Coast, Indonesia.* Bio Web of Conferences, 2023. **70**: p. 05004.
97. Čampara, L., N. Hasanspahić, and S. Vujičić, *Overview of MARPOL ANNEX VI Regulations for Prevention of Air Pollution From Marine Diesel Engines.* SHS Web of Conferences, 2018. **58**: p. 01004.
98. Mansurova, M., et al., *On-Site Analytical Tool Based on Crude Oil Fluorescence and Chemometrics for the Rapid Determination of the Nature and Essential Properties of Oil Spills.* Acs Es&t Water, 2024. **4**(2): p. 621-627.
99. Ajith, P.J., et al., *Energy Efficiency of Offshore Support Vessels.* Iop Conference Series Earth and Environmental Science, 2022. **1044**(1): p. 012005.
100. Jalkanen, J.-P., et al., *Modelling of Discharges From Baltic Sea Shipping.* Ocean Science, 2021. **17**(3): p. 699-728.
101. Сагін, С.В., С.С. Сагін, and V. Madey, *Analysis of Methods of Managing the Environmental Safety of the Navigation Passage of Ships of Maritime Transport.* Technology Audit and Production Reserves, 2023. **4**(3(72)): p. 33-42.
102. Pengky.cn, *Wind Turbine Generator Gearbox (1).* 2025.
103. Zipp, K., *Brakes 101.* 2012, Windpower Engineering % Development.
104. Watanabe, F., et al., *Modelling Passive Yawing Motion of Horizontal Axis Small Wind Turbine: Derivation of New Simplified Equation for Maximum Yaw Rate.* Wind Engineering, 2012. **36**(4): p. 433-441.
105. Klabklay, T. and T. Chitsomboon, *Untitled.* Journal of Engineering and Science Research, 2017. **1**(2).
106. Wanke, G., et al., *Re-Design of an Upwind Rotor for a Downwind Configuration: Design Changes and Cost Evaluation.* 2020.
107. Kress, C., N. Chokani, and R.S. Abhari, *Downwind Wind Turbine Yaw Stability and Performance.* Renewable Energy, 2015. **83**: p. 1157-1165.
108. Saenz-Aguirre, A., et al., *Artificial Neural Network Based Reinforcement Learning for Wind Turbine Yaw Control.* Energies, 2019. **12**(3): p. 436.
109. Karakasis, N., et al., *Active Yaw Control in a Horizontal Axis Wind System Without Requiring Wind Direction Measurement.* Iet Renewable Power Generation, 2016. **10**(9): p. 1441-1449.

110. Bassett, K., R. Carriveau, and D.S.K. Ting, *Vibration Response of a 2.3 MW Wind Turbine to Yaw Motion and Shut Down Events*. Wind Energy, 2011. **14**(8): p. 939-952.

111. Cui, G., et al., *Radial Basis Function Neural Network–based Adaptive Sliding Mode Suspension Control for Maglev Yaw System of Wind Turbines*. Proceedings of the Institution of Mechanical Engineers Part I Journal of Systems and Control Engineering, 2021. **236**(1): p. 66-75.

112. Han, Z., et al., *A 2.5MW Wind Turbine TL-EMPC Yaw Strategy Based on Ideal Wind Measurement by LiDAR*. Ieee Access, 2021. **9**: p. 89866-89877.

113. Zhang, L. and Q. Yang, *A Method for Yaw Error Alignment of Wind Turbine Based on LiDAR*. Ieee Access, 2020. **8**: p. 25052-25059.

114. Jiang, H., Y. Li, and Z. Cheng, *Performances of Ideal Wind Turbine*. Renewable Energy, 2015. **83**: p. 658-662.

115. Fagan, E., et al., *Validation of the Multi-Objective Structural Optimisation of a Composite Wind Turbine Blade*. Composite Structures, 2018. **204**: p. 567-577.

116. Fagan, E., et al., *Physical Experimental Static Testing and Structural Design Optimisation for a Composite Wind Turbine Blade*. Composite Structures, 2017. **164**: p. 90-103.

117. Tenguria, N., N.D. Mittal, and S. Ahmed, *Evaluation of Performance of Horizontal Axis Wind Turbine Blades Based on Optimal Rotor Theory*. Journal of Urban and Environmental Engineering, 2011. **5**(1): p. 15-23.

118. Mahmood, N.M., *Computer Program to Predict Performance of Fast Running Horizontal Axis Wind Turbine to Reaching the Optimum Design*. Diyala Journal of Engineering Sciences, 2011. **4**(2): p. 57-71.

119. Torre, O.d.l., et al., *Accelerated Life Testing Study of a Novel Tidal Turbine Blade Attachment*. International Journal of Fatigue, 2018. **114**: p. 226-237.

120. Lu, Q., et al., *A Study on Design and Aerodynamic Characteristics of a Spiral-Type Wind Turbine Blade*. Journal of the Korean Society of Visualization, 2012. **10**(1): p. 27-33.

121. Liu, Q. and Z. Zhen, *Design of the Blades of Small Horizontal Axis Wind Turbines*. Applied Mechanics and Materials, 2015. **741**: p. 23-27.

122. Huang, Y., H.H. Li, and G.X. Li, *Maximum Wind Power Tracking Strategy of Wind Power Generation System*. Applied Mechanics and Materials, 2013. **313-314**: p. 813-816.

123. Fagan, E., et al., *Design and Testing of a Full-Scale 2 MW Tidal Turbine Blade*. 2020.

124. Grundkötter, E. and J. Melbert, *Precision Blade Deflection Measurement System Using Wireless Inertial Sensor Nodes*. Wind Energy, 2021. **25**(3): p. 432-449.

125. Chen, S., et al., *Transfer Condition Monitoring Across Wind Turbines Using Feature Alignment and Parameter Fine-Tuning*. Measurement Science and Technology, 2023. **34**(12): p. 125901.

126. Wang, W. and H. Cai, *Research on Wind Turbine Maintenance Based on Data-Physical Fusion Modeling*. 2023.

127. Papi, F., et al., *Uncertainty Quantification of the Effects of Blade Damage on the Actual Energy Production of Modern Wind Turbines*. Energies, 2020. **13**(15): p. 3785.
128. Daud, S.Z.M., F. Mustapha, and Z. Adzis, *Lightning Strike Evaluation on Composite and Biocomposite Vertical-Axis Wind Turbine Blade Using Structural Health Monitoring Approach*. Journal of Intelligent Material Systems and Structures, 2018. **29**(17): p. 3444-3455.
129. Galappaththi, U.I.K., et al., *Review of Inspection and Quality Control Techniques for Composite Wind Turbine Blades*. Insight - Non-Destructive Testing and Condition Monitoring, 2012. **54**(2): p. 82-85.
130. Mishnaevsky, L., *How to Repair the Next Generation of Wind Turbine Blades*. Energies, 2023. **16**(23): p. 7694.
131. Asthana, A., et al., *Experimental Validation of the Structural Integrity of Modular Horizontal Axis Wind Turbine Blades*. 2017: p. 694.
132. Zhu, J., et al., *Static and Dynamic Characteristics Study of Wind Turbine Blade*. Advanced Materials Research, 2011. **383-390**: p. 1895-1900.
133. Zhou, F., et al., *Simulation Analysis of Guided Wave Testing on Wind Turbine Blade Shell With Epoxy Resin and Semi-Analytical Finite Element Modeling*. Applied Mechanics and Materials, 2012. **182-183**: p. 1362-1366.
134. Energy Transformer, *Wind Power Transformers: Essential Guide for Renewable Energy Systems*. 2025.
135. Rajňák, M., et al., *Effect of Ferrofluid Magnetization on Transformer Temperature Rise*. Journal of Physics D Applied Physics, 2022. **55**(34): p. 345002.
136. Yao, H., et al., *Numerical Simulation of 35kV Oil-Immersed Transformer Fire and Extinguishing Effects of High-Pressure Fine Water Mist*. International Journal of Heat and Technology, 2023. **41**(3): p. 572-580.
137. Su, P., R. Madan, and S. Purushothaman, *Testing and Approval Standard Development of Online Dissolved Gas Analysis Monitors for Oil-filled Transformers—Part I Hydrogen Monitors*. Process Safety Progress, 2024. **43**(4): p. 686-694.
138. Liu, Y., et al., *Odor Recognition of Deteriorated Mineral Oils Using an Odor-Sensing Array*. Frontiers in Analytical Science, 2022. **2**.
139. Zhang, J., et al., *Effect of Immersion Time on the Combustion Characteristics of Oil-impregnated Transformer Insulating Paperboard*. Fire and Materials, 2022. **47**(3): p. 294-304.
140. Zhang, X., et al., *Numerical Investigation of Influences of Liquid Types on Flow Distribution and Temperature Distribution in Disc Type ON Cooled Transformers*. 2017: p. 1-4.
141. Daghrah, M., et al., *Experimental Study of the Influence of Different Liquids on the Transformer Cooling Performance*. Ieee Transactions on Power Delivery, 2019. **34**(2): p. 588-595.
142. Kaur, K., D. Bhalla, and J. Singh, *Hybridized and Complementary Comparative Analysis of Dissolved Gases in a Power Transformer*. Iop Conference Series Materials Science and Engineering, 2022. **1228**(1): p. 012003.

143. Montero, A., et al., *Dielectric Design of Ester-Filled Power Transformers: AC Stress Analysis*. Ieee Transactions on Power Delivery, 2022. **37**(3): p. 2403-2412.
144. Zelu, Y., et al., *Study on Flow Electrification Hazards With Ester Oils*. 2011: p. 1-4.
145. Xiang, C., et al., *Comparison of Dissolved Gases in Mineral and Vegetable Insulating Oils Under Typical Electrical and Thermal Faults*. Energies, 2016. **9**(5): p. 312.
146. Hossain, M.L., A. Abu-Siada, and S.M. Muyeen, *Methods for Advanced Wind Turbine Condition Monitoring and Early Diagnosis: A Literature Review*. Energies, 2018. **11**(5): p. 1309.
147. Gao, Z. and X. Liu, *An Overview on Fault Diagnosis, Prognosis and Resilient Control for Wind Turbine Systems*. Processes, 2021. **9**(2): p. 300.
148. Kasner, R., et al., *Sustainable Wind Power Plant Modernization*. Energies, 2020. **13**(6): p. 1461.
149. Qawaqzeh, M., et al., *Research of Emergency Modes of Wind Power Plants Using Computer Simulation*. Energies, 2021. **14**(16): p. 4780.
150. Matheswaran, V. and L.S. Miller, *Retrofit Winglets for Wind Turbines*. 2019.
151. Shipurkar, U., et al., *Availability of Wind Turbine Converters With Extreme Modularity*. Ieee Transactions on Sustainable Energy, 2018. **9**(4): p. 1772-1782.
152. Xu, Z., *Adsorption and Sensing Mechanisms of Ni-Doped PtTe₂ Monolayer Upon NO₂ and O₃ in Air-Insulated Switchgears*. RSC Advances, 2023. **13**(26): p. 18129-18137.
153. Han, Y.J., Q. Zhao, and W.Y. Yun, *Optimal Inspection and Replacement Strategy for 145 kV Gas-Insulated Switchgear*. Proceedings of the Institution of Mechanical Engineers Part O Journal of Risk and Reliability, 2019. **236**(2): p. 339-347.
154. Li, Y., et al., *Detection of Ozone and Nitric Oxide in Decomposition Products of Air-Insulated Switchgear Using Ultraviolet Differential Optical Absorption Spectroscopy (UV-DOAS)*. Applied Spectroscopy, 2018. **72**(8): p. 1244-1251.
155. Tan, Q., et al., *Diagnosis of Partial Discharge Based on the Air Components for the 10 kV Air-Insulated Switchgear*. Sensors, 2022. **22**(6): p. 2395.
156. Gu, F.C., *Application of the Convolutional Neural Network in Partial Discharge Spectrum Recognition of Power Apparatus*. Iet Science Measurement & Technology, 2023. **17**(4): p. 137-146.
157. Komiya, G., T. Imai, and Y. Miyauchi, *Development of High Thermal Conductive Epoxy Composite for Large Current Switchgear*. Electrical Engineering in Japan, 2018. **205**(4): p. 9-15.
158. Yamashita, T., et al., *Design of Solid/Gas Composite Insulation System With Embedded Electrode*. Ieee Transactions on Dielectrics and Electrical Insulation, 2019. **26**(1): p. 56-63.
159. Lim, D.-Y. and S. Bae, *Study on Oxygen/Nitrogen Gas Mixtures for the Surface Insulation Performance in Gas Insulated Switchgear*. Ieee Transactions on Dielectrics and Electrical Insulation, 2015. **22**(3): p. 1567-1576.
160. Gui, Y., et al., *Study on the Characteristic Decomposition Components of Air-Insulated Switchgear Cabinet Under Partial Discharge*. Aip Advances, 2016. **6**(7).

161. Kwon, J.-H., et al., *Lightning Impulse Breakdown Characteristic of Dry-Air/Silicone Rubber Hybrid Insulation in Rod-Plane Electrode.* Journal of Electrical Engineering and Technology, 2015. **10**(3): p. 1181-1187.
162. Franck, C.M., A. Chachereau, and J. Pachin, *SF₆-Free Gas-Insulated Switchgear: Current Status and Future Trends.* Ieee Electrical Insulation Magazine, 2021. **37**(1): p. 7-16.
163. Hyrenbach, M., T. Paul, and J. Owens, *Environmental and Safety Aspects of AirPlus Insulated GIS.* Cired - Open Access Proceedings Journal, 2017. **2017**(1): p. 132-135.
164. Saxegaard, M., et al., *Low-Current Interruption in SF6-alternatives.* Cired - Open Access Proceedings Journal, 2017. **2017**(1): p. 286-288.
165. Ceyhuni, S.A., et al., *Investigation of Three-Phase Electrical Endurance Test Results of Vacuum Circuit Breaker.* The European Journal of Research and Development, 2023. **3**(4): p. 186-194.
166. Kafshgari, N.A., N. Ramezani, and H. Nouri, *Effects of High Frequency Modeling &Amp; Grounding System Parameters on Transient Recovery Voltage Across Vacuum Circuit Breakers for Capacitor Switching in Wind Power Plants.* International Journal of Electrical Power & Energy Systems, 2019. **104**: p. 159-168.
167. Yang, Z., et al., *Hardware Design of Online Monitoring Device for the High-Voltage Vacuum Circuit Breaker.* Journal of Physics Conference Series, 2022. **2183**(1): p. 012009.
168. Pei, X., et al., *Fast Operating Moving Coil Actuator for a Vacuum Interrupter.* Ieee Transactions on Energy Conversion, 2017. **32**(3): p. 931-940.
169. Yu, D., et al., *Optimisation of Commutation Circuit Parameters for DC Vacuum Circuit Breakers.* The Journal of Engineering, 2018. **2019**(16): p. 2585-2588.
170. Ahn, H.-M., et al., *Optimal Design of Permanent Magnetic Actuator for Permanent Magnet Reduction and Dynamic Characteristic Improvement Using Response Surface Methodology.* Journal of Electrical Engineering and Technology, 2015. **10**(3): p. 935-943.
171. Florkowski, M., et al., *Overvoltage Impact on Internal Insulation Systems of Transformers in Electrical Networks With Vacuum Circuit Breakers.* Energies, 2020. **13**(23): p. 6380.
172. Nicolae, C., *Scada-supervisory control and data acquisition.* Journal of Sustainable Energy, 2011. **2**(4): p. 1-5.
173. Zhang, F., et al., *Calculation and Analysis of Wind Turbine Health Monitoring Indicators Based on the Relationships With SCADA Data.* Applied Sciences, 2020. **10**(1): p. 410.
174. Yang, L., K. Wang, and J. Zhou, *Multimode High-dimensional Time Series Clustering and Monitoring for Wind Turbine SCADA Data.* Quality and Reliability Engineering International, 2024. **40**(8): p. 4285-4306.
175. Pujana, A., et al., *Hybrid-Model-Based Digital Twin of the Drivetrain of a Wind Turbine and Its Application for Failure Synthetic Data Generation.* Energies, 2023. **16**(2): p. 861.
176. Udo, W. and Y. Muhammad, *Data-Driven Predictive Maintenance of Wind Turbine Based on SCADA Data.* Ieee Access, 2021. **9**: p. 162370-162388.

177. Wei, L., et al., *Comprehensive Aging Assessment of Pitch Systems Combining SCADA and Failure Data.* Iet Renewable Power Generation, 2021. **16**(1): p. 198-210.
178. Delgado, I. and M. Fahim, *Wind Turbine Data Analysis and LSTM-Based Prediction in SCADA System.* Energies, 2020. **14**(1): p. 125.
179. Lin, Z., X. Liu, and M. Collu, *Wind Power Prediction Based on High-Frequency SCADA Data Along With Isolation Forest and Deep Learning Neural Networks.* International Journal of Electrical Power & Energy Systems, 2020. **118**: p. 105835.
180. Yao, Q., et al., *A New Condition-Monitoring Method Based on Multi-Variable Correlation Learning Network for Wind Turbine Fault Detection.* Measurement Science and Technology, 2022. **34**(2): p. 024009.
181. Teng, J. and C.D. Markfort, *A Calibration Procedure for an Analytical Wake Model Using Wind Farm Operational Data.* Energies, 2020. **13**(14): p. 3537.
182. Liu, Y., et al., *Intelligent Wind Turbine Blade Icing Detection Using Supervisory Control and Data Acquisition Data and Ensemble Deep Learning.* Energy Science & Engineering, 2019. **7**(6): p. 2633-2645.
183. Marugán, A.P. and F.P.G.a. Márquez, *Advanced Analytics for Detection and Diagnosis of False Alarms and Faults: A Real Case Study.* Wind Energy, 2019. **22**(11): p. 1622-1635.
184. Martí-Puig, P., et al., *Wind Turbine Prognosis Models Based on SCADA Data and Extreme Learning Machines.* Applied Sciences, 2021. **11**(2): p. 590.
185. Roscher, B. and R. Schelenz, *Usability of SCADA as Predictive Maintenance for Wind Turbines.* Forschung Im Ingenieurwesen, 2021. **85**(2): p. 173-180.
186. Fu, J., et al., *Condition Monitoring of Wind Turbine Gearbox Bearing Based on Deep Learning Model.* Ieee Access, 2019. **7**: p. 57078-57087.
187. Zhang, Z., *Automatic Fault Prediction of Wind Turbine Main Bearing Based on SCADA Data and Artificial Neural Network.* Open Journal of Applied Sciences, 2018. **08**(06): p. 211-225.
188. Abu-Zaid, S., *Performance of Doubly-Fed Induction Generator Based on Wind Turbine.* Journal of Al-Azhar University Engineering Sector, 2018. **13**(48): p. 1110-1116.
189. Wadhwani, M., S. Deshmukh, and H.S. Dhiman, *Digital Twin Framework for Time to Failure Forecasting of Wind Turbine Gearbox: A Concept.* 2022.
190. Ahmed, M.A., et al., *Wireless Network Architecture for Cyber Physical Wind Energy System.* Ieee Access, 2020. **8**: p. 40180-40197.
191. Annoni, J., et al., *A Framework for Autonomous Wind Farms: Wind Direction Consensus.* 2018.
192. Bajracharya, C., R. Grodi, and D.B. Rawat, *Performance Analysis of Wireless Sensor Networks for Wind Turbine Monitoring Systems.* 2015.
193. Wu, H., et al., *Power Grid Reliability Evaluation Considering Wind Farm Cyber Security and Ramping Events.* Applied Sciences, 2019. **9**(15): p. 3003.
194. Qiao, W. and D. Lu, *A Survey on Wind Turbine Condition Monitoring and Fault Diagnosis—Part I: Components and Subsystems.* Ieee Transactions on Industrial Electronics, 2015. **62**(10): p. 6536-6545.

195. Chen, J., *Research on Fault Diagnosis and Health Monitoring of Wind Power Equipment*. 2023.
196. Cisco, *Renewable Generation Wind Farm*. 2022.
197. Troskie, B., R. Toftegaard, and K. Bonnez, *Corrosivity Control Inside Offshore Wind Power Turbines by Combined Dehumidification and Overpressure*. Materials and Corrosion, 2024. **76**(6): p. 851-857.
198. Brijder, R., et al., *Review of Corrosion Monitoring and Prognostics in Offshore Wind Turbine Structures: Current Status and Feasible Approaches*. Frontiers in Energy Research, 2022. **10**.
199. Mei, B.X., *Corrosion Protection Design of Support Structures for Offshore Wind Turbines With Analysis of Material Properties*. Advanced Materials Research, 2013. **703**: p. 336-339.
200. Wang, K. and M. Zhao, *Mathematical Model of Homogeneous Corrosion of Steel Pipe Pile Foundation for Offshore Wind Turbines and Corrosive Action*. Advances in Materials Science and Engineering, 2016. **2016**: p. 1-8.
201. McAuliffe, J., et al., *Corrosion Fatigue Analysis of NREL's 15-Mw Offshore Wind Turbine With Time-Varying Stress Concentration Factors*. Journal of Physics Conference Series, 2024. **2767**(6): p. 062023.
202. Xu, Y., et al., *Optimizing Mechanical Properties and Corrosion Resistance in Core-shell Nanofiber Epoxy Self-healing Coatings: Impact of Shell Material Variation*. Polymer Engineering & Science, 2024. **64**(4): p. 1770-1785.
203. Ntaflos, A., et al., *Enhancing Epoxy Composite Performance With Carbon Nanofillers: A Solution for Moisture Resistance and Extended Durability in Wind Turbine Blade Structures*. Materials, 2024. **17**(2): p. 524.
204. Dong, X.H., T.J. Yuan, and R. Hong, *Corrosion Mechanism on Offshore Wind Turbine Blade in Salt Fog Environment*. Applied Mechanics and Materials, 2013. **432**: p. 258-262.
205. Moghaddam, B.T., et al., *Structural Integrity Assessment of Floating Offshore Wind Turbine Support Structures*. Ocean Engineering, 2020. **208**: p. 107487.
206. Torres, J.F. and F. Petrakopoulou, *A Closer Look at the Environmental Impact of Solar and Wind Energy*. Global Challenges, 2022. **6**(8).
207. Anders Voldsgaard Clausen, *Corrosion protection for offshore turbines requires proper resources and painstaking attention to developing an action plan*. 2013, Wind Systems.
208. Zhu, G., *Lightning and Surge Protection for Wind Turbines*. 2023, LSP.
209. Gaertner, E., et al., *IEA Wind TCP Task 37: Definition of the IEA 15-Megawatt Offshore Reference Wind Turbine*. 2020.
210. Bortolotti, P., et al., *IEA Wind TCP Task 37: Systems Engineering in Wind Energy - WP2.1 Reference Wind Turbines*. 2019.
211. Nogueira, K.d.S., L.C.d.S. Andrade, and P.H.A. Bezerra, *USG Method in the Assessment of Construction Defects in Foundations of Wind Turbines*. 2023.
212. Hermawan, B.M., et al., *Performance Characteristics of Savonius Wind Turbines With Variations Air Gaps in Supporting the Development of Renewable Energy*. Eksergi, 2023. **19**(01): p. 28-31.

213. Zhou, Z.P., Q.Y. Mo, and Z.P. Lei, *Computational Fluid Dynamics Study of a Small Vertical Axis Wind Turbine With Ball-Shaped Blades.* Advanced Materials Research, 2011. **347-353**: p. 319-322.
214. Schubel, P. and R. Crossley, *Wind Turbine Blade Design.* 2014: p. 1-34.
215. Liu, M., et al., *A Knowledge Graph-Based Approach for Assembly Sequence Recommendations for Wind Turbines.* Machines, 2023. **11**(10): p. 930.
216. Mathern, A. and J. Magnusson, *Lessons Learned From the Construction, Inspection, and Defect Assessment of Reinforced Concrete Foundations for Wind Turbines.* Applied Sciences, 2022. **12**(3): p. 1443.
217. Ding, H., et al., *Towing Motion Characteristics of Composite Bucket Foundation for Offshore Wind Turbines.* Energies, 2019. **12**(19): p. 3767.
218. Deb, G.K., *Nature of Wind Turbine Power.* International Journal of Computer and Electrical Engineering, 2012: p. 145-148.
219. Yin, X.J., et al., *Design and Realization of a Pre-Production Platform for Wind Turbine Manufacturing.* Wind Engineering, 2023. **48**(3): p. 436-445.
220. Khan, P.W. and Y.-C. Byun, *Multi-Fault Detection and Classification of Wind Turbines Using Stacking Classifier.* Sensors, 2022. **22**(18): p. 6955.
221. Maldonado-Correa, J., et al., *Using SCADA Data for Wind Turbine Condition Monitoring: A Systematic Literature Review.* Energies, 2020. **13**(12): p. 3132.
222. Pozo, F., Y. Vidal, and Ó. Salgado, *Wind Turbine Condition Monitoring Strategy Through Multiway PCA and Multivariate Inference.* Energies, 2018. **11**(4): p. 749.
223. Wu, Z., X. Wang, and B. Jiang, *Fault Diagnosis for Wind Turbines Based on ReliefF and eXtreme Gradient Boosting.* Applied Sciences, 2020. **10**(9): p. 3258.
224. Cao, M., et al., *Study of Wind Turbine Fault Diagnosis Based on Unscented Kalman Filter and SCADA Data.* Energies, 2016. **9**(10): p. 847.
225. Simani, S., S. Farsoni, and P. Castaldi, *Transfer Learning for Fault Detection With Application to Wind Turbine SCADA Data.* Journal of Energy and Power Technology, 2023. **05**(01): p. 1-12.
226. Liu, Y., et al., *Wind Turbine Spindle Operating State Recognition and Early Warning Driven by SCADA Data.* Energy Engineering, 2023. **120**(5): p. 1223-1237.
227. Yang, H.H., M.L. Huang, and S.W. Yang, *Integrating Auto-Associative Neural Networks With Hotelling T2 Control Charts for Wind Turbine Fault Detection.* Energies, 2015. **8**(10): p. 12100-12115.
228. Kim, B.-S., et al., *Ultimate Load Characteristics of NREL 5-Mw Offshore Wind Turbines With Different Substructures.* International Journal of Energy Research, 2015. **40**(5): p. 639-650.
229. Li, G., et al., *Research on Wind Turbine Power Performance Based on the Operation Data.* E3s Web of Conferences, 2019. **136**: p. 02002.
230. Gebraad, P., et al., *Maximization of the Annual Energy Production of Wind Power Plants by Optimization of Layout and Yaw-Based Wake Control.* Wind Energy, 2016. **20**(1): p. 97-107.
231. Hedevang, E., *Wind Turbine Power Curves Incorporating Turbulence intensity.* Wind Energy, 2012. **17**(2): p. 173-195.
232. Angelou, N. and M. Sjöholm, *Data Reliability Enhancement for Wind-Turbine-Mounted Lidars.* Remote Sensing, 2022. **14**(13): p. 3225.

233. Shin, D. and K. Ko, *Application of the Nacelle Transfer Function by a Nacelle-Mounted Light Detection and Ranging System to Wind Turbine Power Performance Measurement.* Energies, 2019. **12**(6): p. 1087.
234. Honrubia-Escribano, A., et al., *Validation of Generic Models for Variable Speed Operation Wind Turbines Following the Recent Guidelines Issued by IEC 61400-27.* Energies, 2016. **9**(12): p. 1048.
235. Kazda, J. and J. Mann, *Mitigating Impact of Spatial Variance of Turbulence in Wind Energy Applications.* Wind Energy Science, 2020. **5**(2): p. 439-450.
236. Astolfi, D., F. Castellani, and L. Terzi, *Wind Turbine Power Curve Upgrades.* Energies, 2018. **11**(5): p. 1300.
237. Bai, W., et al., *Small Wind Turbine Power Performance Testing With Uncertainty Analysis.* Advanced Materials Research, 2014. **875-877**: p. 1944-1948.
238. Shokrzadeh, S., M.J. Jozani, and E. Bibeau, *Wind Turbine Power Curve Modeling Using Advanced Parametric and Nonparametric Methods.* Ieee Transactions on Sustainable Energy, 2014. **5**(4): p. 1262-1269.
239. Astolfi, D., *A Study of the Impact of Pitch Misalignment on Wind Turbine Performance.* Machines, 2019. **7**(1): p. 8.
240. Wagner, R., et al., *Accounting for the Speed Shear in Wind Turbine Power Performance Measurement.* Wind Energy, 2011. **14**(8): p. 993-1004.
241. Wan, S., L. Cheng, and X. Sheng, *Effects of Yaw Error on Wind Turbine Running Characteristics Based on the Equivalent Wind Speed Model.* Energies, 2015. **8**(7): p. 6286-6301.
242. Han, S., *A complete guide to wind turbine inspection*. 2023, illuminem.
243. Link, H., et al., *Gearbox Reliability Collaborative Project Report: Findings From Phase 1 and Phase 2 Testing.* 2011.
244. Coronado, D. and J. Wenske, *Monitoring the Oil of Wind-Turbine Gearboxes: Main Degradation Indicators and Detection Methods.* Machines, 2018. **6**(2): p. 25.
245. Zhong, X.Y., et al., *Fault Diagnosis for Wind Turbine Gearboxes Based on EMD and the Energy Operator.* Applied Mechanics and Materials, 2013. **281**: p. 10-13.
246. Salem, A., A. Abu-Siada, and S. Islam, *Impact of Gearbox Oil Contamination on the Performance of the Wind Turbine Drivetrain.* Renewable Energy and Power Quality Journal, 2024. **14**(2).
247. Angelis, G.D. and L. Krudde, *Dynamic Reliability Estimation of Wind Turbine Gearboxes Using Real-Time Operational Data.* Journal of Physics Conference Series, 2025. **3025**(1): p. 012013.
248. Bruce, T., H. Long, and R.S. Dwyer-Joyce, *Dynamic Modelling of Wind Turbine Gearbox Bearing Loading During Transient Events.* Iet Renewable Power Generation, 2015. **9**(7): p. 821-830.
249. Cui, Y., et al., *Temperature Prediction for 3 MW Wind-Turbine Gearbox Based on Thermal Network Model.* Machines, 2024. **12**(3): p. 175.
250. Gonçalves, D., et al., *Torque Loss in a Gearbox Lubricated With Wind Turbine Gear Oils.* Lubrication Science, 2013. **25**(4): p. 297-311.
251. Abd-Elwahab, k.t. and A.A. Hassan, *SCADA Data as a Powerful Tool for Early Fault Detection in Wind Turbine Gearboxes.* Wind Engineering, 2020. **45**(5): p. 1317-1326.

252. Kavaz, A.G. and B. Barutçu, *Fault Detection of Wind Turbine Sensors Using Artificial Neural Networks*. Journal of Sensors, 2018. **2018**: p. 1-11.
253. Wei, L., et al., *Wind Turbine Fault Diagnosis by the Approach of SCADA Alarms Analysis*. Applied Sciences, 2021. **12**(1): p. 69.
254. Qiu, Y., et al., *Wind Turbine SCADA Alarm Analysis for Improving Reliability*. Wind Energy, 2011. **15**(8): p. 951-966.
255. Leahy, K., et al., *Diagnosing and PredictingWind Turbine Faults From SCADA Data Using Support Vector Machines*. International Journal of Prognostics and Health Management, 2020. **9**(1).
256. Chen, B., et al., *Wind Turbine SCADA Alarm Pattern Recognition*. 2011: p. 163-163.
257. Milani, A.E., et al., *Boosting Field Data Using Synthetic SCADA Datasets for Wind Turbine Condition Monitoring*. Journal of Physics Conference Series, 2024. **2767**(3): p. 032033.
258. Maseda, F.J., et al., *Sensors Data Analysis in Supervisory Control and Data Acquisition (SCADA) Systems to Foresee Failures With an Undetermined Origin*. Sensors, 2021. **21**(8): p. 2762.
259. Liu, Y., Z. Wu, and X. Wang, *Research on Fault Diagnosis of Wind Turbine Based on SCADA Data*. Ieee Access, 2020. **8**: p. 185557-185569.
260. Revathi, L.V., M. Sudalaimani, and S. Jeyadevi, *An ABC-SVM Based Fault Prognosis of Wind Turbines Using SCADA Data*. Scientia Iranica, 2024. **0**(0): p. 0-0.
261. Li, J.B., et al., *Determination of Energy Storage Capacity for a Hybrid Renewable Power Generation System*. Applied Mechanics and Materials, 2014. **528**: p. 371-379.
262. Kan, H.J. and L.N. Tan, *The Influence of Wind–Solar Hybrid Generation System on Transmission Service Rate*. Advanced Materials Research, 2014. **1070-1072**: p. 1472-1476.
263. Sohoni, V., S. Gupta, and R. Nema, *Design of Wind-Pv Based Hybrid Standalone Energy Systems for Three Sites in Central India*. Ecti Transactions on Electrical Engineering Electronics and Communications, 2019. **17**(1): p. 24-34.
264. Heidary, B., et al., *Exergy of a Hybrid Solar-Wind Reverse Osmosis-MSF Desalination System*. Resource-Efficient Technologies, 2019(1): p. 8-19.
265. Nair, A.R., et al., *Modelling and Optimising the Value of a Hybrid Solar-Wind System*. Iop Conference Series Materials Science and Engineering, 2017. **197**: p. 012035.
266. Jha, N., et al., *Energy-Efficient Hybrid Power System Model Based on Solar and Wind Energy for Integrated Grids*. Mathematical Problems in Engineering, 2022. **2022**: p. 1-12.
267. Chong, W.T., et al., *Urban Eco-Greenergy™ Hybrid Wind-Solar Photovoltaic Energy System and Its Applications*. International Journal of Precision Engineering and Manufacturing, 2015. **16**(7): p. 1263-1268.
268. Zhang, S.J., B. Li, and S. Liu, *The Design for Zero Emission Greenhouse System*. Applied Mechanics and Materials, 2014. **631-632**: p. 310-313.
269. Mohammed, K.G., A.Q. Ramli, and U.A.U. Amirulddin, *A Novel Design of DC-AC Electrical Machine Rotary Converter for Hybrid Solar and Wind Energy*

Applications. Iop Conference Series Earth and Environmental Science, 2013. **16**: p. 012052.

270. Xie, Y. and B. Jiao, *Research of a Kind of Wind-Solar Hybrid Power Experiment System.* Applied Mechanics and Materials, 2011. **121-126**: p. 3667-3671.
271. Tummala, S.R. and S. Pendyala, *A Novel Proposal of Using Solar Panels Mounted on Wind Turbines for Enhancement of Power Generation.* Wind Engineering, 2024. **49**(3): p. 622-631.
272. Vick, B. and B.A. Neal, *Analysis of Off-Grid Hybrid Wind Turbine/Solar PV Water Pumping Systems.* Solar Energy, 2012. **86**(5): p. 1197-1207.
273. Akpan, U.V., et al., *Design, Sizing and Optimization of a Solar- Wind Hybrid Power System.* Journal of Energy Research and Reviews, 2022: p. 50-71.
274. Liu, X., *An in-Depth Study of the Principles and Technologies of Wind-Solar Complementary Systems: Optimization Strategies and Future Development Trends.* Applied and Computational Engineering, 2024. **70**(1): p. 133-138.
275. Rehman, S., et al., *Feasibility Study of a Wind–pv–diesel Hybrid Power System for a Village.* Renewable Energy, 2012. **38**(1): p. 258-268.
276. Al-Areqi, T.S.A., et al., *Development of an Integrated Solar-Wind Power Generator.* 2022: p. 196-204.
277. Bhandari, B., et al., *Optimization of Hybrid Renewable Energy Power Systems: A Review.* International Journal of Precision Engineering and Manufacturing-Green Technology, 2015. **2**(1): p. 99-112.
278. Diemuodeke, O.E., et al., *Optimal Mapping of Hybrid Renewable Energy Systems for Locations Using Multi-Criteria Decision-Making Algorithm.* Renewable Energy, 2019. **134**: p. 461-477.
279. Muralidharan, S., et al., *A Comparative Study of Machine Learning Techniques for Wind Turbine Performance Prediction.* E3s Web of Conferences, 2023. **387**: p. 04011.
280. Garan, M., K. Tidriri, and I. Kovalenko, *A Data-Centric Machine Learning Methodology: Application on Predictive Maintenance of Wind Turbines.* Energies, 2022. **15**(3): p. 826.
281. Vives, J., *Monitoring and Detection of Wind Turbine Vibration With KNN-Algorithm.* Journal of Computer and Communications, 2022. **10**(07): p. 1-12.
282. Guo, L., et al., *An Improved Feature Selection Method Based on Random Forest Algorithm for Wind Turbine Condition Monitoring.* Sensors, 2021. **21**(16): p. 5654.
283. Santolamazza, A., D. Dadi, and V. Introna, *A Data-Mining Approach for Wind Turbine Fault Detection Based on SCADA Data Analysis Using Artificial Neural Networks.* Energies, 2021. **14**(7): p. 1845.
284. Kestel, K., et al., *Offshore Field Experimentation for Novel Hybrid Condition Monitoring Approaches.* Journal of Physics Conference Series, 2024. **2745**(1): p. 012009.
285. Matsui, T., K. Yamamoto, and J. Ogata, *Anomaly Detection Using a <scp>SCADA</Scp> Feature Extractor and Machine Learning to Detect Lightning Damage on Wind Turbine Blades.* Ieej Transactions on Electrical and Electronic Engineering, 2022. **17**(6): p. 945-951.

286. Zhi, S. and H. Shen, *A Self-Supervised Learning Method for Fault Detection of Wind Turbines*. Measurement Science and Technology, 2024. **35**(11): p. 116118.
287. Kim, J.-H., et al., *A Design of the Real-Time Simulation for Wind Turbine Modeling With Machine Learning*. Journal of Electrical Engineering and Technology, 2023. **18**(4): p. 3277-3285.
288. Kong, L., et al., *Research on Wind Turbine Fault Detection Based on the Fusion of ASL-CatBoost and TtRSA*. Sensors, 2023. **23**(15): p. 6741.
289. Korkos, P., et al., *Fault Detection in a Wind Turbine Hydraulic Pitch System Using Deep Autoencoder Extracted Features*. PHM Society European Conference, 2022. **7**(1): p. 261-268.
290. Lebranchu, A., et al., *A Combined Mono- And Multi-Turbine Approach for Fault Indicator Synthesis and Wind Turbine Monitoring Using SCADA Data*. Isa Transactions, 2019. **87**: p. 272-281.
291. Desai, A., et al., *Prognosis of Wind Turbine Gearbox Bearing Failures Using SCADA and Modeled Data*. Annual Conference of the PHM Society, 2020. **12**(1): p. 10.
292. Pandit, R. and D. Infield, *SCADA-based Wind Turbine Anomaly Detection Using Gaussian Process Models for Wind Turbine Condition Monitoring Purposes*. Iet Renewable Power Generation, 2018. **12**(11): p. 1249-1255.
293. Abbas, N., et al., *A Reference Open-Source Controller for Fixed and Floating Offshore Wind Turbines*. Wind Energy Science, 2022. **7**(1): p. 53-73.
294. Asmuth, H. and H. Korb, *WakeNet 0.1 - A Simple Three-Dimensional Wake Model Based on Convolutional Neural Networks*. Journal of Physics Conference Series, 2022. **2265**(2): p. 022066.
295. Kaleli̇, M.S., A. Kaleli, and C.Ç. Yalçıner, *Torque Prediction Based Performance Analysis of Small-Scale Wind Turbines Using Data Driven Modelling Methods*. Erzincan Üniversitesi Fen Bilimleri Enstitüsü Dergisi, 2021. **14**(1): p. 260-269.
296. Zhang, K.W., *Design for Enterprise Asset Management Evaluation System of Power Generation Enterprises Based on KPI*. Advanced Materials Research, 2013. **694-697**: p. 3401-3405.
297. Antwi, B.O. and E.K. Avickson, *Integrating Sap, AI, and Data Analytics for Advanced Enterprise Management*. International Journal of Research Publication and Reviews, 2024. **5**(10): p. 621-636.
298. Elete, T.Y., et al., *Digital Transformation in the Oil and Gas Industry: A Comprehensive Review of Operational Efficiencies and Case Studies*. International Journal of Applied Research in Social Sciences, 2024. **6**(11): p. 2611-2643.
299. Oyeniyi, L.D., C.E. Ugochukwu, and N.Z. Mhlongo, *IoT Applications in Asset Management: A Review of Accounting and Tracking Techniques*. International Journal of Science and Research Archive, 2024. **11**(2): p. 1510-1525.

Index

A

B

C

D

E

F

G

H

I

J

L

M

N

O

P

Q

R

S

T

U

V

W

Y

Z

Made in the USA
Las Vegas, NV
31 October 2026

57067591R00293